Malgorzata Lekka, Daniel Navajas, Manfred Radmacher and Alessai.... . ... (Lu..)
Mechanics of Cells and Tissues in Diseases

Also of Interest

Mechanics of Cells and Tissues in Diseases.
Biomedical Applications
Edited by Malgorzata Lekka, Daniel Navajas, Manfred Radmacher,
Alessandro Podestà, 2022
ISBN 978-3-11-099972-3, e-ISBN (PDF) 978-3-11-098938-0

Medical Physics.
Volume 1: Physical Aspects of Organs and Imaging
Hartmut Zabel, 2017
ISBN 978-3-11-037281-6, e-ISBN (PDF) 978-3-11-037283-0

Medical Physics.
Volume 2: Radiology, Lasers, Nanoparticles and Prosthetics
Hartmut Zabel, 2017
ISBN 978-3-11-055310-9, e-ISBN (PDF) 978-3-11-055311-6

Biomimetics.
A Molecular Perspective
Raz Jelinek, 2021
ISBN 978-3-11-070944-5, e-ISBN (PDF) 978-3-11-070949-0

Quantum Electrodynamics of Photosynthesis.
Mathematical Description of Light, Life and Matter
Artur Braun, 2020
ISBN 978-3-11-062692-6, e-ISBN (PDF) 978-3-11-062994-1

Biosensors.
Fundamentals and Applications
Chandra Mouli Pandey and Bansi Dhar Malhotra, 2019
ISBN 978-3-11-063780-9, e-ISBN (PDF) 978-3-11-064108-0

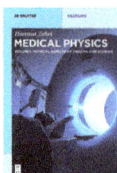

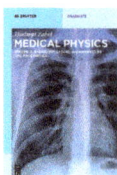

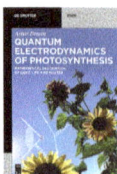

Mechanics of Cells and Tissues in Diseases

Biomedical Methods

Edited by
Malgorzata Lekka, Daniel Navajas, Manfred Radmacher
and Alessandro Podestà

Volume 1

DE GRUYTER

Editors

Prof. Dr. Malgorzata Lekka
Department of Biophysical Microstructures
Institute of Nuclear Physics
Polish Academy of Sciences
ul. Radzikowskiego 152
31-342 Kraków
Poland
malgorzata.lekka@ifj.edu.pl

Prof. Dr. Daniel Navajas
Unitat de Biofísica i Bioenginyeria
Facultat de Medicina i Ciències de la Salut
Universitat de Barcelona
Institute for Bioengineering of Catalonia
Barcelona. Spain
Spain
dnavajas@ub.edu

Prof. Dr. Manfred Radmacher
Institute of Biophysics
University of Bremen
Otto-Hahn-Allee 1
28359 Bremen
radmacher@uni-bremen.de

Prof. Dr. Alessandro Podestà
Department of Physics "Aldo Pontremoli" and
CIMaINa
Università degli Studi di Milano
via Celoria 16
20133 Milano
Italy
alessandro.podesta@mi.infn.it

ISBN 978-3-11-064059-5
e-ISBN (PDF) 978-3-11-064063-2
e-ISBN (EPUB) 978-3-11-064068-7

Library of Congress Control Number: 2022936609

Bibliographic information published by the Deutsche Nationalbibliothek
The Deutsche Nationalbibliothek lists this publication in the Deutsche Nationalbibliografie;
detailed bibliographic data are available on the Internet at http://dnb.dnb.de.

© 2023 Walter de Gruyter GmbH, Berlin/Boston
Cover image: background: Olga Kurbatova/iStock/Getty Images Plus, drawing: Malgorzata Lekka
Typesetting: Integra Software Services Pvt. Ltd.
Printing and binding: CPI books GmbH, Leck

www.degruyter.com

Preface

The mechanical properties of cells can be used to distinguish pathological from normal cells and tissues in many diseases, not only those where the relation between mechanics and physiology of the disease is obvious, like infarcted heart tissue, but also those where this relation is less obvious or still unknown, like cancer. This book outlines the physics behind cell and tissue mechanics, describes the methods, which can be used to determine the mechanical properties of single cells and tissues, and presents various diseases, in which a mechanical fingerprint could be established. Cell mechanics has the potential to serve as an assay, which could be widely used in the future. This book aims to introduce this topic to researchers from backgrounds as varied as biophysics, biomedical engineering, biotechnology, as well as graduate students from biology to medicine to introduce this novel and exciting concept to the community. In this book, we introduce to several aspects of cell biology, emphasizing the importance of the cytoskeleton, the cell membrane and glycocalyx, and the extracellular matrix. One chapter introduces the physics of continuum mechanics and its application to cells, including viscoelastic measurements. Then, various methods for measuring the mechanical properties of cells and tissues are discussed. Finally, evidence on the mechanical fingerprint of diseases is presented, discussing the properties of pathological cells from cancer, muscular dystrophy to diabetes, to name just a few here.

The first volume presents a comprehensive description of the basic concepts of soft matter mechanics and of the nano- and microscale *biomedical methods* that characterize the mechanical properties of cells and tissues.

The second volume is dedicated to discussing several *biomedical applications* of the mechanical phenotyping of cells and tissues to specific disease models. The topical chapters on mechanics in disease are preceded by chapters describing cell and tissue structure and their relationship with the biomechanical properties, as well as by describing dedicated sample preparation methods for the nano- and microscale mechanical measurements.

This book has been written for the primary benefit of young researchers but also of senior scientists, involved in interdisciplinary studies at the boundary of Physics, Biology and Medicine, and committed to transforming academic scientific and technological knowledge into useful diagnostic tools in the clinic.

We like to thank all authors of the various chapters for their valuable contributions. We appreciate very much your efforts and your continuing support over the time needed to create this work.

https://doi.org/10.1515/9783110640632-202

Acknowledgment

We acknowledge the support of the European Union's Horizon 2020 research and innovation program under the Marie Skłodowska-Curie grant agreement no. 812772, project Phys2Biomed.

https://doi.org/10.1515/9783110640632-203

Table of Contents Volume 1 – Biomedical Methods

Table of Contents of Volume 2 – Biomedical Applications

Mechanics in Diseased Cells and Tissues

Contributing Authors

Yara Abidine
University Grenoble Alpes, CNRS, LIPhy
Grenoble, France
Current address: Department of Clinical
Microbiology, Faculty of Medicine
Wallenberg Centre for Molecular Medicine
Umeå University, Sweden
yara.abidine@umu.se

Charles T. Anderson
Department of Biology and Center for
Lignocellulose Structure and Formation
The Pennsylvania State University
University Park, PA, USA
cta3@psu.edu

Massimo Alfano
Division of Experimental Oncology/Unit of
Urology, URI, IRCCS Ospedale San Raffaele
Milan, Italy
alfano.massimo@hsr.it

Nelda Antonovaite
Optics 11 Life,
Amsterdam, Netherlands
nelda373@gmail.com

Manuela Brás
i3S Instituto de Investigação e
Inovação em Saúde, Universidade do
Porto, Portugal
INEB – Instituto de Engenharia Biomédica
Porto, Portugal
FEUP – Faculdade de Engenharia da
Universidade do Porto, Portugal
mbras@i3s.up.pt

Kristian Brat
Department of Respiratory Diseases
University Hospital Brno, Brno
Czech Republic
Brat.Kristian@fnbrno.cz

Massimiliano Berardi
Optics 11 Life,
Amsterdam, Netherlands
massimiliano.berardi@optics11.com

Kevin Bielawski
Optics 11 Life,
Amsterdam, Netherlands
kevin.bielawski@optics11.com

Ignacio Casuso
Aix-Marseille Univ, CNRS, INSERM, LAI, Turing
centre for biological systems, Marseille
France

Shu-wen W. Chen
Institut de Biologie Structurale, Univ.
Grenoble Alpes, CEA, CNRS, F-38000
Grenoble, France

Matteo Chighizola
CIMaINa and Dipartimento di Fisica "Aldo
Pontremoli", Università degli Studi di Milano
Milan, Italy
matteo.chighizola@unimi.it

Thomas Decaens
Institute for Advanced Biosciences, Grenoble-
Alpes University, Inserm U1209 – CNRS UMR
5309, and Hepatology Department, University
Hospital of Grenoble Alpes, Grenoble, France
tdecaens@chu-grenoble.fr

Thierry Desnos
Aix Marseille Université, CNRS, CEA, Institut
de Biosciences et Biotechnologies Aix-
Marseille, Equipe Bioénergies et
Microalgues, CEA Cadarache, Saint-Paul-lez-
Durance, France
thierry.desnos@cea.fr

https://doi.org/10.1515/9783110640632-205

Simone Dinarelli
Institute for the Structure of Matter, CNR
Rome, Italy
simone.dinarelli@ism.cnr.it

Peter Dvorak
Department of Biology, Faculty of Medicine
Masaryk University, Brno, Czech Republic
ICRC, St. Anne's University Hospital, Brno
Czech Republic
dvorak.josefov@icloud.com

Vincent Dupres
Cellular Microbiology and Physics of Infection
Group, Univ. Lille, CNRS, Inserm, CHU Lille
Institut Pasteur Lille, Center for Infection and
Immunity of Lille, Lille, France
vincent.dupres@ibl.cnrs.fr

Allen Ehrlicher
Department of Bioengineering, McGill
University, Montreal, Canada
aje.mcgill@gmail.com

Ramon Farré
Unitat de Biofísica i Bioenginyeria
Facultat de Medicina i Ciències de la Salut
Universitat de Barcelona, Barcelona
Spain CIBER de Enfermedades Respiratorias
Madrid, Spain
Institut d'Investigacions Biomediques August
Pi Sunyer, Barcelona, Spain
rfarre@ub.edu

Conor Fields
School of Chemistry and Conway Institute for
Biomolecular and Biomedical Science
University College Dublin, Dublin
Republic of Ireland

Dorota Gil
Chair of Medical Biochemistry, Jagiellonian
University Medical College, Kraków, Poland
dorotabeata.gil@uj.edu.pl

Marco Girasole
Institute for the Structure of Matter, CNR,
Rome, Italy
marco.girasole@ism.cnr.it

Christian Godon
Aix Marseille Université, CNRS, CEA, Institut
de Biosciences et Biotechnologies Aix-
Marseille, Laboratoire de Signalisation pour
l'adaptation des végétaux à leur
environnement, CEA Cadarache, Saint-Paul-
lez-Durance, France
godon@cea.fr

Wolfgang Goldmann
Department of Physics, Biophysics Group
Friedrich-Alexander-University Erlangen-
Nuremberg, Erlangen, Germany
wgoldmann@biomed.uni-erlangen.de

Hatice Holuigue
CIMaINa and Dipartimento di Fisica "Aldo
Pontremoli", Università degli Studi di Milano
Milan, Italy
hatice.holuigue@unimi.it

Sébastien Janel
Cellular Microbiology and Physics of Infection
Group, Univ. Lille, CNRS, Inserm, CHU Lille,
Institut Pasteur Lille, Center for Infection and
Immunity of Lille, Lille, France
sebastien.janel@cnrs.fr

Tae-Hyung Kim
University of California, Los Angeles, CA, USA
Current address: Department of Pathology at
the University of New Mexico, Albuquerque
NM, USA
takim@salud.unm.edu

Harinderbir Kaur
Univ. Grenoble Alpes, CEA, CNRS, Institut de
Biologie Structurale, Grenoble, France
harinderbir.kaur@ibs.fr

Prem Kumar Viji Babu
Institute of Biophysics, University of Bremen
Bremen, Germany
Current address: NanoLSI, Kanazawa
University, Kanazawa, Japan
pk@biophysik.uni-bremen.de

Leda Lacaria
Aix-Marseille Univ, INSERM, CNRS, LAI, Turing
Centre for Living Systems, Marseille, France
leda.lacaria@inserm.fr

Frank Lafont
Cellular Microbiology and Physics of Infection
Group, Univ. Lille, CNRS, Inserm, CHU Lille
Institut Pasteur Lille, Center for Infection and
Immunity of Lille, Lille, France
frank.lafont@pasteur-lille.fr

Piotr Laidler
Chair of Medical Biochemistry, Jagiellonian
University Medical College, Kraków, Poland
piotr.laidler@uj.edu.pl

Gil Lee
School of Chemistry and Conway Institute for
Biomolecular and Biomedical Science
University College Dublin, Dublin
Republic of Ireland
gil.lee@ucd.ie

Peng Li
School of Chemistry and Conway Institute for
Biomolecular and Biomedical Science,
University College Dublin, Dublin, Republic of
Ireland

Malgorzata Lekka
Department of Biophysical Microstructures
Institute of Nuclear Physics, Polish Academy
of Sciences, Kraków, Poland
malgorzata.lekka@ifj.edu.pl

Ewelina Lorenc
CIMaINa and Dipartimento di Fisica "Aldo
Pontremoli", Università degli Studi di Milano
Milan, Italy
ewelina.lorenc@unimi.it

Chau Ly
Department of Bioengineering, University of
California, Los Angeles, CA, USA

Arnaud Millet
Institute for Advanced Biosciences, Grenoble-
Alpes University, Inserm and Research
Department University Hospital of Grenoble
Alpes, Grenoble, France
arnaud.millet@inserm.fr

Daniel Navajas
Unitat de Biofísica i Bioenginyeria
Facultat de Medicina i Ciències de la Salut
Universitat de Barcelona
Institute for Bioengineering of Catalonia
Barcelona. Spain
dnavajas@ub.edu

Hans Oberleithner
Thaur, Austria
oberlei@gmx.at

Jordi Otero
Unitat de Biofísica i Bioenginyeria, Facultat
de Medicina i Ciències de la Salut
Universitat de Barcelona, Barcelona, Spain
CIBER de Enfermedades Respiratorias
Madrid, Spain
jorge.otero@ub.edu

Jean-Luc Pellequer
Univ. Grenoble Alpes, CEA, CNRS, Institut de
Biologie Structurale, Grenoble, France
jean-luc.pellequer@ibs.fr

Martin Pesl
Department of Biology, Faculty of Medicine
Masaryk University, Brno, Czech Republic
ICRC, St. Anne's University Hospital, Brno
Czech Republic
First Department of Internal Medicine, Cardio-
Angiology, Faculty of Medicine, Masaryk
University, Brno, Czech Republic
martin.pesl@fnusa.cz

Alessandro Podestà
Dept. of Physics "Aldo Pontremoli" and
CIMAINA, University of Milano, Milan, Italy
alessandro.podesta@mi.infn.it

Jan Pribyl
CEITEC, Masaryk University, Brno, Czech
Republic

Manfred Radmacher
Institute of Biophysics, University of Bremen
Bremen, Germany
radmacher@uni-bremen.de

Lorena Redondo-Morata
Cellular Microbiology and Physics of Infection
Group, Univ. Lille, CNRS, Inserm, CHU Lille
Institut Pasteur Lille, Center for Infection and
Immunity of Lille, Lille, France
lorena.redondo-morata@inserm.fr

Carmela Rianna
Institute of Biophysics, University of Bremen
Bremen, Germany
Current address: Institute of Applied Physics
University of Tübingen, Tübingen, Germany
carmela.rianna@uni-tuebingen.de

Felix Rico
Aix-Marseille Univ, INSERM, CNRS, LAI, Turing
Centre for Living Systems, Marseille, France
felix.rico@inserm.fr

Jorge Rodriguez-Ramos
Aix-Marseille Univ, CNRS, INSERM, LAI, Turing
Centre for Living Systems, Marseille, France
jorge.rodriguez-ramos@inserm.fr

Vladimir Rotrekl
Department of Biology, Faculty of Medicine
Masaryk University, Brno, Czech Republic
ICRC, St. Anne's University Hospital, Brno
Czech Republic
vrotrekl@med.muni.cz

Niek Rijnveld
Optics 11 Life,
Amsterdam, Netherlands
niek.rijnveld@optics11.com

Amy C. Rowat
Department of Integrative Biology &
Physiology, University of California
Los Angeles, CA, USA
rowat@ucla.edu

Carsten Schulte
CIMaINa and Dipartimento di Fisica "Aldo
Pontremoli", Università degli Studi di Milano
Milan, Italy
carsten.schulte@unimi.it

Petr Skladal
Department of Biochemistry, Faculty of Science
Masaryk University, Brno, Czech Republic
petr.skladal@ceitec.muni.cz

Zdenek Starek
ICRC, St. Anne's University Hospital, Brno
Czech Republic
44278@mail.muni.cz

Marta Targosz-Korecka
Center for Nanometer-Scale Science and
Advanced Materials, NANOSAM, Faculty of
Physics, Astronomy and Applied Computer
Science, Jagiellonian University, Kraków
Poland
marta.targosz-korecka@uj.edu.pl

Jean-Marie Teulon
Univ. Grenoble Alpes, CEA, CNRS, Institut de
Biologie Structurale, Grenoble, France
jean-marie.teulon@cea.fr

Pouria Tirgar
Department of Bioengineering, McGill
University, Montreal, QC, Canada
pouria.tirgarbahnamiri@mail.mcgill.ca

Anita Wdowicz
School of Chemistry and Conway Institute for
Biomolecular and Biomedical Science
University College Dublin, Dublin
Republic of Ireland

Claude Verdier
University Grenoble Alpes, CNRS, LIPhy
Grenoble, France
claude.verdier@univ-grenoble-alpes.fr

Ellen Zelinsky
Department of Biology and Center for
Lignocellulose Structure and Formation, The
Pennsylvania State University, University
Park, PA, USA

Joanna Zemla
Institute of Nuclear Physics, Polish Academy
of Sciences, Kraków, Poland
Joanna.Zemla@ifj.edu.pl

Marta Zarzycka
Chair of Medical Biochemistry, Jagiellonian
University Medical College, Kraków, Poland
marta.lydka@uj.edu.pl

M. Lekka, D. Navajas, A. Podestà, M. Radmacher

1 Introduction

Europe faces crucial health challenges due to demographic aging, an increase of sedentary- and nutrition-linked problems and the emergence of infectious diseases. The population of Europeans aged 65 and more is expected to double over the next 50 years, with the subsequent increase of chronic diseases, including age-related co-morbidities, like neurodegenerative diseases and cancer. *Early detection and diagnosis of diseases* provide essential tools for a better quality of life for aged people by allowing improved prognosis and personalized healthcare to optimize therapeutic strategies.

Nanotechnology allows the manipulation and inspection of matter at the nanometer scale with unprecedented sensitivity and spatial resolution, also providing new perspectives for the investigation of biological systems. The nanoscale investigation of cells and tissues revealed that *specific diseases have well-pronounced mechanical fingerprints*. For example, cancerous cells have shown to be typically significantly softer than normal cells, while the extracellular matrix (ECM) of cancerous tissue is typically stiffer than in normal cases. It is not surprising since anatomopathologists and histologists in the clinics are used to observe and often to sense directly by palpation, significant mechanical changes occurring in diseased tissues and organs as a consequence of the progression of the pathology.

Several nanoscale experimental techniques such as atomic force microscopy (AFM), nanoindenters, magnetic or optical tweezers, super-resolution photonic microscopy, the use of micropatterns and micropillars, and microfluidics have allowed to reliably probe the mechanical properties of cells and tissues within the research conducted at the laboratory level. Quantitative cell and tissue (nano)mechanobiology offer the possibility of adding to the growing list of molecular markers a new class of markers based on physical properties, specifically on suitable elastic and viscoelastic parameters (a short historical perspective on mechanobiology is presented at the end of this section). Therefore, it is appropriate to say that *a new paradigm arose – the use of the mechanical fingerprint of cells and tissues to detect and diagnose diseases*. Crucial, for this purpose, is paving the road to bring nanomechanical tests to the clinic.

Among several approaches that have been developed, AFM presents some unique advantages.

It combines topographical information with force mapping allowing to measure adhesive and mechanical (rheological) properties of cells and tissues. In addition, when using functionalized probes, specific molecular interactions can be probed with high spatial and force resolution. On the other hand, it is fair to note that AFM is still not popularized within the biology and medical community and is not yet included in the toolbox of cell mechanobiologists. Despite the reported successes of AFM in the study of biological systems and its potentialities, still to be fully developed in the

https://doi.org/10.1515/9783110640632-001

research milieu, the mechanical phenotyping of clinically relevant samples aimed at producing diagnostic cues requires *novel, clinic-oriented tools, featuring ease of use and high throughput*, which are still to be developed, although interesting products are appearing on the market. At the same time, the *standardization of procedures for both preparation and mechanical testing of clinical samples* must be implemented; when framed into the clinical environment, these objectives represent far greater challenges compared to when they are pursued among different research laboratories, an already complex task that has been only partially fulfilled to date.

Given that innovation in the health field proceeds through the development of advanced knowledge, novel methodologies and technologies, and their application to human diseases in the places where diseases are confronted, the clinics, the tools to support this challenging strategy, are interdisciplinary and intersectoral research, cooperation through networking, and effective communication and sharing of methodologies. A recently funded EU project, Phys2BioMed,[1] aims at accomplishing these tasks through the interdisciplinary and cross-sectoral training of a team of early-stage researchers, taking advantage of a network of research institutions, companies, and hospitals.

This book is aimed at collecting and summarizing, to the primary benefit of young researchers but also of senior scientists involved in interdisciplinary studies, the present knowledge on the fundamental biomechanical markers of diseases, like cancer, and on the experimental techniques that are used to characterize the mechanical properties of cells and tissues; we emphasized the application of these techniques to the study of clinically relevant samples.

1.1 History of Biomechanical Investigations of Cells and Tissues

Or The Long Route from Biomechanics via Mechanobiology to Mechanics in Diseases

Biomechanics, that is, employing principles from mechanics to describe biological systems has a very long tradition. The possibly first, clear, and prominent example was the physics behind blood circulation. Even the concept that blood circulates in the first place was, a (then controversially discussed) consequence of the simple fact that the heart is pumping more than 200 kg of blood per hour, raised by Harvey (1628). The shear amount suggests that there has to be a circular network, which at that time was

---

1 Phys2BioMed, https://www.phys2biomed.eu/, a EU MSCA-ITN-ETN project, has received funding from the European Union's Horizon 2020 research and Innovation program under the H2020-MSCA-ITN-2018 grant agreement no. 812772.

not apparent, since the connection between the arterial and the venous system could not be seen: capillaries are just too small to be detected by the naked eye and optical microscopy was in its infancy then. This controversy could only be solved once optical microscopy became available. This early example shows how important the connection between developing new concepts and the availability of appropriate techniques is. Harvey can be considered as a disciple of Galileo, who is well known for his contributions to astronomy, but actually started studying medicine (Fung 1993) or, as we would call it nowadays, physiology. Other early examples of biomechanics are the work of Young (1800), understanding the generation of sound, which relies strongly on the elastic properties of the vocal cords, or understanding the function of the lung, which is an interplay of the mechanical properties of lung tissue (most importantly its very high extensibility and the hydrodynamics of airflow). Another more modern example is the biomechanics of our locomotor system, that is, the functional interplay of muscles, tendons, and the skeleton. This – rather macroscopic – application of mechanics in biology can be understood as applying the laws of mechanics, as a part of physics, and the concepts from engineering (envisioning the heart as a pump with valves, describing our locomotor system as a combination of forces and levers) to better understand how biological systems work.

To get a better insight into how mechanical aspects of biological systems are actually generated or created a more microscopic or even molecular understanding of mechanics in biology is needed. This more modern interpretation of the theme, which now is called mechanobiology, describes how active molecules (like the motor enzymes myosin or flagella in single-celled organisms) or structural molecules (like the actin cytoskeleton or polymeric proteins in the ECM) organize themselves in cells or tissues, generate forces, or are affected by external forces (deform) or even sense external forces (mechanosensation). These modern concepts require the availability of microscopic and structural techniques to characterize tissues, cells, and macromolecules like polymers down to nanometer dimensions. It also requires tools to sense and apply mechanical forces at increasingly smaller length and force scales.

Thus, the field profited and progressed from various techniques, which became available to study mechanical properties of cells and tissues. To highlight this idea (not giving a concise overview here), we just present key pioneering concepts here: (1) micropipette aspiration (Hochmuth 2000), first developed by Hochmuth (1993) to characterize red blood cells and later applied by Evans (1989) to measure the membrane properties of granulocytes. This technique, which was the first to probe viscoelastic properties of cells, and thus is conceptionally very important. However, its applicability was limited since it could not apply to adherent cells and tissues. (2) Scanning acoustic microscopy overcomes these limitations since it could probe mechanical properties of three-dimensional samples of cells and tissues, as shown by Bereiter-Hahn and his group (Kundu 1991). However, since the shear modulus has been derived from the speed of sound, which mainly reflects the properties of water, only affected to some degree by the polymeric network within and between cells, this technique has not found widespread applications. (3) Finally, the cell poker, developed by Eliot Elson (1988), allowed to locally probe the mechanical properties of cell samples with a cylindrical millimeter-sized probe. This setup resembles the current AFMs to a large extent, albeit being more macroscopic in scales. Elson also coined the notion that the mechanical properties of cells reflect the state and composition of the cytoskeletal network. From there, it was only a short step to the notion that mechanical properties of cells and tissues reflect the state of cells and anything, which will change the state of cells, including diseases, which always will affect the metabolisms of cells, could also be detected by changes in the mechanical fingerprint of cells. This final jump in our way how to think about cell mechanics was largely due to the availability of the AFM, which allowed for the first time to measure viscoelastic properties of living cells at subcellular resolution without interfering with the function and activity of cells. So, it basically combines all features from our above three examples: AFM allows to characterize viscoelastic properties of cells and tissues as micropipette suction does; it is a microscopy technique with even higher spatial resolution than the scanning acoustic microscope; and it employs the direct interaction between a probe in physical contact with the sample as the cell poker does.

The AFM has been invented in 1986 by Binnig et al. (1986) and could be understood as a variant of the scanning tunneling microscope just shortly before being invented by Binnig and Gerber. Initially, only serving the purpose to extend the use of scanning probe techniques beyond electrically conducting samples, it was soon understood by Paul Hansma that this opens the route to visualize biological samples in physiological conditions (Drake 1989). In a very early application (Weisenhorn 1993), a height anomaly (due to compression) in soft samples was discovered, which was soon understood and used to determine the elastic properties of living cells with a high spatial resolution (Radmacher 1995). In 1996, in a short note, Goldmann et al. (1996) demonstrated that viscoelastic properties of cells could be related to the presence of vinculin, a protein participating in the focal adhesions being the link between a cell and the ECM. The vinculin-deficient cells were softer, which was attributed to remodeling of the cell cytoskeleton. As a consequence of increasing knowledge on the viscoelastic properties of cells and tissues (Alcaraz 2003), the idea of mechanics in diseases evolved. The earliest implementation of this idea was the evidence that cancer cells tend to be softer than normal cells by Lekka et al. (1999).

Most important is to envision cells as active matter, which not only react to external mechanical stimuli but also generate forces and stresses within cells and tissues (Kasza et al. 2007). The modern – molecular-based – understanding of cellular function and metabolisms, the knowledge on the complex molecular interactions in active cellular polymeric networks (intracellularly the cytoskeleton and extracellularly the ECM), the central role of this network in processes like migration and force generation, but also in responding to external mechanical cues make it conceivable that diseases, which could be most generally be defined as abnormalities in the metabolism of cells, will result in a specific mechanical fingerprint. Thus, mechanics could serve as an assay to define and determine the state of a cell, or even the state of a disease, which generalizes Elliot Elson's view that mechanics could serve to define the state of the cytoskeleton. Usually, assays used for detecting diseases will be molecular-based and highly specific in detecting certain molecular markers of a particular disease. The specificity and the sensitivity of these molecular-based assays are a clear advantage. However, these assays can only detect what they are aiming for. A non-specific assay, like the mechanical fingerprint, has the advantage of an unbiased, generally applicable test, which can give hints for diseases, which you were not looking for in the first place. This new paradigm in designing a mechanical assay for detecting diseases can and will add to our knowledge on diseases in general and may hopefully also find its way from research to biomedical applications in the clinic in the not so far future.

References

Alcaraz, J., L. Buscemi, M. Grabulosa, X. Trepat, B. Fabry, R. Farré and D. Navajas (2003). "Microrheology of human lung epithelial cells measured by atomic force microscopy." Biophysical Journal **84**: 2071–2079.

Binnig, G., C. F. Quate and C. Gerber (1986). "Atomic force microscope." Physical Review Letters **56**(9): 930–933.

Drake, B., C. B. Prater, A. L. Weisenhorn, S. A. C. Gould, T. R. Albrecht, C. F. Quate, D. S. Cannell, H. G. Hansma and P. K. Hansma (1989). "Imaging crystals, polymers and biological processes in water with AFM." Science **243**: 1586–1589.

Elson, E. L. (1988). "Cellular mechanics as an indicator of cytoskeletal structure and function." Annual Review of Biophysics and Biophysical Chemistry **17**: 397–430.

Evans, E. and A. Yeung (1989). "Apparent viscosity and cortical tension of blood granulocytes determined by micropipet aspiration." Biophysical Journal **56**(1): 151.

Fung, Y. C. (1993). Biomechanics – mechanical properties of living tissues. New York, Springer.

Goldmann, W. H. and R. M. Ezzell (1996). "Viscoelasticity in wild-type and vinculin-deficient (5.51) Mouse F9 embryonic carcinoma cells examined by atomic force microscopy and rheology." Experimental Cell Research **266**: 234–237.

Harvey, W. (1628). Exercitatio anatomica de motu cordis et sanguinis in animalibus. Founders of experimental physiology. J. B. W. Blasius, K. Kramer. München.

Hochmuth, R. M. (1993). "Measuring the mechanical properties of individual human blood cells." Journal of Biomechanical Engineering **115**: 515–519.

Hochmuth, R. M. (2000). "Micropipette aspiration of living cells." Journal of Biomechanics **33**(1): 15–22.

Kasza, K. E., A. C. Rowat, J. Liu, T. E. Angelini, C. P. Brangwynne, G. H. Koenderink and D. A. Weitz (2007). "The cell as a material." Current Opinion in Cell Biology **19**: 101–107.

Kundu, T., J. Bereiter-Hahn and K. Hillmann (1991). "Measuring elastic properties of cells by evaluation of scanning acoustic microscopy V(z) values using simplex algorithm." Biophysical Journal **59**(5): 1194–1207.

Lekka, M., P. Laidler, D. Gil, J. Lekki, Z. Stachura and A. Z. Hrynmiewicz (1999). "Elasticity of normal and cancerous human bladder cells studied by scanning force microscopy." European Biophysics Journal: EBJ **28**(4): 312–316.

Radmacher, M., M. Fritz and P. K. Hansma (1995). "Measuring the elastic properties of biological materials with the atomic force microscope." Biophysical Journal **68**(2): A139.

Weisenhorn, A. L., M. Khorsandi, S. Kasas, V. Gotozos, M. R. Celio and H. J. Butt (1993). "Deformation and height anomaly of soft surfaces studied with the AFM." Nanotechnology **4**: 106–113.

Young, T. (1800). "Outlines of experiments and inquiries respecting sound and light. By Thomas Young, M. D. F. R. S. In a Letter to Edward Whitaker Gray, M. D. Sec. R. S." Philosophical Transactions of the Royal Society of London **90**: 106–150.

Soft Matter Mechanics

Daniel Navajas, Manfred Radmacher

2.1 Introduction to Viscoelasticity

2.1.1 Introduction

Mechanical properties of cells are a major determinant of critical cell functions including motility, contraction, gene expression, division, and differentiation. Moreover, cells sense and actively respond to the mechanical features of their microenvironment. Therefore, precise knowledge of the mechanical properties of cells and tissues is needed to understand their function in health and disease.

Elastic materials deform instantaneously in response to external forces and immediately recover their initial shape when unloaded. Moreover, these materials show time-independent mechanical behavior with no energy dissipation. On the other hand, the application of a constant shear force to a fluid induces constant velocity flow and energy dissipation. Although cell and tissue mechanics are usually studied assuming that they are elastic materials, they exhibit many viscoelastic features. Materials exhibiting both solid- and liquid-like features are known as viscoelastic materials.

This chapter describes the general mechanical properties of linear viscoelastic materials, common tests used to characterize their mechanical behavior, and simple models to interpret cell and tissue viscoelasticity.

2.1.2 Hookean Elastic Materials

When a normal force F is applied to a face of a linear elastic parallelepiped of area A with relaxed length L_0, it deforms to length L (Figure 2.1.1). The force per unit of area is defined as normal stress, σ:

$$\sigma = \frac{F}{A} \tag{2.1.1}$$

Deformation can be characterized by the stretch ratio, λ:

Acknowledgments: This work was supported in part by the Spanish Ministry of Economy and Competitiveness (DPI2017-83721-P), Generalitat de Catalunya (CERCA Program), the Marie Sklodowska-Curie Action, Innovative Training Networks 2018, EU grant agreement no. 812772.

Daniel Navajas, Unitat de Biofísica i Bioenginyeria, Facultat de Medicina i Ciències de la Salut, Universitat de Barcelona, Institute for Bioengineering of Catalonia, Barcelona, Spain.
dnavajas@ub.edu
Manfred Radmacher, Institute of Biophysics, University of Bremen, Bremen, Germany

https://doi.org/10.1515/9783110640632-002

$$\lambda = \frac{L}{L_o} \tag{2.1.2}$$

or by the relative strain ε:

$$\varepsilon = \frac{\Delta L}{L_o} \tag{2.1.3}$$

Material properties are described by constitutive equations defined as stress–strain relationships. A Hookean elastic solid defines a material with a proportional stress–strain relationship when subjected to uniaxial stretching as follows:

$$\sigma = E\varepsilon \tag{2.1.4}$$

where the constant of proportionality E is Young's modulus of the material.

When a Hookean elastic solid is subjected to shear stress, $\tau = F/A$, the deformation is characterized by the angle α (Figure 2.1.1) with the constitutive equation:

$$\tau = G \tan \alpha \tag{2.1.5}$$

where G is the shear modulus of elasticity. Both elastic moduli, E and G, are related as follows:

$$G = \frac{E}{2(1+v)} \tag{2.1.6}$$

where v is Poisson's ratio, which is 0.5 in incompressible materials.

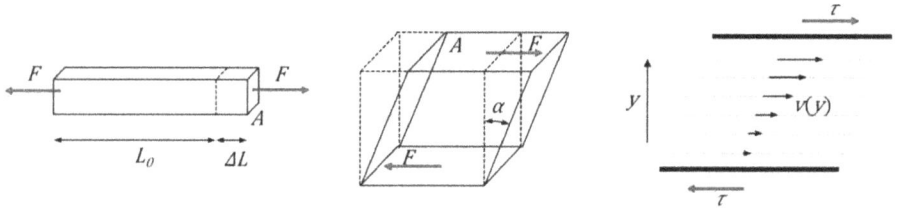

Figure 2.1.1: Deformation of materials in response to force. Left: Hookean elastic solid under normal force. Centre: Hookean elastic solid under shear force. Right: Newtonian viscous liquid. F: force, A: area, L_0: relaxed length, ΔL: length increase, τ: shear stress, v: velocity.

2.1.3 Newtonian Viscous Fluids

When a Newtonian viscous liquid is placed between a stationary plate and a parallel plate subjected to shear stress, the velocity of the parallel layers of fluid increases linearly with its distance to the stationary plate (Figure 2.1.1). Newtonian viscous liquids obey the following constitutive equation:

$$\tau = \eta \frac{dv}{dy} \tag{2.1.7}$$

where η is the coefficient of viscosity.

2.1.4 Viscoelastic Materials

Hookean elastic solids and Newtonian viscous fluids are ideal materials. However, most biological materials are viscoelastic, that is, they exhibit both solid- and fluid-like features. In contrast with elastic solids, the stress–strain relationships of viscoelastic materials depend on time. When a viscoelastic body is suddenly stretched, and the strain is maintained constant afterward, the resulting stress decreases with time. This behavior is called stress relaxation. If a constant force is suddenly applied, the strain increases with time, a behavior known as creep. Moreover, when a viscoelastic material is subjected to cyclic strain, it exhibits a stress–strain hysteresis loop with different σ–ε plots during loading and unloading. The area within the loop is the energy dissipated per cycle.

2.1.5 Spring–Dashpot Mechanical Models of Viscoelastic Materials

The behavior of linear viscoelastic materials can be described by simple mechanical models consisting of a combination of linear springs and dashpots (Figure 2.1.2). The deformation of a linear spring follows Hooke's law:

$$\sigma = E\varepsilon \tag{2.1.8}$$

where ε is the strain of the spring and E is the spring constant. The strain rate of the dashpot is proportional to the applied force:

$$\sigma = \eta \dot{\varepsilon} \tag{2.1.9}$$

where η is dashpot viscosity. Note that for infinitely slow deformations, $\dot{\varepsilon} \to 0$, and the dashpot does not generate viscous force ($\sigma \to 0$). By contrast, if the deformation is applied infinitely fast, $\dot{\varepsilon} \to \infty$ and the dashpot behaves as a rigid element.

Figure 2.1.2 depicts creep and stress relaxation responses of commonly used spring–dashpot viscoelastic models. The Maxwell model is defined as a spring arranged in series with a dashpot. A sudden application of a force step of amplitude σ_0 results in a continuous increase in spring strain (Fung, 1993):

$$\varepsilon(t) = \sigma_0 \left(\frac{1}{E} + \frac{1}{\eta} t \right) H(t) \tag{2.1.10}$$

where $H(t)$ is the Heaviside or unit step function, which is 0 for $t < 0$, 1/2 for $t = 0$ and 1 at $t > 0$. Note that a force step generates an immediate strain of the spring, σ_0/E, which is followed by creep of the dashpot with constant speed σ_0/η.

On the other hand, a sudden application of a strain step of amplitude ε_0 generates a peak in stress ($\varepsilon_0 E$) followed by an exponential decay with time constant $\tau = \eta/E$:

$$\sigma(t) = \varepsilon_0 E e^{-t/\tau} H(t) \tag{2.1.11}$$

The Voight model is composed of a parallel combination of a spring and a dashpot. Application of a force step results in an exponential increase in elongation with time constant, $\tau = \eta/E$, until reaching a constant deformation σ_0/E:

$$\varepsilon(t) = \frac{\sigma_0}{E} \left(1 - e^{-t/\tau} \right) H(t) \tag{2.1.12}$$

The stress relaxation in response to an elongation step is

$$\sigma(t) = \varepsilon_0 \eta \delta(t) + \varepsilon_0 E H(t) \tag{2.1.13}$$

where $\delta(t)$ is the Dirac-delta or unit-impulse function. According to eq. (2.1.13), a strain step induces an instantaneous sharp peak in force followed by a constant stress, $\varepsilon_0 E$. In practice, however, the application of a sharp strain step is not possible, since the dashpot cannot be instantaneously stretched.

The standard linear solid is a three-element model consisting of a spring in parallel with a Maxwell body. The creep response of the standard linear solid to a force step is

$$\varepsilon(t) = \frac{\sigma_0}{E_0} \left[1 - \left(1 - \frac{\tau_r}{\tau_c} \right) e^{-t/\tau_c} \right] H(t) \tag{2.1.14}$$

where τ_r is the relaxation time for a strain step defined as

$$\tau_r = \frac{\eta}{E_1} \tag{2.1.15}$$

and τ_c is the retardation time for a stress step defined as

$$\tau_c = \tau_r \left(1 + \frac{E_1}{E_0} \right) \tag{2.1.16}$$

Note that $\tau_c > \tau_r$.

Application of a force step induces an immediate strain, $\sigma_0/(E_0 + E_1)$, followed by an exponential strain increase with a retardation time τ_c up to σ_0/E_0. On the other hand, the stress relaxation response to a strain step is

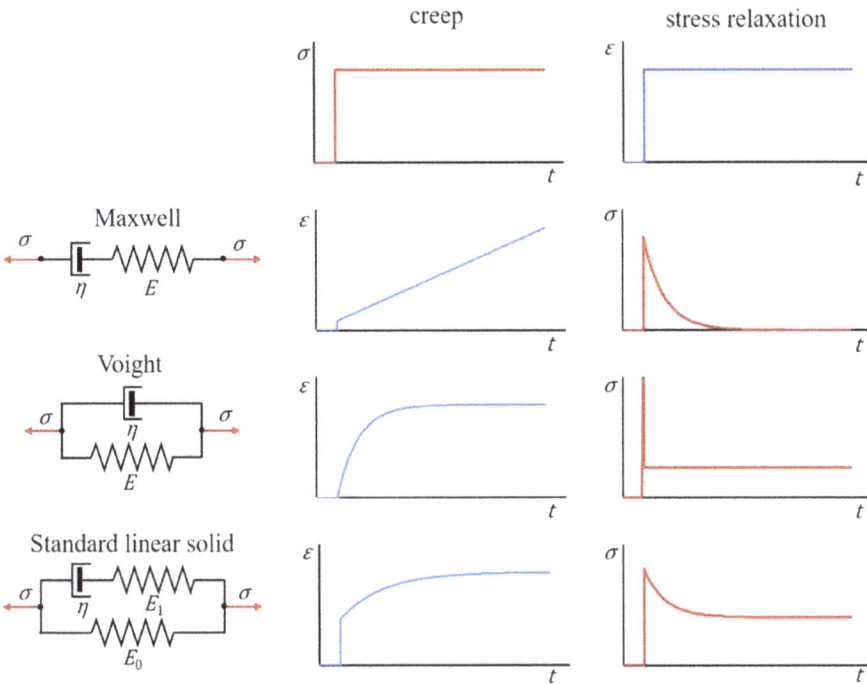

Figure 2.1.2: Creep and stress relaxation in spring–dashpot models of viscoelastic materials. Left: Commonly used models consisting of simple combinations of springs with elastic modulus, E, and dashpots with viscosity, η. Centre: Strain (ε) in response to a stress step (σ). Right: Stress in response to a strain step (stress relaxation). η: dashpot viscosity, E: spring constant.

$$\sigma(t) = \varepsilon_0 E_0 \left[1 - \left(1 - \frac{\tau_c}{\tau_r} \right) e^{-t/\tau_r} \right] H(t) \tag{2.1.17}$$

Immediately after the application of the strain step, the instantaneous elastic modulus of the standard linear solid is $E_0 + E_1$ inducing a sudden peak of stress of $\varepsilon_0(E_0 + E_1)$. Then, the stress decays exponentially with a relaxation time, τ_r, to a value of $E_0\varepsilon_0$, E_0 being the elastic modulus of the relaxed state.

2.1.6 Response to Sinusoidal Loading

When a linear viscoelastic material is subjected to sinusoidal strain:

$$\varepsilon(t) = \varepsilon_o \sin(\omega t) \tag{2.1.18}$$

ω being the angular frequency and ε_o the amplitude (Figure 2.1.3), the induced stress is also a sinusoidal function with the same frequency but out of phase with the stress:

$$\sigma(t) = \sigma_o \sin(\omega t + \phi) \tag{2.1.19}$$

where σ_o is the amplitude of the sinusoidal stress and ϕ is the phase shift between stress and strain:

$$\phi = 2\pi\Delta t/T \tag{2.1.20}$$

T being the period of the waveform. Therefore, the stress–strain plot is an elliptical hysteresis loop (Lakes, 2009). The area inside the loop is the energy dissipated in each cycle per unit volume of the material and is called damping energy ΔW:

$$\Delta W = \pi\sigma_a\varepsilon_a \sin\phi \tag{2.1.21}$$

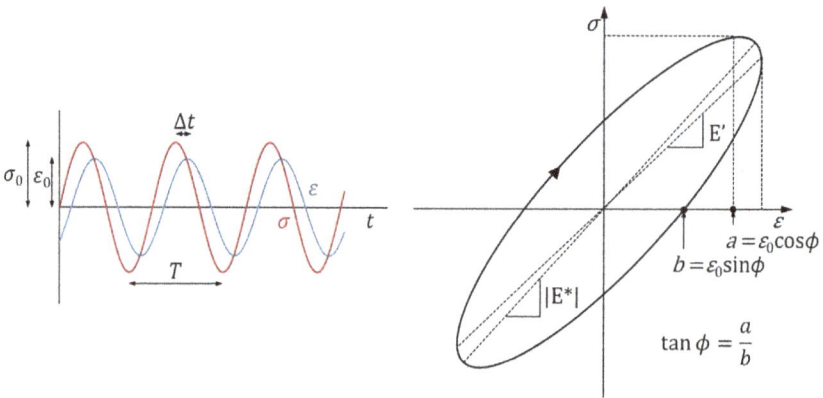

Figure 2.1.3: Response of a linear viscoelastic solid to sinusoidal load. Left: Sinusoidal stress σ with σ_0 amplitude in response to a sinusoidal strain ε with ε_0 amplitude. T is the period of the oscillation. Δt is time shift. Right: Stress–strain elliptical loop. $|E^*|$ and E' are the modulus and real part of the complex shear modulus E^*, respectively.

In an elastic material, $\phi = 0$ and $\Delta W = 0$. The stress–strain relationship can be characterized in the frequency domain with the complex elastic modulus defined as

$$E^*(\omega) = \frac{\sigma(\omega)}{\varepsilon(\omega)} \tag{2.1.22}$$

where $\sigma(\omega)$ and $\varepsilon(\omega)$ are Fourier transforms of stress and strain, respectively. $E^*(\omega)$ is a complex number:

$$E^*(\omega) = E'(\omega) + iE''(\omega) \tag{2.1.23}$$

where i is the imaginary unit $i = \sqrt{-1}$. The real part, $E'(\omega)$, is called elastic or storage modulus, and the imaginary part, $E''(\omega)$, is viscous or loss modulus. The ratio between imaginary and real parts of $E^*(\omega)$:

$$\tan\phi(\omega) = \frac{E''(\omega)}{E'(\omega)} \qquad (2.1.24)$$

is called loss tangent, mechanical damping, or hysteresivity and is a measure of the internal friction of the material. The loss tangent is taken as an index of the degree of solid- or liquid-like behavior. In a purely elastic material, $E''(\omega) = 0$ and $\tan\phi(\omega) = 0$. On the other hand, in a purely viscous material, $E'(\omega) = 0$ and $\tan\phi(\omega) = \infty$.

$E^*(\omega)$ can be measured with atomic force microscopy (AFM) by applying low amplitude sinusoidal oscillation around an operating indentation, δ_0. Taking the first two terms of the Taylor expansion of the tip–sample contact model and transforming into the frequency domain, the complex elastic modulus for a pyramidal tip of semi-included angle θ is (Rico et al., 2005)

$$E^*(\omega) = \frac{\sqrt{2}(1-\upsilon)}{\delta_0 \tan\theta} \frac{F(\omega)}{\delta(\omega)} \qquad (2.1.25)$$

and for a spherical tip with a radius R

$$E^*(\omega) = \frac{1-\upsilon}{4(R\delta_0)^{1/2}} \frac{F(\omega)}{\delta(\omega)} \qquad (2.1.26)$$

$E^*(\omega)$ of a spring is frequency independent with $E'(\omega) = E$ and $E''(\omega) = 0$. On the other hand, a dashpot has $E'(\omega) = 0$ and $E''(\omega) = \eta\omega$.

Since a Voight model is composed of a spring and dashpot arranged in parallel, its complex elastic modulus is

$$E^*(\omega) = E + i\eta\omega \qquad (2.1.27)$$

The complex elastic modulus of a Maxwell model is

$$E^*(\omega) = \frac{E\tau^2\omega^2 + iE\tau\omega}{1+\tau^2\omega^2} \qquad (2.1.28)$$

The standard linear solid has a complex elastic modulus:

$$E^*(\omega) = E_0 + \frac{E_1\tau_r^2\omega^2 + iE_1\tau_r\omega}{1+\tau_r^2\omega^2} \qquad (2.1.29)$$

The loss tangent presents a peak at a characteristic frequency, $\omega_0 = 1/\sqrt{\tau_r\tau_c}$ (Figure 2.1.4). The loss modulus exhibits a peak slightly shifted to higher frequencies. Both the storage modulus and the modulus of $E^*(\omega)$ ($|E^*(\omega)|$) increase from E_0 when $\omega \to 0$ to $E_0 + E_1$ when $\omega \to \infty$ with the faster growth around ω_0.

Viscoelastic behavior of materials can be also expressed in terms of the complex shear modulus, $G^*(\omega)$, defined as the ratio between shear stress and angular

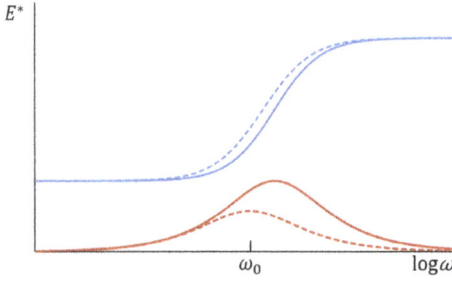

Figure 2.1.4: Complex elastic modulus $E^*(\omega)$ of a standard linear solid. Solid blue line: elastic modulus $E'(\omega)$. Solid red line: viscous modulus $E''(\omega)$. Dashed blue line: modulus of the complex elastic modulus $|E^*(\omega)|$. Dashed red line: loss tangent $\tan\phi$. ω_0 is the characteristic frequency.

deformation in the frequency domain. Transforming eq. (2.1.6) to the frequency domain $G^*(\omega)$ and $E^*(\omega)$ are related as

$$G^*(\omega) = \frac{E^*(\omega)}{2(1+\upsilon)} \tag{2.1.30}$$

2.1.7 Power-Law Rheology

Spring–dashpot viscoelastic models assume that the material can be characterized by discrete time constants or characteristic frequencies. However, oscillatory measurements performed in cells and tissues over a wide frequency range revealed scale-free behavior, with the storage and loss moduli increasing with frequency as a weak power law of frequency (Fabry et al., 2001, Alcaraz et al., 2003, Luque et al., 2013). This rheological behavior can be characterized by a frequency dependence of the complex elastic modulus as

$$E^*(\omega) = A^*(i\omega)^\alpha \tag{2.1.31}$$

where $i^\alpha = \cos(\pi\alpha/2) + i\sin(\pi\alpha/2)$. Thus, the storage and loss moduli increase with frequency as a power law with the same exponent:

$$E'(\omega) = A\cos(\pi\alpha/2)^*\omega^\alpha \tag{2.1.32}$$

$$E''(\omega) = A\sin(\pi\alpha/2)^*\omega^\alpha \tag{2.1.33}$$

where A is a scale factor for the storage and loss moduli. Equation (2.1.31) implies a frequency-independent loss tangent and coupled with the power-law exponent as:

$$\frac{E''(\omega)}{E'(\omega)} = \tan\frac{\pi\alpha}{2} \tag{2.1.34}$$

The loss modulus of cells and tissues has been reported to exhibit a steeper frequency dependence at high frequencies (Fabry et al., 2001, Alcaraz et al., 2003). A Newtonian viscous term with apparent viscosity η could be added to eq. (2.1.31) to account for the stronger frequency dependence of the loss modulus:

$$E^*(\omega) = A^*(i\omega)^\alpha + i\eta\omega \tag{2.1.35}$$

Alternatively, cell and tissue rheology can be described over a broad frequency range with a linear superposition of two power laws (Luque et al., 2013, Andreu et al., 2014):

$$E^*(\omega) = A^*(i\omega)^\alpha + B^*(i\omega)^{3/4} \tag{2.1.36}$$

Accordingly, the rheological behavior at low frequencies can be characterized as a weak power law with exponent α consistent with a soft glassy regime. The steeper frequency dependence of the loss modulus at high frequencies can be accounted for by a power-law regime with an exponent of 3/4, suggesting an increasing contribution of entropic elasticity.

Noteworthy, a power-law frequency behavior with exponent α (eq. (2.1.31)) corresponds in the time domain to a creep response with strain increasing over time as a power law (Desprat et al., 2005):

$$\varepsilon(t) = \varepsilon_0 t^\alpha \tag{2.1.37}$$

where ε_0 is a scale factor. The corresponding stress relaxation response also follows a power law with stress decaying with time as (Efremov et al., 2017)

$$\sigma(t) = \sigma_0 t^{-\alpha} \tag{2.1.38}$$

σ_0 being a scale factor for stress relaxation.

2.1.8 Discussion

Cell and tissue rheology are usually interpreted in terms of models consisting of a few springs and dashpots (Darling et al., 2006, Rianna and Radmacher, 2017). These simple models exhibit a single relaxation time constant and a single characteristic frequency. Therefore, most of the change in creep and stress relaxation occurs over one decade of time (Figure 2.1.2). Roughly speaking, the scales of experimental time and frequency windows can be related as $t = 1/\omega$. Thus, most of the change in the complex elastic modulus is limited to one decade of frequency (Figure 2.1.4). Models with a single time constant can fit creep and stress relaxation data recorded in an experimental time window expanding a time decade or oscillatory data recorded in a frequency band of a frequency decade. The standard linear solid model can be generalized by including additional Maxwell elements arranged in parallel, with a spring to account for more time constants and characteristic frequencies (Moreno-Flores et al., 2010). It should be noted that identification of multiple time constants or characteristic frequencies requires measurement over a wide time or frequency window. However, complex elastic modulus measurements performed in cells and tissues with AFM and magnetic tweezers over a broad frequency band of several decades revealed scale-free power-law viscoelasticity corresponding to a continuous distribution of relaxation times (Fabry et al., 2001, Alcaraz et al., 2003, Luque et al., 2013, Rebelo et al., 2014, Rother et al., 2014, Hecht et al., 2015, Rigato et al., 2017).

Creep and stress relaxation functions are defined as the response to an instantaneous change in stress and strain, respectively. Although small sharp changes in strain have been applied to measure the stress relaxation response of cells (Yango et al., 2016), a linear strain ramp followed by a constant level is usually employed (Darling et al., 2006, Moreno-Flores et al., 2010, Hecht et al., 2015). As a result of the deviation from the ideal strain step, the stress also deviates from the ideal response. Therefore, linear strain loading allows us to accurately compute creep and stress relaxation function only for times one decade longer than the loading rise time. It should be also noted that a constant strain level after the ramp is not easily maintained with conventional AFM devices. Stress relaxation during constant cantilever displacement progressively reduces cantilever deflection, which results in a slight decrease in indentation. By contrast, a constant force level can be readily obtained by operating the AFM in force control mode. Interestingly, complex elastic modulus measured in the frequency domain as the ratio of stress–strain Fourier transforms does not require pure sinusoidal oscillation. In fact, the complex elastic modulus can be measured simultaneously over a broad bandwidth by applying an oscillatory signal composed of several sinusoidal oscillations. However, it should be noted that stress relaxation data at short times and complex elastic modulus data at high frequencies measured with AFM must be corrected for the hydrodynamic drag force in the cantilever (Alcaraz et al., 2002).

Cell and tissue rheology can be characterized in the time domain by measuring creep or stress relaxation functions. The parameters of simple spring–dashpot models can be estimated by fitting creep or stress relaxation recordings with the theoretical function of the model. Alternatively, the rheological behavior can be characterized in the frequency domain by applying sinusoidal stress or strain loading. In linear materials, time-domain response can be computed from the frequency-domain response using Laplace analysis. Therefore, assuming linearity, creep, stress relaxation, and sinusoidal loading assays can provide equivalent information of cell and tissue viscoelasticity. Moreover, methods to compute viscoelastic parameters from conventional force–displacement curves measured with AFM have been developed (Rebelo et al., 2013, Efremov et al., 2017). However, cells and tissues are not pure linear materials (Kollmannsberger and Fabry, 2011, Jorba et al., 2019). Assuming quasilinear mechanics (Fung, 1993), nonlinear viscoelasticity can be readily characterized by superimposing small-amplitude strain and stress steps or sinusoidal oscillation at different levels of a given operating strain or stress (Alcaraz et al., 2003). Additional assessment of nonlinearity can be obtained by measuring cell or tissue viscoelasticity under different levels of global sample stretch (Kollmannsberger and Fabry, 2011, Jorba et al., 2019).

References

Alcaraz, J., L. Buscemi, M. Grabulosa, X. Trepat, B. Fabry, R. Farre and D. Navajas (2003). "Microrheology of human lung epithelial cells measured by atomic force microscopy." Biophysical Journal **84**: 2071–2079.

Alcaraz, J., L. Buscemi, M. Puig-de-morales, J. Colchero, A. Baro and D. Navajas (2002). "Correction of microrheological measurements of soft samples with atomic force microscopy for the hydrodynamic drag on the cantilever." Langmuir **18**: 716–721.

Andreu, I., T. Luque, A. Sancho, B. Pelacho, O. Iglesias-Garcia, E. Melo, R. Farre, F. Prosper, M. R. Elizalde and D. Navajas (2014). "Heterogeneous micromechanical properties of the extracellular matrix in healthy and infarcted hearts." Acta Biomaterialia **10**(7): 3235–3242.

Darling, E. M., S. Zauscher and F. Guilak (2006). "Viscoelastic properties of zonal articular chondrocytes measured by atomic force microscopy." Osteoarthritis and Cartilage **14**(6): 571–579.

Desprat, N., A. Richert, J. Simeon and A. Asnacios (2005). "Creep function of a single living cell." Biophysical Journal **88**: 2224–2233.

Efremov, Y. M., W.-H. Wang, S. D. Hardy, R. L. Geahlen and A. Raman (2017). "Measuring nanoscale viscoelastic parameters of cells directly from AFM force-displacement curves." Scientific Reports **7**: 1541.

Fabry, B., G. N. Maksym, J. P. Butler, M. Glogauer, D. Navajas and J. J. Fredberg (2001). "Scaling the microrheology of living cells." Physical Review Letters **87**(14): 148102.

Fung, Y. C. (1993). Biomechanics: Mechanical properties of living tissues. New York, Springer.

Hecht, F. M., J. Rheinlaender, N. Schierbaum, W. H. Goldmann, B. Fabry and T. E. Schaffer (2015). "Imaging viscoelastic properties of live cells by AFM: Power-law rheology on the nanoscale." Soft Matter **11**(23): 4584–4591.

Jorba, I., G. Beltran, B. Falcones, B. Suki, R. Farre, J. M. Garcia-Aznar and D. Navajas (2019). "Nonlinear elasticity of the lung extracellular microenvironment is regulated by macroscale tissue strain." Acta Biomaterialia **92**: 265–276.

Kollmannsberger, P. and B. Fabry (2011). "Linear and nonlinear rheology of living cells." Annual Review of Materials Research **41**: 75–97.

Lakes, R. (2009). Viscoelastic materials. Cambridge, Cambridge University Press.

Luque, T., E. Melo, E. Garreta, J. Cortiella, J. Nichols, R. Farre and D. Navajas (2013). "Local micromechanical properties of decellularized lung scaffolds measured with atomic force microscopy." Acta Biomaterialia **9**(6): 6852–6859.

Moreno-Flores, S., R. Benitez, M. Vivanco and J. L. Toca-Herrera (2010). "Stress relaxation and creep on living cells with the atomic force microscope: A means to calculate elastic moduli and viscosities of cell components." Nanotechnology **21**(44): 445101.

Rebelo, L. M., J. S. de Sousa, J. M. Filho, J. Schape, H. Doschke and M. Radmacher (2014). "Microrheology of cells with magnetic force modulation atomic force microscopy." Soft Matter **10**(13): 2141–2149.

Rebelo, L. M., J. S. de Sousa, J. Mendes Filho and M. Radmacher (2013). "Comparison of the viscoelastic properties of cells from different kidney cancer phenotypes measured with atomic force microscopy." Nanotechnology **24**(5): 055102.

Rianna, C. and M. Radmacher (2017). "Comparison of viscoelastic properties of cancer and normal thyroid cells on different stiffness substrates." European Biophysics Journal **46**(4): 309–324.

Rico, F., P. Roca-Cusachs, N. Gavara, R. Farre, M. Rotger and D. Navajas (2005). "Probing mechanical properties of living cells by atomic force microscopy with blunted pyramidal cantilever tips." Physical Review E **72**: 021914.

Rigato, A., A. Miyagi, S. Scheuring and F. Rico (2017). "High-frequency microrheology reveals cytoskeleton dynamics in living cells." Nature Physics **13**: 771–775.

Rother, J., H. Nöding, I. Mey and A. Janshoff (2014). "Atomic force microscopy-based microrheology reveals significant differences in the viscoelastic response between malign and benign cell lines." Open Biology **4**(5): 140046.

Yango, A., J. Schape, C. Rianna, H. Doschke and M. Radmacher (2016). "Measuring the viscoelastic creep of soft samples by step response AFM." Soft Matter **12**(40): 8297–8306.

Leda Lacaria, Alessandro Podestà, Manfred Radmacher, Felix Rico

2.2 Contact Mechanics

The study of the deformation between two bodies that are in contact, upon application of force, can reveal relevant information about the mechanical properties of the two objects and their composition. This problem, first considered by Heinrich Hertz for two contacting spheres (Hertz, 1881, 1882), forms the basis of the mechanical tests that employ suitable indenters, including those performed by the atomic force microscopy (AFM) and commercial nanoindenters. To simplify the problem, one of the bodies can be considered as a rigid nondeformable probe and the other as an infinite half space, the sample, of which we want to study the elastic properties. The underlying formalism to determine the pressure, upon a given deformation, was generalized by Boussinesq, from which it takes its name: the Boussinesq's problem (Boussinesq, 1885, Love, 1939). The scientific field that studies the properties of materials through the force–deformation interaction between two bodies in contact is commonly called contact mechanics (Johnson, 1985).

In a simple configuration, relevant to mechanical measurements using AFM or nanoindenters, the approaching probe will apply a pressure on the sample, and due to the repulsion of atoms, which is basically a consequence of the Pauli exclusion principle, will cause its deformation. The ratio between the applied force and the deformation, or the applied stress and the resulting strain, is a quantitative description of the mechanical properties of the sample. Traditionally, theoretical models from contact mechanics are used to fit experimental data and obtain the mechanical properties of materials at macroscopic or mesoscopic scales.

Nowadays, the development of advanced techniques to manipulate rigid probes of nanoscale dimensions, such as AFM, nanoindenters, and optical and magnetic

Acknowledgments: The authors thank Andreas Janshoff and Javier López for the insightful discussions. We acknowledge the support of the European Union's Horizon 2020 research and innovation program under the Marie Skłodowska-Curie grant agreement no. 713750, project DOC2AMU. Also, this work has been carried out with the financial support of the Regional Council of Provence-Alpes-Côte d'Azur and the A*MIDEX (no. ANR- 11-IDEX-0001-02), funded by the Investissements d'Avenir project funded by the French Government and managed by the French National Research Agency (ANR). In addition, this project has received funding from the European Research Council (ERC, grant agreement no. 772257) and from the Human Frontier Science Program (HFSP, grant no. RGP0056/2018). The authors thank Elsevier for the permission of the free reuse of figure 3 in Wu et al for Figure 2.2.7.

Leda Lacaria, Felix Rico, Aix-Marseille Univ, INSERM, CNRS, LAI, Turing Centre for Living Systems, Marseille, France
Alessandro Podestà, Dipartimento di Fisica "Aldo Pontremoli" and C.I.Ma.I.Na, Università degli Studi di Milano, Milan, Italy
Manfred Radmacher, Institute of Biophysics, University of Bremen, Bremen, Germany

https://doi.org/10.1515/9783110640632-003

tweezers, has opened a new door for the study of the mechanics of cells and tissues, from the macroscale to the micro and nanoscale. Indeed, while traditional models of contact mechanics were principally used to describe the macroscopic mechanical properties of materials in mechanical engineering and material sciences, currently, contact mechanics has been rediscovered as a powerful tool to investigate the local, nanoscale mechanical properties of soft living materials such as cells and tissues, making contact mechanics well known also in the field of biophysics and mechanobiology (Radmacher et al., 1992).

In this chapter, we describe the most relevant contact models to analyze AFM force–deformation measurements to determine the mechanics of soft matter, including cells and tissues. We illustrate the different contact elastic and viscoelastic models, specifying the most common tip geometries and sample configurations. We provide a brief introduction to the theoretical tools required to obtain the described models, but leave the full development by referring to the original texts.

2.2.1 Elasticity and AFM Force–Deformation Measurements

Contact mechanics involves the determination of the pressure-deformation relationship between two bodies in contact. In conventional AFM measurements, an external force (F) is exerted by the rigid probe that is in contact with the sample, and a deformation or indentation (δ) is induced due to the pressure on the sample's surface.

A typical AFM measurement, therefore, is represented by a force-distance curve, where the distance is the z position of the probe relative to the sample, as shown in Figure 2.2.1.

When the probe is far away from the sample, the cantilever deflection, and thus the force, is zero. After the point of contact between the probe and the sample, the pressure of the probe starts to deform a soft sample and the force increases with the sample indentation.

The contact region starts from the point of contact between the probe and the sample, and ends at the point of maximum force and maximum indentation of the sample.

The precise determination of the contact point allows the conversion of the force–distance curve in a force–indentation curve that can be fitted with the expressions from contact mechanics models $F(\delta)$.

A typical force curve on a rigid material will be linear, where deflection and distance are identical. On a soft sample, like for example, a cell or a biological tissue, the slope will be smaller than 1 and often nonlinear, depending on the probe geometry, as a consequence of increasing contact area between the probe and the sample surface.

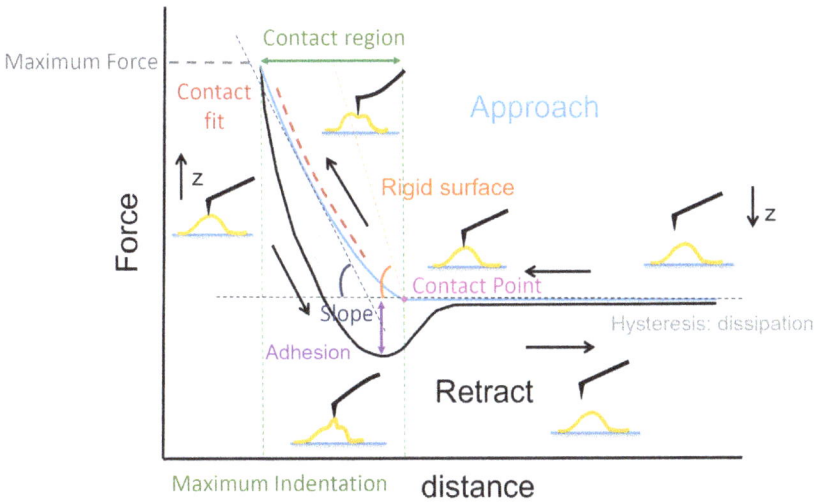

Figure 2.2.1: Schematic typical AFM measurement (force–distance curve) on a soft material, for example, a cell. In blue, the approach curve and, in black, the retract. In green is shown the contact region, starting from the contact point between the probe and the soft sample, and delimited by the point of maximum force and maximum indentation. The black and orange dashed lines show the typical difference in slope for a rigid surface (orange) and a soft surface (black). In violet, the adhesion effect on the retract curve; in gray, the hysteresis in the curve due to dissipation of energy.

Moreover, in biological samples and soft materials, in general, we can have energy dissipation or viscous effects, which will result in a difference, often termed hysteresis, between the approach and the retract curve.

In soft materials, adhesion between the probe and the sample is often present, which becomes visible as negative forces during retract.

Determining an analytical (or numerical) relation between the local stress and strain distribution in the contact region is the common route to derive a relation between the experimentally accessible force and indentation, allowing the deduction of the materials' properties. This interplay between the theoretical derivations and experimental data is the essence of contact mechanics.

The elasticity of a material is its tendency to resist deformation when subjected to an external force, combined with its ability to return to its original size and shape when the force is removed. This property of the material can be described by the Young's modulus (E), which is related to the shear modulus (G) and the bulk modulus (K), defined as the ratio between the stress (σ) and the strain (ε) in the uniaxial compression: $\sigma = E\varepsilon$, as already explained in detail in Chapter 2.1. A material is defined as elastic when its response to the application of an external force, in terms of deformation and recovery, is instantaneous and there is no energy dissipated. In the general theory of elasticity, the sample is solid and regarded as a continuous material. Therefore, when the body is not deformed, it is in thermal equilibrium. During the deformation of the body, there is a change in the shape of the boundary surface, the arrangement of the components of the solid is modified, and small internal forces are generated because the system tends to return to the state of equilibrium. These local forces are due to the repulsion of the molecules and components within the body and are termed internal stresses. The total force applied on the body is identical to the continuous sum, the integral, over all internal stresses developed inside the body upon deformation.

2.2.2 Contact Elastic Models

In this section we describe some of the most common contact elastic models, according to the geometry of the contact. In the context of AFM, contact models have been developed for those geometries that correspond to commercially available probes. All models in this section assume that a nondeformable probe of known geometry indents an infinite pure elastic, isotropic, linear half space of known Young's modulus E and Poisson's ratio v. The Poisson's ratio is the amount of transversal elongation divided by the amount of axial compression. When studying cells and tissues, the sample is often assumed to be incompressible (a good approximation for most elastomers and gels, as well as for cells and tissues), which leads to $v = 0.5$.

2.2.2.1 The Hertz Model

The application of mechanics to contact problems first began with Heinrich Hertz (1881), who, in his pioneering paper entitled "On the contact of elastic bodies," solved the problem of elastic contact between two spheres (two lenses). This model can be extended to a sphere in contact with an infinite half space, which is a common configuration of AFM measurements.

Hertz's approach was based on the following assumptions, which define Hertzian contact:

1) The surfaces are frictionless and perfectly smooth.
2) The surfaces are continuous.
3) The probe and sample are isotropic, axisymmetric, and show a linear elastic response.
4) The strains are very small in comparison to any relevant dimension of the body.
5) The bodies are considered as infinite half-spaces.
6) The surfaces do not show adhesion.

The assumption 1) allows us to state that only a normal pressure acts between the parts in contact. When the assumption 2) is held, the area of contact is much smaller than the characteristic dimensions of the contacting bodies, that is, the area of contact between the two bodies is close to zero, which implies that the pressures are finite. Indeed, even if the forces due to the repulsion of the atoms of the two bodies in contact become very large locally, the integral of these forces through an infinitely small area is always finite. According to the assumption 3), since the two surfaces are isotropic and axisymmetric, their common normal is parallel to the direction of the pressure that each body exerts on the other. Thus, the surface of contact, and also the surface of pressure, lies on the tangent plane xy of the two surfaces, and the normal is parallel to the z-direction. In this condition, the distance of any point of either surface from the common tangent plane is in the neighborhood of the point of contact, and can be described by a homogeneous quadratic function of x and y. This last assumption is important, since a sphere can only be described as a paraboloid of revolution for small deformations relative to the radius of curvature. Assumption 4) allows us to apply the linear theory of elasticity. Indeed, when the strains[1] in a continuum body are smaller than the body dimensions, the displacement gradient is smaller than 1, and the strain tensor can be linearized, according to the infinitesimal strain theory (Slaughter, 2002); instead, if the strains are comparable to the body dimensions, the finite strain theory or the plasticity theory should be used. Assumption 5) assures that the influence of the boundaries of the two bodies can be ignored. Finally, assumption 6) states that the contact is adhesionless, which allows to exclude forces other than the elastic ones present inside or outside the contact area.

The analysis presented further is based on the approach used by Dimitriadis et al. (2002) following Landau and Lifshitz (1986), as will be seen to become useful later on. In an analogy to AFM measurements, a spherical probe of radius R is assumed to apply a total force F in the z direction to a half-space, making contact at $z = 0$, as shown in Figure 2.2.2. The force is distributed over the contact circle of radius a centered at $r = 0$ (Figure 2.2.2), where r is the in-plane radial distance in the z-plane. The axial deformation field u_z can be obtained by integrating the product of the displacement profile for a point-concentrated force, that is, the Green's function G and the pressure distribution P_z over the contact area A:

$$u_z(r,z) = \iint_A P_z(r_s)G(s)dA \qquad (2.2.1)$$

---

[1] The strain, in general, is a tensor and it is defined as the change in an element of length when the body is deformed, that is, the gradient of the displacement vector ∇u. The displacement u of a point in a body due to the deformation is the distance between the coordinate of the point before and after the deformation: $\mathbf{u} = x_i' - x_i$; more details are described in Landau and Lifshitz (1986)).

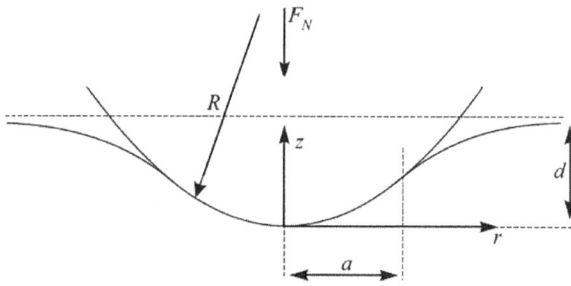

Figure 2.2.2: Schematic parabolic probe indenting an elastic infinite half space.

where the Green's function is given by

$$G(s) = -\frac{1+v}{2\pi Es}\left(\frac{z^2}{s^2} + 2(1-v)\right)$$ (2.2.2)

The distribution of pressure applied by the sphere was obtained by Hertz as

$$P_z(r_s) = p_0\sqrt{a^2 - r_s^2}$$ (2.2.3)

where $s = |\mathbf{r}| = (r^2 + r_s^2 - 2rr_s \cos\phi + z^2)^{1/2}$ is the distance between the source $(r_s, 0, 0)$ and the observation point (r, ϕ, z); being ϕ the azimuthal angle of the cylindrical coordinates, and $p_0 = (2E/\pi R)(1/(1-v^2))$.

If the deformation field in eq. (2.2.1) is assumed to conform with a paraboloid of revolution

$$u_z = \delta - \frac{1}{2R}r^2$$ (2.2.4)

the contact radius depends only on the indentation of the sample (δ) and the radius of curvature of the probe, and can be written in terms of the applied force and the mechanical parameters of the half-space,

$$a^2 = \delta R = \left(\frac{3(1-v^2)}{4E}FR\right)^{2/3}$$ (2.2.5)

Equation (2.2.5) can be rearranged in terms of the normal applied force as a function of the deformation or indentation $F(\delta)$:

$$F = \frac{4}{3}\frac{E}{1-v^2}\sqrt{R\delta^3}$$ (2.2.6)

Thus, according to the Hertz model, the relation between force and indentation depends only on the intrinsic properties of the material, the compressibility expressed by the Poisson's ratio v, the elasticity expressed by the Young's modulus E, and on the radius of curvature R.

The Hertz model can also be used, and was originally developed to describe the contact between two elastic spheres (with respective radii R_1 and R_2, Poisson's ratio v_1 and v_2 and Young's moduli E_1 and E_2) compressed by a force F, resulting in a contact area of radius a, as shown in Figure 2.2.3. In that case, an effective Young's modulus E_{eff} takes into account the elastic properties of both bodies and can be defined as follows:

$$\frac{1}{E_{eff}} = \frac{1-v_1{}^2}{E_1} + \frac{1-v_2{}^2}{E_2} \tag{2.2.7}$$

The effective radius R_{eff} is defined as follows:

$$\frac{1}{R_{eff}} = \frac{1}{R_1} + \frac{1}{R_2} \tag{2.2.8}$$

Assuming that one of the two bodies has infinite curvature radius $R_2 \to \infty$, the problem of the two spheres reduces to the problem of a half infinite elastic space, as shown in Figure 2.2.3, and the only geometrical variable becomes the radius R of the spherical probe. Moreover, if we consider the probe as infinitely rigid, $E_1 \gg E_2$, so $E_1 \to \infty$, the solution will depend only on the Poisson's ratio and the Young's modulus of the half infinite space.

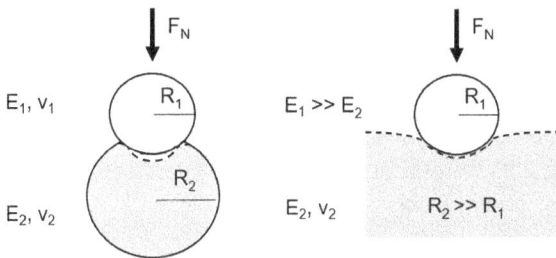

Figure 2.2.3: Schematic drawing of the Hertz contact problem: the general case of two spheres in contact and the particular case of a rigid spherical probe indenting an infinite half-space.

The Hertz model is largely used to measure the mechanical properties of materials, but to obtain reliable values of elasticity, it is important to carefully evaluate if the assumptions described above are valid. The distribution of stresses transmitted inside the deformed material depends on the area of contact between the rigid probe and the sample. Indeed, the normal force is the result of the local pressure distributed over the area of contact, which changes during the penetration of the probe inside the sample. For that reason, the expression of the function $F(\delta)$ depends strictly on the geometry of the problem, and in particular, on the shape of the probe. It is important to note that the derivation of the Hertz model assumes that the indenter is a paraboloid of

revolution. Thus, its application to spheres and similar geometries is limited, in principle, to small indentations compared to the radius of curvature.

Since the publication of the Hertz solution for the spherical probe contact problem,
several models have been formulated to obtain a force–indentation relation for probes
of different geometries. Boussinesq derived a solution to the problem corresponding to
the case of a solid of revolution indenting a half space, whose axis is normal to the
original boundary of the space (Boussinesq, 1885). Thanks to the Boussinesq solution,
subsequently, it was possible to derivate the model for a flat-ended cylindrical probe
(Love, 1929) and a conical probe (Love, 1939). Sneddon (1965) formulated a general solution for a probe of arbitrary profile, from which he derived, as particular cases, other
geometries, including a sphere not approximated by a paraboloid of revolution, which
presents the advantage of not being limited by the constraint $\delta \ll R$.

Sneddon derived the general integrals required to calculate the indentation and
the applied force as a function of the shape profile function of the punch, as well as
the corresponding pressure distributions for different geometries.

For most of them, he showed that the force–indentation relationship can be
generally written as

$$F = CE_{\text{eff}}\delta^{m} \qquad (2.2.9)$$

where F is the normal applied force, C reflects the indenter geometry constants, E_{eff}
refers to the effective modulus of the body, δ is the indentation, and m is an exponent that depends on how the area of contact changes during the indentation of the
probe, in turn depending on the shape profile function of the indenter. In the following, we provide a list of solutions derived by Sneddon for different geometries,
together with more recent contact problem solutions for geometries closer to the actual shapes of the available AFM probes used in experiments. A review of the different models is provided in Table 2.2.1 and the force-indentation relationships for
different geometries and the corresponding curves are shown in Figure 2.2.4.

2.2.2.2 Paraboloid of Revolution

As solved by Hertz, the solution for a paraboloid of revolution of curvature $1/2\,R$ is

$$F = \frac{4}{3}\frac{E}{(1-v^2)}\sqrt{R}\delta^{3/2} \qquad (2.2.10)$$

This is the well-known Hertz model derived in (eq. 2.2.6). The contact radius is
$a = \sqrt{R\delta}$, and the force–contact radius relation is

$$F = \frac{4}{3}\frac{E}{(1-v^2)R}a^{3} \qquad (2.2.11)$$

2.2.2.3 Flat-Ended Cylinder

The expression for a flat-ended cylinder of radius a is

$$F = 2a \frac{E}{1-v^2} \delta \tag{2.2.12}$$

Notice that unlike the model for the paraboloid of revolution (eq. (2.2.10)), the relationship is linear, as the area of contact does not change with indentation (the contact radius a is constant). Cylindrical punches are thus ideal to probe the nonlinearity of the sample.

2.2.2.4 Sphere

Sneddon provided the exact solution for a spherical probe of radius R in terms of the radius of contact a (which is now different from $\sqrt{R\delta}$)

$$F = \frac{E_s}{2(1-v_s^2)} \left[(a^2 + R^2) \ln \left(\frac{R+a}{R-a} \right) - 2Ra \right] \tag{2.2.13}$$

after correction for a typographical error (Heuberger et al., 1996), where the indentation is a function of the radius of contact

$$\delta = \frac{1}{2} a \ln \left(\frac{R+a}{R-a} \right) \tag{2.2.14}$$

The advantage of this solution compared to the Hertz model is that it is valid for any applied indentation, provided $a \leq R$. The disadvantage is that it requires numerical computations (Chyasnavichyus et al., 2016) or approximations (Kontomaris and Malamou, 2021a). The corrections for the case of large indentations, obtained by Kontomaris and Malamou (2021) and Müller et al. (2019), are as follows:

$$\Omega_{\text{Kontomaris}} = c_1 + \sum_{k=2}^{6} \frac{3}{2k} c_k y^{k-3/2} \tag{2.2.15}$$

with coefficients:

$$c_1 = 1.0100000$$
$$c_2 = -0.0730300$$
$$c_3 = -0.1357000$$
$$c_4 = 0.0359800$$
$$c_5 = -0.0040240$$
$$c_6 = 0.0001653$$

and

$$\Omega_{\text{Muller}} = 1 - \frac{1}{10}\gamma - \frac{1}{840}\gamma^2 - \frac{1}{15,120}\gamma^3 + \frac{1,357}{6,652,800}\gamma^4 \tag{2.2.16}$$

where $\gamma = \delta/R$ is the ratio between the indentation and the radius of the probe, and $\Omega(\gamma)$ is the polynomial correction factor that can be multiplied for the Hertz equation (eq. (2.2.6)) in the case of large indentation in the following form:

$$F = F_{\text{Hertz}}\Omega(\gamma(\delta, R)) \tag{2.2.17}$$

Even if the formulation of the two correction factors is different (the Müller correction is a fourth-order power law expansion, while the Kontomaris correction is not; indeed, the second term is of power 1/2), the two corrections are equivalent, compared to the Sneddon formula.

As expected, the spherical solution by Sneddon approximates the Hertz model for small indentations ($\delta \ll R$). The general solution for an ellipsoid of revolution was also derived in the Sneddon work.

2.2.2.5 Other Geometries

Sneddon provided the general exact solution for a punch with a shape profile described by the polynomial $z = \sum c_n r^n$. For more complex geometries, approximations are often necessary. A general development that leads to some of the solutions derived by the Sneddon approach was proposed by Barber and Billings to derive the approximate solution for a punch of arbitrary shape, and they applied it to an n-sided regular pyramid with semi-included angle θ (Barber and Billings, 1990). This powerful approach is briefly described in the following and has been recently used to derive more complex problems that will be listed later.

The pressure distribution $P(r)$ for a probe of projected area A can be obtained by knowing the expression of the work $F\delta$ for a flat cylinder probe, using the Betti–Rayleigh reciprocal theorem (Betti, 1872, Rayleigh, 1873), which relates the two properties (Gavara and Chadwick, 2012).

Betti's reciprocal theorem states that for a linear elastic material, for two displacement systems at points A and B, due to the two respective forces F and F^*, the work done by F applied in A through the relative deformation produced by F^* in A is equal to the work done by F^* applied in B through the relative deformation produced by F in B:

$$F_A d_A^* = F_B^* d_B \tag{2.2.18}$$

In general, we can write the relative deformation as the integral over all strains of all elements of the total surface $d = \int_S u_i dS$ and the total force as $F = \int_S T_i \, ds$, thus equation 2.2.18 becomes:

$$\int_S T_i u^*_i dS = \int_S T^*_i u_i dS \qquad (2.2.19)$$

where S is the surface of the projected area of the indenting punch, and T_i and T^*_i are the normal internal stresses due to respectively the forces F and F^*, while u_i and u^*_i are the corresponding displacements.

In our AFM problem, the force is applied just in the z-direction and the area corresponds to the projected area of a probe of arbitrary shape. Accordingly, a probe of flat-ended cylindrical shape can be considered in the point A (left part of the equation 2.2.19) and an arbitrary probe shape in the point B (right part of the equation 2.2.19), of which we want to obtain the force-indentation expression.

The force F and the corresponding displacements u_i are valid only for a frictionless contact, since we consider only the component of the stress tensor normal to the surface. For the arbitrary probe shape, the surface displacements u_i can be described by the shape profile function of the probe $f(x,y)$ or $f(r,\phi)$, while the stress T^* is the pressure distribution $P^*(r,\phi)$. For the flat-endend cylindrical probe displacement u^*_i corresponding to the force F, are known (see equation 2.2.12) and the total deformation is the uniform indentation δ^* induced by F.

Accordingly, the equation 2.2.19 becomes:

$$\iint_S P(r,\phi)\delta^* \ r\, dr\, d\phi = \iint_S P^*(r,\phi)f(r,\phi)r\, dr\, d\phi$$

Since the indentation δ^* is a constant, it can be taken outside of the integral:

$$F = \iint_S P(r,\phi) \ r\, dr\, d\phi = 1/\delta^* \iint_S P^*(r,\phi)f(r,\phi)r\, dr\, d\phi$$

This leads to the following form of the total force:

$$F = \int_S P^*(r,\phi)\frac{f(r,\phi)}{\delta^*}r\, dr\, d\phi \qquad (2.2.20)$$

The analytical form of $P^*(r,\phi)$ is only known for an elliptical contact area and is given by

$$P^*(r,\phi) = \frac{E}{\pi(1-v^2)}\frac{\delta^*}{\sqrt{a^2-r^2}} \qquad (2.2.21)$$

Thus, the pressure distribution can be approximated by that of a flat cylindrical punch, with an area corresponding to the best ellipse approximation to the actual contact area. The radius of contact is then determined by imposing that the derivative of the force over the contact radius is maximal $((\partial F/\partial a) = 0)$ (Shield, 1967). More refined approaches for the flat punch pressure distribution are available in the literature (Barber and Billings, 1990, Fabrikant, 1986).

This approach is thus a valid approximation for an indenter of arbitrary shape, provided the shape, the profile function is known. For example, in the case of a conical probe, the corresponding displacement under the punch is $f(r, \phi) = u_z = \delta - r\cot\theta$, where r is the radial distance from the vertex. At this point, knowing that $\int d\phi = 2\pi$, the eq. (2.2.10) can be written as

$$F = 2\frac{E}{1-v^2}\int_0^a \frac{\delta - r\cot\theta}{\sqrt{a^2 - r^2}}rdr = 2\frac{E}{1-v^2}\left(\delta a - \frac{\pi}{4}a^2\cot\theta\right) \tag{2.2.22}$$

By imposing $(\partial F/\partial a) = 0$, we find the radius of contact $a = ((2\tan\theta)/\pi)\delta$, and eq. (2.2.22) can be written as

$$F = \frac{2\tan\theta}{\pi}\frac{E}{1-v^2}\delta^2 \tag{2.2.23}$$

In the case of a cone, the solution is exact, as the best ellipse approximation for the flat punch is actually a circle of radius a.

2.2.2.6 Cone

The solution for a conical probe of semi-included angle θ, often called the Sneddon model, as obtained in the previous section,

$$F = \frac{2\tan\theta}{\pi}\frac{E}{1-v^2}\delta^2 \tag{2.2.24}$$

It is important to note that the semi-included angle θ is defined from the vertical axis (z) to the face of the cone, and not from the plane tangent to the face, as some other works do.[2] In this case, the contact radius is given by $a = ((2\tan\theta)/\pi)\delta$.

2.2.2.7 *n*-Sided Regular Pyramid

The approach described above has been used originally by Barbers and Billing to provide a solution for a regular *n*-sided pyramid of semi-included angle θ. The general solution is

$$F \approx \frac{2\tan\theta}{n\,\sin\frac{\pi}{n}}\frac{E}{1-v^2}\delta^2 \tag{2.2.25}$$

For $n \to \infty$, we recover the solution for a cone. For the common four-sided pyramidal tip, often used in AFM, the expression is

---

2 We opted for using semi-included angle, instead of semi-open, as it is more explicit the definition we use here.

$$F \approx \frac{\tan \theta}{\sqrt{2}} \frac{E}{1 - v^2} \delta^2 \qquad\qquad (2.2.26)$$

This result is within 6% of the numerical solution provided by Bilodeau (1992). The contact radius is given by $a = \left(\delta \tan \theta / \sqrt{2} \right)$.

2.2.2.8 Blunted Cone and Pyramid

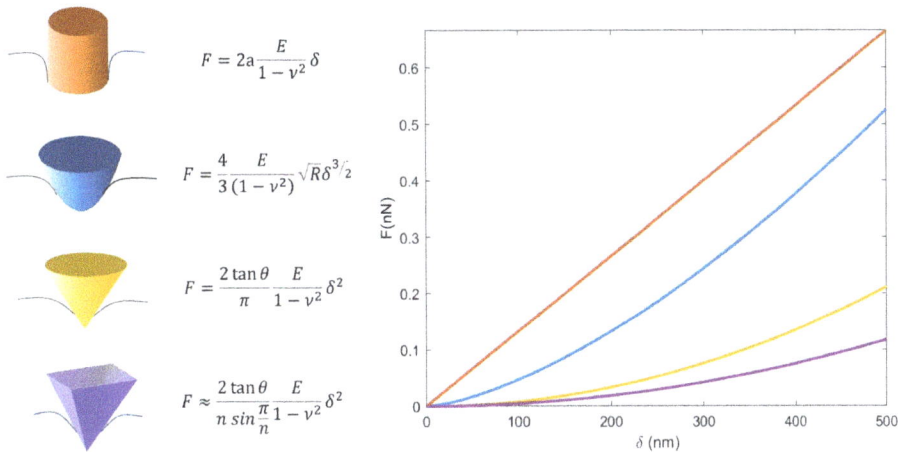

$$F = 2a \frac{E}{1 - v^2} \delta$$

$$F = \frac{4}{3} \frac{E}{(1 - v^2)} \sqrt{R} \delta^{3/2}$$

$$F = \frac{2 \tan \theta}{\pi} \frac{E}{1 - v^2} \delta^2$$

$$F \approx \frac{2 \tan \theta}{n \sin \frac{\pi}{n}} \frac{E}{1 - v^2} \delta^2$$

Figure 2.2.4: Force-indentation relationships for different probe geometries: a cylindrical indenter (Sneddon model), a paraboloidal indenter (Hertz model), a conical indenter (Sneddon model), and a pyramidal indenter (Barber and Billings model). The parameters used for the figures are: $E = 1{,}000$ Pa, $v = 0.5$, $R = 700$ nm, $a = 500$ nm, $\theta = 45$.

While the above equations are satisfactory and provide good approximations to several tip shapes and experimental conditions, the ideal geometry of a cone or a pyramid is often not found in experimental AFM probes. In particular, for small indentations of some tens of nanometers, the ideally sharp apex appears blunted. For this reason, solutions for more realistic probe shape profiles have been proposed. We provide here the general solution for an n-sided regular pyramid described above, but with a blunted tip of radius R_c at the apex, as shown in figure (Figure 2.2.5). The solution was found following the above described method using Betti's reciprocal theorem and the Rayleigh–Ritz approximation, which is shown in Table 2.2.1 (Rico et al., 2005). As shown for the case of a sharp indenter, for $n \to \infty$, we recover the solution for a blunted cone, which was proposed earlier by Briscoe et al. (1994). It is interesting to comment on the original definition of the cap width, which considered the case of a sphere that emerged tangential with the pyramid faces. The original work defined this parameter as $b = R_c \cos \theta$. Nevertheless, to solve the problem, the actual

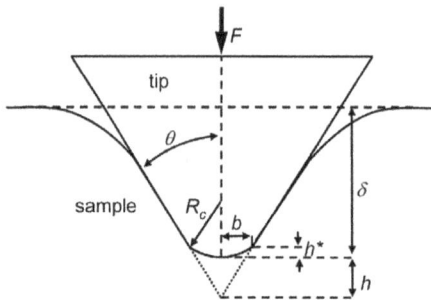

Figure 2.2.5: Schematic drawing of the the blunted pyramidal probe model, where θ in the angle of the pyramid, R_c the blunt tip radius of curvature, δ the indentation, b the point of transition between the pyramidal and the parabolical shape, h and h^* the characteristic height of the blunted tip. Reproduced with permission from Rico et al. (2005).

integration was performed for a paraboloid of radius R_c. This introduced a slight discontinuity in the numerical solutions for the contact radius. To avoid this discontinuity, the cap width is better defined by $b = R_c/\tan \theta$, i.e. by assuming a paraboloid, and not a sphere, emerging tangential to the pyramidal faces. This new definition does not invalidate the original approximation but makes the solution self-consistent.

The emergence of new tip geometries becoming commercially available, such as blunted cylinders or hemispheres, makes the use of the approximate methods proposed above increasingly useful. Moreover, the complexity of the biological samples often requires further corrections. In some cases, the Hertzian requirements are not fulfilled on some systems or are too restrictive to describe the complex materials like biological samples, among others. Several additional theoretical approaches have been formulated to overcome the limitations of Hertzian mechanics in describing more realistic systems, and the most relevant to describe the biological samples are summarized in the following paragraphs.

Table 2.2.1: Principal force–indentation equations for different probe geometries.[3]

Probe geometry	Force–indentation function, F(δ)	Reference	Limits of validity
Paraboloid of curvature $1/2\,R$ (radius R)	$F = \dfrac{4\sqrt{R}}{3} \dfrac{E}{(1-v^2)} \delta^{3/2}$	Hertz (1881)	$\delta \ll R$

---

3 F_{sphere} has to be identified with Eq. 2.2.13 , $F_{cylinder}$ has to be identified with Eq. 2.2.12 and F_{cone} has to be identified with Eq. 2.2.24. and F_{Hertz} has to be identified with Eq. 2.2.20.

Table 2.2.1 (continued)

Probe geometry	Force–indentation function, F(δ)	Reference	Limits of validity
Sphere of radius R and radius of contact a	$$F = \frac{E_s}{2(1-v^2)}\left[(a^2+R^2)\ln\left(\frac{R+a}{R-a}\right)-2Ra\right]$$ $$\delta = \frac{1}{2}a\ln\left(\frac{R+a}{R-a}\right)$$	Sneddon (1965)	$a \le R$
Flat-ended cylinder of radius a	$$F = 2a\frac{E}{1-v^2}\delta$$	Love (1929), Sneddon (1965), Rico (2007)	$\delta \ll a$
Cone of semi-included angle θ	$$F = \frac{2\tan\theta}{\pi}\frac{E}{1-v^2}\delta^2$$	Love (1939), Sneddon (1965)	$\delta \ll \tan\theta$
Blunted n-sided pyramidal probe[4]	$$F = \begin{cases} \frac{4\sqrt{R}}{3}\frac{E}{(1-v^2)}\delta^{3/2} & a < b \text{ spherical cap} \\ & a > b \\ f(n,\delta,a,b)\frac{E}{(1-v^2)} & \text{blunted } n\text{-sided regular pyramid} \end{cases}$$ In the limit $b \to 0$: n-sided regular pyramid $$F = \frac{2\tan\theta}{\left(n\sin\left(\frac{\pi}{n}\right)\right)}\frac{E}{1-v}\delta^2$$ $$\delta = \frac{2}{n\sin\left(\frac{\pi}{n}\right)}\tan(\theta)\delta$$	Rico (2005), Barber and Billings (1990)	$h^* < \delta \ll 1$ $F'_{sc} = F'_p$ when $a = b$
Needle	$$F = \begin{cases} F_{sphere}(\delta) & \delta \le \delta_c \\ F_{sphere}(\delta_c)+F_{cylinder}(\delta-\delta_c) & \delta > \delta_c \end{cases}$$	Garcia and Garcia (2018)	$\delta \ll R$ $\delta \ll a$
Nanowire	$$F = \begin{cases} F_{cone}(\delta) & \delta \le \delta_c \\ F_{cone}(\delta_c)+F_{cylinder}(\delta-\delta_c) & \delta > \delta_c \end{cases}$$	Garcia and Garcia (2018a)	$\delta \ll \tan\theta$ $\delta \ll a$
Sphere of radius R and radius of contact a suitable also for large deformations	$F = F_{Hertz}\Omega(\gamma(\delta,R))$ with $$\Omega_{Kontomaris} = c_1 + \sum_{k=2}^{6}\frac{3}{2k}c_k\gamma^{k-\frac{3}{2}} \text{ or}$$ $$\Omega_{M\ddot{u}ller} = 1-\frac{1}{2}\gamma-\frac{1}{840}\gamma^2-\frac{1}{15120}\gamma^3$$ $$+\frac{1357}{6652800}\gamma^4$$	(Kontomaris and Malamou, 2021a; Müller et al., 2019)	$\gamma = \delta/R$

---

4 The function $f(n,\delta,a,b) = 2(\delta a - \frac{n}{\pi}\sin\left(\frac{\pi}{n}\right)\frac{a^2}{2\tan\theta}\left(\frac{\pi}{2}-\arcsin\frac{b}{a}\right)-\frac{a^3}{3R_c}+(a^2-b^2)^{\frac{1}{2}}(\frac{n}{\pi}\sin\left(\frac{\pi}{n}\right)\frac{b}{2\tan\theta}+\frac{a^2-b^2}{3R_c}))$,
$\delta-\frac{a}{\tan\theta}\frac{n}{\pi}\sin\left(\frac{\pi}{n}\right)\left(\frac{\pi}{2}-\arcsin\frac{b}{a}\right)+\frac{a}{R_c}\left[(a^2-b^2)^{1/2}-a\right]=0$ and $b = R_c\cos\theta$ for an emerging sphere or $b = \frac{R_c}{\tan\theta}$ for an emerging paraboloid, $h^* = \frac{b^2}{2R_c}$, the condition $F'_{sc}(a=b)=F'_p$ is valid when there is a smooth transition between spherical cap and pyramidal tip shape.

2.2.3 Models Considering Finite Sample Thickness (Bottom Effect)

One of the Hertzian model's assumptions, as previously explained, is the infinite sample thickness (assumption number 5). In the case of biological samples, such as a thin lipid layer or a cell, this assumption does not always hold. Indeed, often in experimental conditions, cells featuring thin sections are grown on rigid substrates. In this case, when the indentation δ cannot be considered as being much smaller than the sample thickness h, the stress distribution is influenced by the presence of a stiffer substrate below the sample.[5] To correct for this bottom effect, advanced force–indentation relationships for different tip geometries and sample configurations have been proposed.

2.2.3.1 Bottom Effect Correction for a Paraboloid

A solution was proposed by Dimitriadis et al. (2002) for this problem in 2002 for the case of a spherical probe or, more formally, of a paraboloid of revolution. Before Dimitriadis' model, other solutions for the bottom effect correction were formulated by Popov (2013), Tu and Gazis (1964), Dhaliwal and Rau (1970), Chen and Engel (1972), and Aleksandrov (1968, 1969), but these calculations required extensive numerical computation, and the Aleksandrov analytic solution is not valid for incompressible materials with Poissson's ratio $v = 0.5$. Thus, it is not suitable for biological samples that contain a large amount of water. Indeed, the difficulty in formulating a convenient approach for routine use, to correct the bottom effect, is due to the intrinsic nonlinearity of the problem whereby the applied total force in the z direction depends on the distribution of pressure, and thus on the contact area. Dimitriadis et al. solved this issue by dividing the problem of integral equations into a hierarchy of simplified subproblems; simpler integral equations that can be solved analytically using the method of images.

In Dimitriadis' work, two different cases are studied:
– the sample is not bonded to the supporting substrate
– the sample is bonded to the supporting substrate

In the first case, the authors used the method of images to construct an approximate solution, while in the second case, they first derived the Green's function and then used it to compute an explicit expression of the force versus indentation relationship, following the same procedure as in the first case.

---

5 As a rule of thumb, $\delta < 0.2\,h$ (Dimitriadis et al., 2002) is often used as a threshold for ignoring the contribution of the underlying hard substrate.

The method of images is based on the idea that any solution of an integral equation that satisfies the appropriate boundary conditions, like eq. (2.2.4), is the unique solution. The interface between the sample and the rigid substrate is considered a singularity where the solution of the problem is unknown.

The authors assumed a sample of thickness h extending in the x-y plane and supported on a rigid substrate located at $z = -h$. The force application is the same as in the Hertz model described earlier, but it is assumed that the rigid boundary modifies the pressure distribution. The authors combine multiple images of the probe that apply normal forces on planes located at distances corresponding to multiples of $2h$ from the surface of the force application. Thus, an infinite number of images emerge, each representing an infinite half-space with a known solution. This procedure is carried out on an infinite sequence of images.

As in the Hertz model, s is the distance between the source and the observation point, $s = |\mathbf{r}| = r^2 + r_s^2 - 2rr_s \cos\phi$. For a general image at $z = 2nh$, $n = 0, 1, 2, \ldots$, the Green's function from eq. (2.2.2) becomes:

$$G_n(s) = (-1)^n \frac{1+v}{2\pi E} \left\{ \frac{z^2}{(s^2 + z^2)^{\frac{3}{2}}} + \frac{2(1-v)}{(s^2 + z^2)^{\frac{1}{2}}} \right\}\Bigg|_{z = 2nh} \qquad n = 0, 1, 2, \ldots \qquad (2.2.27)$$

The total Green's function of all the images is then:

$$G_{tot}(s) = \sum_{n = 0, \pm 1, \pm 2, \ldots} G_n(s) \qquad (2.2.28)$$

If we assume that the sample is not bonded to the supporting substrate, the surface in contact with the rigid substrate is free to slide horizontally, that is, there is no friction and no adhesive contact between the sample and the support.

The region of interest of the original problem is only the probe contact region at the surface of the sample. Accordingly we can assume that $s \leq h$, especially if the thickness of the sample is larger than the radius of the probe, and that the strain is small. The strain is defined as the ratio between the deformation and the original height or thickness of the body, in a direction parallel to the applied force. Thus, in our case, we can define the parameter $\varepsilon = \delta/h$, which is small and can be regarded as a strain. At this point, we can expand the Green's function in a Taylor series, in terms of ε:

$$G_{tot}(s) = G_\infty(s)\left(1 + \varepsilon\alpha(s) + \varepsilon^3\beta(s) + \varepsilon^5\gamma(s) + \cdots\right) \qquad (2.2.29)$$

where

$$G_\infty(s) = \frac{(1-v^2)}{\pi E}\left(\frac{1}{s}\right)$$

is the Green's function for the surface indentation of an infinite half-space and the higher order terms correct for the bottom effect. The coefficients of this series depend on the Poisson's ratio v as well:

$$\alpha(s) = \alpha_0(v)\frac{s}{\delta}, \ \beta(s) = \beta_0(v)\left(\frac{s}{\delta}\right)^3, \ \gamma(s) = \gamma_0(v)\left(\frac{s}{\delta}\right)^5, \ \cdots \qquad (2.2.30)$$

The expressions of the coefficients $\alpha_0(v)$, $\beta_0(v)$, and $\gamma_0(v)$ can be found in the original work by Dimitriadis et al. (2002).

Since the probe is spherical and we assume negligible long-range interactions, the displacement field will follow the shape of the probe and eq. (2.2.4) becomes:

$$\delta - \frac{r^2}{2R} = \iint\limits_A P(r_s)G_{tot}(s)dA \qquad (2.2.31)$$

where the Green's function $G_{tot}(s)$ is given by eq. (2.2.29). The contact area and, accordingly, the contact radius are assumed to be independent of h, because the radius of the probe is smaller than the sample thickness. Thus, the presence of the rigid substrate modifies the pressure profile without affecting the contact area. It is then reasonable that the pressure profile depends on the small parameter ε. Expanding it in Taylor series as before for G_{tot}:

$$P(r_s) = P_\infty(r_s)\left(1 + \varepsilon\alpha(s) + \varepsilon^3\beta(s) + \varepsilon^5\gamma(s) + \cdots\right) \qquad (2.2.32)$$

Through the substitution of eq. (2.2.32) into eq. (2.2.31), we obtain a series of integral equations of different orders for P_∞, P_1, P_2, P_3, etc. Each order problem can be solved separately. The first order problem solution is exactly the Hertz model of a rigid spherical probe indenting an infinite half-space (eq. (2.2.6)). Calculating the solutions of the integral equations until the fourth order, the expression of the force is a function of the indentation of a spherical probe indenting an infinite half-space multiplied by a series of terms correcting for the finite thickness or bottom effect:

$$F = \frac{4\sqrt{R}}{3}\frac{E}{(1-v^2)}\delta^{3/2}\left[1 - \frac{2\alpha_0}{\pi}\chi + \frac{4\alpha_0^2}{\pi^2}\chi^2 - \frac{8}{\pi^3}\left(\alpha_0^3 + \frac{4\pi^2}{15}\beta_0\right)\chi^3 + \frac{16}{\pi^4}\left(\alpha_0^4 + \frac{3\pi^2}{5}\beta_0\alpha_0\right)\chi^4\right]$$

$$(2.2.33)$$

where $\chi = (a/h) = \sqrt{R\delta}/h$.

Equation (2.2.33) is the finite thickness solution for a parabolic probe (closely approximates a spherical probe), valid whether the sample is bonded to the substrate or not. The difference between the two cases consists in the parameters α_0 and β_0, which depend on the Poisson's ratio v.

Noticeably, the bottom effect correction does not depend trivially on the ratio of the vertical lengths δ and h, but on the ratio of the horizontal dimension of the contact – the contact radius a to the sample height h. The correction, therefore, considers the propagation of the strain and stress fields into the bulk volume of the sample, and not only their vertical extension. Therefore, we must expect stronger finite-thickness effects for large spherical-parabolic probe with respect to sharp ones.

For most biological samples, we can assume $v = 0.5$. In this case, the equations for bonded and not bonded samples become:

Sample not bonded:

$$F = \frac{16\sqrt{R}E}{9} \delta^{3/2} \left[1 + 0.884\chi + 0.781\chi^2 + 0.386\chi^3 + 0.0048\chi^4\right] \qquad (2.2.34)$$

Sample bonded:

$$F = \frac{16\sqrt{R}E}{9} \delta^{3/2} \left[1 + 1.133\chi + 1.283\chi^2 + 0.769\chi^3 + 0.0975\chi^4\right] \qquad (2.2.35)$$

The curves corresponding to these bottom-corrected equations are shown in Figure 2.2.6 and compared with the Hertz model. When comparing these two equations, the apparent stiffness of the sample is larger for the bonded case, which corresponds with intuition, since the sample is not allowed to slide laterally.

For the intermediate case, in which only some parts of the sample are bonded, we can replace eqs. (2.2.34) and (2.2.35) with a similar equation in which the numerical coefficients are the average of the corresponding coefficients in the two cases; this can be appropriate for cells that adhere locally to a substrate through dynamically forming focal adhesion complexes (Gavara and Chadwick, 2012).

Clearly, for any given tip radius, there is a limited range of thickness and indentation for which eqs. (2.2.34) and (2.2.35) are valid. In order to respect the assumption of linear elasticity of the material, the maximum total strain should never exceed 10%, or $\delta \leq 0.1\, h$. We can consider that the bottom effect correction is required if the first term of the series adds at least to 10% of the force. For the bonded case, this means that

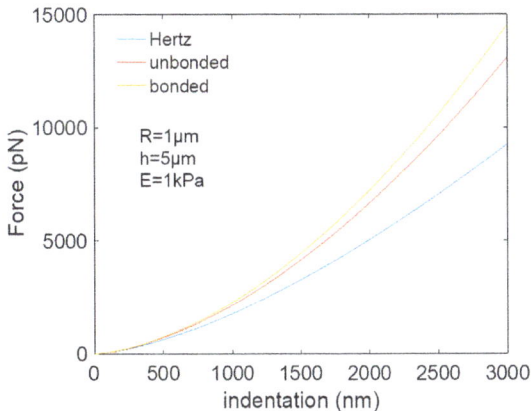

Figure 2.2.6: Force–indentation curve of bottom effect model in the case of sample bonded and unbonded in comparison to the Hertz model for a parabolic intender with curvature radius $R = 1\,\mu m$, sample thickness $h = 5\,\mu m$, and Young's modulus $E = 1\,kPa$.

$1.133\chi \geq 0.1$, with $\chi = \sqrt{R\delta}/h$, which implies $h \leq 12.83R$. Thus, for sample thickness with $h > 13R$, the infinite half-space assumption can be considered a good approximation.

At the same time, to safely use this model, h cannot be much smaller than R. Indeed, the parameter χ of this series has to be small enough. If χ is too large, the series expansion may lose accuracy and could diverge. The series converges if $\chi \leq 1$, which implies $h \geq 0.1\ R$.

The previous formulation assumes Hertzian contact area $(a = \sqrt{R\delta})$, which is not necessarily given. Considering that the contact area at each indentation point follows $((\partial F(\delta, a))/\partial a) = 0$, Garcia and Garcia (2018b) found that the equivalent of eq. (2.2.33) for a bonded sample presents slightly different correction terms:

$$F = \frac{16\sqrt{RE}}{9}\delta^{3/2}\left[1 + 1.133\chi + 1.497\chi^2 + 1.469\chi^3 + 0.755\chi^4\right]. \tag{2.2.36}$$

which imposes a stronger correction.

If the sample is very thin, $h \leq 0.1R$, these models do not describe the physics of the problem, and the following formulas obtained in Chadwick (2002) should be used.

Sample not bonded:

$$F = \left(\frac{2\pi}{3}\right)E\sqrt{R\delta^3}\chi \tag{2.2.37}$$

Sample bonded:

$$F = \left(\frac{2\pi}{3}\right)E\sqrt{R\delta^3}\chi^3 \tag{2.2.38}$$

In conclusion, the bottom effect correction is necessary to avoid an overestimation of the Young's Modulus when the sample thickness is not infinite. If the sample is too thin compared to the radius, approximately $h \leq 13\ R$ (Dimitriadis et al., 2002), the rigid substrate blocks the propagation of the stress field induced by the tip in the sample. This phenomenon affects the distribution of the pressure and decreases the deformation of the sample, with a consequent increase in the apparent Young's Modulus E.

It should also be noted that the bottom effect correction allows to consider in the analysis, those thinner regions of the sample (like the peripheral extensions and lamellipodia in the cells), where usually the estimation of the Young's modulus is inaccurate.

Besides, the vertical spatial constraint represented by the substrate supporting the system under study, it can also be assumed that lateral confinement and boundaries produce similar effects, for which analytical corrections are not available, but only FEM simulations are available (Garcia and Garcia, 2018, supplement material). This can be relevant, for example, for the nanomechanical characterization of cells in a confluent layer, where strong cell–cell interactions determine lateral boundaries, which can interfere with the strain field induced by the indenter, especially when large spherical probes are used.

2.2.3.2 Bottom Effect Correction for Other Geometries

In the case of other tip geometries, like conical and pyramidal, the pressure distribution $P(r)$ for the bottom effect correction can be approximated, as explained before, knowing the expression of the work $F\delta$ for a flat cylinder probe and using Betti's reciprocal theorem, and then the Rayleigh-Ritz approximation. The pressure distribution can then be computed using the integral equation:

$$\delta = \iint_A P(r)G_{tot}(s)dA \qquad (2.2.39)$$

Using the Green's function described above for finally finding the force as a function of the contact radius. Finally, the contact radius can be found by imposing $\frac{\partial F}{\partial a} = 0$. The full development to correct the bottom effect of a bonded sample of finite thickness for a conical probe was formulated in the work of Gavara et al. (2016). However, according to later works, the solution provided was not convergent, probably due to a typographical error (Garcia and Garcia, 2018b; Managuli and Roy, 2018). Following the same approach, slightly different factors were found, which lead to a convergent series of the force F (more details in the supplementary materials of Garcia and Garcia, 2018a). We provide here the solution reported by Garcia and Garcia for a cone:

$$F = \frac{8\tan\theta}{3\pi}E\delta^2\left[1 + 0.721\tan\theta\frac{\delta}{h} + 0.650\tan^2\theta\left(\frac{\delta}{h}\right)^2 + 0.491\tan^3\theta\left(\frac{\delta}{h}\right)^3 + \right.$$
$$\left. + 0.225\tan^4\theta\left(\frac{\delta}{h}\right)^4 + O\left(\left(\frac{\delta}{h}\right)^5\right)\right] \qquad (2.2.40)$$

Following the same approach, Garcia and Garcia also reported the solutions for a paraboloid, extending the solution for the case of non-Hertzian contact radius (as described above), and for a flat-ended cylinder of radius a. Interestingly, for a flat-ended cylinder, the force–indentation relationship remains linear for finite sample thicknesses. The principal force–indentation solutions for bottom effect correction for different geometrical probe models are shown in Table 2.2.2. Other solutions for the indentation of different probe geometries indenting thin samples, with their own limitations, can be found in the literature (Akhremitchev and Walker, 1999, Long et al., 2011, Yang, 1998).

Recent developments have established the use of soft substrates for cell culture. In that case, a bottom effect correction may be needed to prevent underestimation of cell mechanics. Solutions have been provided for the case of two-layered elastic substrates, which might be important for the accurate estimation of the Young's modulus of living cells growing on soft hydrogels (Doss et al., 2019). In the case of a thin sample with Young's modulus E_1 placed on a softer substrate with Young's modulus E_2, where $E_1 > E_2$, the force–indentation relation is described by the following equation, obtained by Doss et al.:

Table 2.2.2: Principal force–indentation equations for the bottom effect corrected models for different probe geometries.[6]

Probe geometry	Force–indentation function, F(δ)	Reference	Limits of validity
Paraboloid of curvature $1/2\,R$ (radius R) Sample not bonded	$F = \dfrac{16\sqrt{R}E}{9}\delta^{3/2}$ $[1+0.884\chi+0.781\chi^2+0.386\chi^3+0.0048\chi^4]$	Dimitriadis et al. (2002)	$\chi=\sqrt{R\delta}/h \ll 1$
Paraboloid of curvature $1/2\,R$ (radius R) Sample bonded	$F = \dfrac{16\sqrt{R}E}{9}\delta^{3/2}$ $[1+1.133\chi+1.283\chi^2+0.769\chi^3+0.0975\chi^4]$	Dimitriadis et al. (2002)	$\chi=\sqrt{R\delta}/h \ll 1$
Cone of semi-included angle θ or pyramid	$F = \dfrac{8\tan\theta E}{3\pi}\delta^2[1+0.721\tan\theta\chi+0.650\tan^2\theta\chi^2$ $+0.491\tan^3\theta\chi^3++0.225\tan^4\theta\chi^4+O(\chi^5)]$	Gavara et al. (2016), Garcia and Garcia (2018a)	$\chi=\delta/h \ll 1$
Sphere (paraboloid very close to a sphere of radius R)	$F = \dfrac{16\sqrt{R}E}{9}\delta^{3/2}$ $[1+1.133\chi+1.497\chi^2+1.469\chi^3+0.755\chi^4]$	Garcia and Garcia (2018a)	$\chi=\dfrac{\sqrt{R\delta}}{h} \ll 1$ $\delta \ll R$
Flat-ended cylinder of radius a	$F = \dfrac{8aE}{3}\delta$ $[1+1.133\chi+1.283\chi^2+0.598\chi^3+0.291\chi^4]$	Garcia and Garcia (2018a)	$\chi=\dfrac{a}{h} \ll 1$
Needle	$F = \begin{cases} F_{sphere}(\delta) & \delta \leq \delta_c \\ F_{sphere}(\delta_c)+F_{cylinder}(\delta-\delta_c) & \delta > \delta_c \end{cases}$	Garcia and Garcia (2018a)	$\dfrac{\sqrt{R\delta}}{h} \ll 1$ $\dfrac{a}{h} \ll 1$ $\delta \ll R$
Nanowire	$F = \begin{cases} F_{cone}(\delta) & \delta \leq \delta_c \\ F_{cone}(\delta_c)+F_{cylinder}(\delta-\delta_c) & \delta > \delta_c \end{cases}$	Garcia and Garcia (2018a)	$\dfrac{a}{h} \ll 1$ $\delta/h \ll 1$
Paraboloid of curvature $1/2\,R$ (radius R) Layered soft elastic substrate	$F = \dfrac{16\sqrt{R}E_1}{9}\delta^{3/2}$ $\left(\dfrac{0.85\chi+3.36\chi^2+1}{(0.85\chi+3.36\chi^2)\left(\frac{E_1}{E_2}\right)^{0.72-0.34\chi+0.51\chi^2}+1}\right)$	Doss et al. (2019)	$\chi=\sqrt{R\delta}/h \ll 1$ $E_1 > E_2$

---

6 The difference of the needle and nanowire models between Tables 2.2.1 and 2.2.2 is that in Table 2.2.1 F_{sphere}, $F_{cylinder}$, and F_{cone} are not bottom effect corrected, while in Table 2.2.2 they are

$$F = \frac{16\sqrt{R}E_1}{9}\delta^{3/2}\left(\frac{0.85\chi + 3.36\chi^2 + 1}{(0.85\chi + 3.36\chi^2)\left(\frac{E_1}{E_2}\right)^{0.72 - 0.34\chi + 0.51\chi^2} + 1}\right) \qquad (2.2.41)$$

Recently, Rheinlaender et al. proposed a model for the indentation of a living cell on a soft substrate. The model assumes a uniform deformation of the cell contact area, and was validated for the case of large spherical tips, which causes not only an indentation of the cell but also a coupled indentation of the soft substrate ($R \gg h_{cell}$) (Rheinlaender et al., 2020).

2.2.4 Viscoelastic Models

Living cells have a viscous, liquid-like component, and an elastic, solid-like component, both coupled and arising from the complex filament network of the cytoskeleton within the cytosol (Fabry et al., 2001, Rigato et al., 2017). Thus, cells are viscoelastic, and such are many other systems that can be studied by AFM or other indenters. When a force is applied on a purely elastic material, there is no dissipation of energy and response to the force is instantaneous. On the contrary, when the sample is viscoelastic, there is a loss of energy inside the material and the response of the sample to the external stimulus is delayed. Therefore, viscoelasticity is a time-dependent anelastic behavior of materials.

The viscoelastic response of complex systems, such as cells or tissues, formed of different types of polymers, often occurs over a wide range of time scales and comprises a continuum of relaxation times. Dissipative stresses inside the material can be due to the structure and mechanical properties of the polymeric network, but may also be an effect of the flow of liquid through the porous matrix (Moeendarbary et al., 2013, Kalcioglu et al., 2012).

Unlike purely elastic materials that recover their shape after the applied load is removed, and unlike purely viscous materials that remain in the deformed state after the applied load is removed, viscoelastic materials present a superposition of these two properties. Such a behavior may be linear (stress and strain are proportional) or nonlinear. We will only consider the linear viscoelastic regime. Viscoelasticity is observed as a combination of both recoverable elastic and permanent

---

corrected for the bottom effect, thanks to the multiplication with the Taylor series coefficient. Moreover, we added to this table two models for a paraboloidal probe by Dimitriadis et al. and Garcia et al. because both are valid models, but they present some differences: the Garcia et al. model is valid when the paraboloid is closer to a sphere and the force values are higher than in the Dimitriadis et al. model, if the same parameters are used. This is well explained in the supplementary materials of Garcia and Garcia (2018).

viscous deformation. Nearly all biopolymer solutions, cells, and biological tissues exhibit viscoelasticity.

So far, we have been focusing on the approach curve to fit the contact elastic models. This may be partially due to a historical choice motivated by the fact that the retraction curve often features adhesion events that may affect the fitting procedure and the final results (Radmacher et al., 1996). In addition to the presence of adhesive features, as observed in Figure 2.2.1, the approach and retract curve may not overlap, the force in the retract trace typically being lower. Excluding the presence of plastic deformations, which is reasonable for relatively small indentations, this reflects some kind of dissipation during the whole cycle, likely due to the viscoelastic nature of cells. Indeed, the opening of the approaching-retracting force curve loop typically increases as the ramping frequency increases. Different types of approaches can be applied to determine the dynamic response of materials. Mainly, two types of experiments are carried out: oscillatory experiments (an oscillatory strain or stress is applied and the resulting stress or strain, respectively, is measured) at a constant frequency or sweeping over a frequency range, and creep experiments (stress is applied and kept constant, while strain over time is measured) or relaxation experiments (strain is applied and kept constant, while the stress over time is measured) (Alcaraz et al., 2003, Mahaffy et al., 2000). In this section, we will focus on the recent solutions describing the shape of the force–distance curves on a viscoelastic material. Recent AFM applications of theoretical developments have led to a set of equations that allow a complete fit of the loading and unloading force curves (or approach and retract), taking into account a viscoelastic response of the sample.

There are various approaches to determine the viscoelastic force–indentation relationship of a viscoelastic sample – some using numerical integration and others using analytical approximations – all based on the seminal works by Graham (1965), Lee and Radok (1960), and Ting (1966) and more recent work by Brückner et al. (2017), Garcia et al. (2020), Efremov et al. (2017, 2019). We will follow here the approach recently reported by Brückner and co-workers, which provides an analytical solution assuming constant indentation velocity and power law rheology. The general approach requires solving two problems, one for the loading part and another for the unloading part. Brückner and coworkers assumed a linear indentation ramp that leads to a maximum in the contact radius upon loading, and then decreases upon unloading. The solution uses equivalent elastic contact mechanics, with the elastic modulus being a function of the loading history. Thus, it requires the definition of the viscoelastic response of the sample. As shown earlier for a number of cell types and tissues, using AFM and other techniques, the viscoelasticity of living cells and extracellular matrices is well described by a power law relaxation function, $E(t) = E_0 \Psi(t) = E_0 (t/t_0)^{-\beta}$, both at low and at high frequencies (Balland et al., 2006, Fabry et al., 2001, Jorba et al., 2017, 2019, Rigato et al., 2017). Thus, we will limit our description to the solution of a material exhibiting such a power law response. While solutions have been proposed for other viscoelastic relaxation functions $\Psi(t)$, such as standard linear solid or Kelvin–Voigt models, their application to living cells is limited (Garcia et al., 2020, Garcia and Garcia, 2018c, Greenwood, 2010).

All proposed approaches depart from the definition of time-dependent force ($F(t)$) that in turn depend on the time-dependent Young's modulus ($E(t)$):

$$F(t) = \tilde{C}^{-1} \int_0^t E(t-\tau) \frac{\partial \delta(t)^n}{\partial \tau} d\tau \tag{2.2.42}$$

where $E(t) = E_0 \psi(t-\tau)$, with ψ being the relaxation function describing the viscoelastic response of the material, δ is the indentation, and the pre-factor $\tilde{C}$ and the exponent n are constants that depend on the tip geometry as defined below. Actually, they are given by the force and indentation in the equivalent contact elastic model (e.g., $n = 2$ for a cone, $n = 3/2$ for a paraboloid of revolution).

To determine the prefactor $\tilde{C}$ and the indentation, Brückner et al. (2017) use the approach by Popov for an arbitrary axisymmetric punch of profile $f(r)$ that relates the indentation up to a time $t = t_m$ (time of maximum contact radius $a(t_m) = a_{max}$) with the time dependent on the contact radius (Popov, 2010):

$$\delta(t) = a(t) \int_0^{a(t)} \frac{f'(r)dr}{\sqrt{a^2(t) - r^2}} \tag{2.2.43}$$

This assumes a monotonically increasing contact radius, which is achieved for the case of a linear indentation ramp $\delta(t) = v_0 t$. The force is then given by

$$F(t) = 2E_0 \int_0^t \psi(t-\tau) \frac{\partial}{\partial \tau} \left(\int_0^{a(\tau)} \frac{f'(r)r^2 dr}{\sqrt{a^2(\tau) - r^2}} \right) d\tau \tag{2.2.44}$$

The indentation profiles for a cone, a paraboloid, and a flat cylinder have been defined above.

For a cone

$$\delta(t) = \left(\frac{\pi}{2}\right) a(t)\cot\alpha, \; \tilde{C}_c = \frac{\pi(1-v^2)}{2\tan\alpha}, \text{ and } n = 2 \tag{2.2.45}$$

For a paraboloid of revolution

$$\delta(t) = \left(\frac{a^2(t)}{R}\right), \; \tilde{C}_p = \frac{3(1-v^2)}{4\sqrt{R}}, \text{ and } n = \frac{3}{2} \tag{2.2.46}$$

and for a flat-ended cylinder of radius a_{cp}, the indentation does not depend on the contact radius and, therefore, we can simply use $\tilde{C}_{cp} = (1-v^2)/2a_{cp}$.

In principle, the loading trace can be generalized for a nonaxisymmetric punch by defining the appropriate punch profile, depending on the azimuthal angle $f(r)$

and using the Rayleigh–Ritz approximation of the contact area. Although not rigorously developed, the symmetry of the problem suggests a possible approximation using such an approach. Indeed, in the work of Brückner and coworkers, the authors assumed that for a regular four-sided pyramid, the final solution will be the same as for a cone, but with a different geometrical prefactor $\delta(t) = \sqrt{2}a(t)\cot\alpha$, and $C = 1.342\left((1-v^2)/\tan\alpha\right)$ is obtained from the numerical solution by Bilodeau (1992), or $\tilde{C}_{pyr} = \left((\sqrt{2}(1-v^2))/\tan\alpha\right)$ from the analytical approximation by Barber and Billings (1990).

As mentioned earlier, the relaxation function is assumed to follow a power law response with $\psi(t) = (t/t_0)^{-\beta}$, that is, $E(t) = E_0(t/t_0)^{-\beta}$. Thus, from eq. (2.2.41), for the loading or approach trace[7] for a cone, they obtain

$$F_a(t) = \frac{v_0^2}{\tilde{C}_c} \int_0^t E_0 \left(\frac{t-\tau}{t_0}\right)^{-\beta} \tau \, d\tau = 2\frac{v_0^2}{\tilde{C}_c} E_0 \frac{t_0^\beta}{(2-3\beta+\beta^2)} t^{2-\beta}$$

$$= 2\frac{v_0^2}{\tilde{C}_c} E_0 \frac{t_0^\beta \Gamma[2]\Gamma[1-\beta]}{\Gamma[3-\beta]} t^{2-\beta} = 2\frac{E(t)}{\tilde{C}_c(2-3\beta+\beta^2)} \delta^2(t) \qquad (2.2.47)$$

For a paraboloid

$$F_a(t) = \frac{v_0^{\frac{3}{2}}}{\tilde{C}_p} \int_0^t E_0 \left(\frac{t-\tau}{t_0}\right)^{-\beta} \tau^{1/2} d\tau = \frac{v_0^{\frac{3}{2}}}{\tilde{C}_p} E_0 \frac{t_0^\beta 3\sqrt{\pi}\Gamma(1-\beta)}{4\Gamma\left(\frac{5}{2}-\beta\right)} t^{\frac{3}{2}-\beta} = \frac{3E(t)\sqrt{\pi}\Gamma(1-\beta)}{\tilde{C}_p 4\Gamma\left(\frac{5}{2}-\beta\right)} \delta^{3/2}(t)$$

$$(2.2.48)$$

And for a flat-ended cylinder

$$F_a(t) = \frac{v_0}{\tilde{C}_{cp}} \int_0^t E_0 \left(\frac{t-\tau}{t_0}\right)^{-\beta} d\tau = \frac{v_0}{\tilde{C}_{cp}} E_0 \frac{t_0^\beta}{(1-\beta)} t^{1-\beta} = \frac{E(t)}{\tilde{C}_{cp}(1-\beta)} \delta(t) \qquad (2.2.49)$$

Calculation of the unloading trace requires knowledge of the time $t_1 < t_m$ at which $a(t) = a(t_1)$ for $t > t_m$. Again, assuming unloading with a linear indentation ramp at the same rate, than during loading, going from t_m until $\delta(t) = 0$. Thus, the indentation follows $\delta(t) = v_0(2t_m - t)$ and the contact radius decreases monotonically for $t > t_m$ from the maximum at $t = t_m$ with the condition to find $t_1(t)$ being

$$\int_{t_1(t)}^t \psi(t-\tau) \frac{\partial \delta(t)}{\partial \tau} d\tau = 0 \qquad (2.2.50)$$

Substituting the form of $\delta(t)$ and solving $\Psi(t) = (t/t_0)^{-\beta}$ leads to

---

7 In the loading or approach trace equations, the expression $\dfrac{\Gamma[2]\Gamma[1-\beta]}{\Gamma[3-\beta]} = 1/(2-3\beta+\beta^2)$, the same also for the unloading or retract trace equations.

$$t_1(t) = t - 2^{\frac{1}{1-\beta}}(t - t_m) \tag{2.2.51}$$

A relevant assumption of this approach is that the contact area increases monotonically with time, which is important for using Lee and Radok's (1960) viscoelastic correspondence principle. Thus, as developed by Ting, the retract trace is derived from the approach elastic solution, with an increasing contact area equation (2.2.43), but with the integration going from 0 to $t_1(t)$. Thus, for a cone, we obtain

$$F_r(t) = 2\frac{v_0{}^2 E_0 t_0{}^\beta}{\tilde{C}_c(2 - 3\beta + \beta^2)}\left(2\left[t(2 - \beta) + 2^{\frac{1}{1-\beta}}(1 - \beta)(t - t_m)\right](t - t_m)^{1-\beta} - t^{2-\beta}\right) \tag{2.2.52}$$

For a paraboloid, the retract expression would involve ordinary hypergeometric functions ($_2F_1$):

$$F_r(t) = \frac{3v_0{}^{\frac{3}{2}}E_0 t_0{}^\beta}{\tilde{C}_p(3 + 4(\beta - 2)\beta)}\, t_1^{-1/2}(t - t_1)^{1-\beta}\left((2\beta - 1)t_1 - t + {}_2F_1\left(1, \frac{1}{2} - \beta, \frac{1}{2}, \frac{t_1}{t}\right)\right) \tag{2.2.53}$$

And for a flat-ended cylinder

$$F_r(t) = \frac{v_0}{\tilde{C}_{cp}}E_0\frac{t_0{}^\beta}{(1 - \beta)}\left(t^{1-\beta} - 2(t - t_m)^{1-\beta}\right) \tag{2.2.54}$$

As noticed by Ting, the contact radius is given by the indenter shape during loading, but it also depends on the material properties during unloading.

The loading and unloading trace equations for the viscoelastic model obtained by Brückner et al. are recapitulated in Table 2.2.3. In principle, the above approach should be also valid for the case of different loading and unloading velocities. As mentioned earlier, the developed models assume a linear indentation ramp. This is an important assumption. This condition was verified by the authors in AFM measurements on cells, concluding that it will be valid when applying a linear piezo ramp using relatively stiff cantilevers to indent a soft sample. More practically, the assumption is valid if the deflection of the cantilever is negligible compared to the applied indentation. If the linear indentation rate assumption does not hold, numerical approaches have been proposed by Efremov et al. (2017, 2019), including corrections for the bottom effect, based on the above equations for $F(t)$ and t_1 (eqs. (2.2.47) and (2.2.46)). An analytical approximation for a viscoelastic sample of finite thickness was also recently reported by Garcia et al. (2020), as introduced in the next section.

In Table 2.2.3, $\tilde{C}_p = (3(1 - v^2)/4\sqrt{R})$, for a paraboloid, $\tilde{C}_c = (\pi(1 - v^2)/2\tan\alpha)$ for a cone $\tilde{C}_{pyr} = \left(\sqrt{2}(1 - v^2)/\tan\alpha\right)$ for a pyramid, $\tilde{C}_{cp} = ((1 - v^2)/2a_{cp})$ for a flat cylinder.

Table 2.2.3: Principal force–time equations for the viscoelastic models for different probe geometries, where $F(t)$ is the approach curve and $F_b(t)$ is the retract.[8]

Probe geometry	Force–time function, F(t)	Reference	Limits of validity
Cone of semi-included angle θ or pyramid	$F_a(t) = \dfrac{2}{\tilde{C}_c \text{ or } \tilde{C}_{pyr}} \dfrac{v_0{}^2 E_0 t_0{}^\beta}{(2-3\beta+\beta^2)} t^{2-\beta}$ $F_r(t) = \dfrac{2}{\tilde{C}_c \text{ or } \tilde{C}_{pyr}} \dfrac{v_0{}^2 E_0 t_0{}^\beta}{(2-3\beta+\beta^2)}$ $\left(2\left[t(2-\beta) + 2^{\frac{1}{1-\beta}}(1-\beta)(t-t_m) \right] \right.$ $\left. (t-t_m)^{1-\beta} - t^{2-\beta} \right)$	Brückner et al. (2017)	Linear, equal approach and retract $\dot{\delta}(t)$
Paraboloid of curvature 1/2 R (radius R)	$F_a(t) = \dfrac{v_0{}^{3/2}}{\tilde{C}_p} E_0 \dfrac{t_0{}^\beta 3\sqrt{\pi}\Gamma(1-\beta)}{4\Gamma(\frac{5}{2}-\beta)} t^{\frac{3}{2}-\beta}$ $F_r(t) = \dfrac{3v_0{}^{\frac{3}{2}} E_0 t_0{}^\beta}{\tilde{C}_p(3+4(\beta-2)\beta)} t_1{}^{-1/2}(t-t_1)^{1-\beta}$ $\left((2\beta-1)t_1 - t + {}_2F_1\left(1, \frac{1}{2}-\beta, \frac{1}{2}, \frac{t_1}{t}\right) \right)$	Brückner et al. (2017) and this chapter	Linear, equal approach and retract $\dot{\delta}(t)$
Flat-ended cylinder of radius a	$F_a(t) = \dfrac{v_0}{\tilde{C}_{cp}} E_0 \dfrac{t_0{}^\beta}{(1-\beta)} t^{1-\beta}$ $F_r(t) = \dfrac{v_0}{\tilde{C}_{cp}} E_0 \dfrac{t_0{}^\beta}{(1-\beta)} \left(t^{1-\beta} - 2(t-t_m)^{1-\beta} \right)$	Brückner et al. (2017) and this chapter	Linear, equal approach and retract $\dot{\delta}(t)$

2.2.5 Bottom Effect Correction for Viscoelastic Models

In the previous sections we have seen the importance of probe geometry, sample thickness, and sample viscoelasticity when mechanical properties of soft materials are to be obtained from force–distance curves. The contact models explained above consider different probe geometries, sample finite thickness, and sample viscoelasticity. Both living tissues and cells are viscoelastic, as mentioned earlier, but while tissues are generally thick compared to the indentation achieved during nanomechanical measurements, living cells can be very thin and a bottom effect correction is often

---

8 We prefer to present in this table the force–time instead of the force–indentation equations because the viscoelastic model depends on time, as explained before, so it is easier to obtain the analytical expression of force in function of time from the theory. The conversion from a force–time curve to a force–indentation curve is straightforward when it is known the waveform that describes the indentation $\delta(t)$.

needed. Therefore, an accurate determination of cell mechanics across the whole cell surface requires combining a viscoelastic model with a bottom-effect correction for different probe geometries, to consider at the same time the finite thickness of the cell and its complex viscoelastic response. Different approaches have been proposed (Darling et al., 2007).

A simple and practical analytical solution was proposed by Garcia and Garcia (2018c). In this model, the cell was considered as an incompressible material with a linear viscoelastic response described by a Kelvin–Voigt model. The authors later realized that the Kelvin–Voigt model leads to an artifactual jump in the retraction curve, not observed experimentally in living cells, and proposed the solution for a power law model (Garcia et al., 2020). Thus, only the model derived from the power law relaxation function is valid in both loading and unloading traces for living cells. The universal power law response observed in living cells further justifies this choice (Fabry et al., 2001).

The proposed development is based on:
1) Betti's reciprocal theorem and Rayleigh–Ritz approximation to relate pressures and deformations for different geometries
2) The equivalence principle between elastic and viscoelastic deformation
3) Ting's method to obtain the force as a function of deformation history
4) Boundary conditions involving a cell adherent on a rigid support

The authors provided analytical solutions for a conical indenter and we refer to the original article for the specific equations (Garcia et al., 2020).

We have provided relevant contact models to quantify the mechanics of soft, complex systems from AFM force–distance curves, focusing on the importance of probe geometry, sample thickness, and energy dissipation. One of the parameters that we have ignored and that might be relevant is adhesion. This will be briefly addressed in the following section.

2.2.6 Contact Models Considering Adhesion

As mentioned above, the dissipative response of biomaterials, reflected by hysteresis between loading and unloading curves, is often due to the viscoelastic response of the sample. However, adhesion between the probe and the sample may also be at the origin. Indeed, when working with cells or tissues, adhesion between the AFM tip and the sample is often observed and tip passivation strategies can be used to minimize it.

On some occasions, probes are functionalized with adhesion proteins to measure adhesion or non-specific interactions that occur between the tip and the sample. In that case, the force–distance curves will feature pronounced negative forces upon unloading, due to the stretching of the sample through the formed bonds. The analysis of this type of curves is generally not carried out using continuum contact elastic models considering adhesion and might not be convenient for measurements on cells or tissues, as adhesion is often mediated by discrete adhesion complexes that are not well described by the formalism required in continuum mechanics. Instead, non-specific adhesion on macroscopic objects is more prone to this type of analysis. Nevertheless, it might be useful to know the available models developed in the context of classical non-specific adhesion on macroscopic objects. Moreover, some commercial software use such models. We thus briefly describe here the most known JKR and DMT contact models, since they may help in better estimating the sample elastic modulus under conditions of adhesion.

After the initial work by Derjaguin (1934), the first approaches to adhesion between elastic bodies in contact were developed by Johnson, Kendall, and Roberts (the JKR model), and by Derjaguin, Muller, and Toporov (the DMT model) (Derjaguin et al., 1975, Johnson et al., 1971). Tabor discussed the transition from the DMT to the JKR regime (Tabor, 1977), while Maugis developed a generalized multiparametric model (Maugis, 1992, Popov, 2010).

When considering adhesion between two elastic bodies, unlike the developments explained earlier, in addition to compression stresses, we should consider tensile traction, mainly generated both outside or inside the contact area, depending on the adhesive properties of the interacting surfaces and on the size of the contact region. Thus, at zero applied force, the contact area is not zero, but finite.

In this section, we will present two of the most used models: JKR and DMT (Derjaguin et al., 1975, Johnson et al., 1971). In general, JKR is valid for large, flexible spheres, and short-range adhesive interactions. Under these conditions, the leading contribution to adhesion comes from inside the contact area, and it therefore depends on the applied load. DMT is typically valid for small, rigid surfaces in contact (or at least one of the two), and/or long-ranged adhesive interactions. Under these conditions, the contribution to adhesion from outside the contact area is dominating, and nearly constant (Popov, 2017). Both models consider a spherical indenter of radius R in contact with a planar surface. Often, the models are presented with the sphere being the elastic body, while the surface is rigid; but the formulation is valid for the opposite case too.

The JKR assumes that the energy of the interaction is given by an elastic, storage term, described by the Young's modulus (E), and an attractive, dissipative term, described by a surface energy (γ),[9] acting only within the contact area (Johnson, 1985). The pressure distribution in the contact area is then assumed to be a superposition of the Hertzian pressure due to compressive stresses of the elastic body around the center

---

9 In the following formulas, the same surface energy γ is used for both contacting surfaces so that the work of adhesion W per unit area required to separate the two surfaces is $W = -2\gamma$. This definition may vary for different references. In the general case of different surface energies γ_1 and γ_2, the work of adhesion is $W = -(\gamma_1 + \gamma_2 - \gamma_{12})$, where γ_{12} is the interfacial energy.

and the flat cylinder pressure due to tensile stresses (diverging at the rim of contact) given by the presence of adhesion forces. By considering the work done in compression by the pressure and minimizing the total energy at equilibrium, it was found that in the JKR model, there still exists a formal Hertzian relationship between the contact radius a and an effective force F_{JKR}, which takes adhesion into account:

$$F_{JKR} = \frac{4}{3}\frac{E^*}{R}a^3 \qquad (2.2.55)$$

with

$$F_{JKR} = F_n + 6\pi\gamma R + \sqrt{12\pi\gamma RF_n + (6\pi\gamma R)^2} \qquad (2.2.56)$$

where F_n is the external applied force and $E^* = (E/(1-v^2))$ is the reduced Young's modulus.

Solving for F_n provides a relation between F_n and the contact radius a:

$$F_n = E^* \left(\frac{4}{3}\frac{a^3}{R} - 4\sqrt{\frac{\pi\gamma a^3}{E^*}} \right) \qquad (2.2.57)$$

Notice that the first term in eq. (2.2.57) reminds the Hertz model, but with a larger contact radius. The second term accounts for the force due to adhesion and depends on the rigidity of the sample (the softer the sample, the larger is the contact area and the higher is the adhesive force). Because of the adhesion forces, the same contact radius can be obtained with a smaller external applied force F_n.

In the limit of zero applied force ($F_n = 0$), the JKR model predicts a finite contact radius a_0:

$$a_0 = \left(\frac{9\pi\gamma R^2}{E^*} \right)^{\frac{1}{3}} \qquad (2.2.58)$$

It is possible to apply a negative external force to overcome adhesion. The contact radius reduces to a critical value, after which the contact is broken and the stress is suddenly released (pull-off, PO). The critical pull-off force $F_{JKR,PO}$ can be obtained from eq. (2.2.56) by noticing that the term under the square root must be nonnegative, that is, $12\pi\gamma RF_n + (6\pi\gamma R)^2 \geq 0$, and finding the force for which the equality strictly holds:

$$F_{JKR, PO} = -3\pi\gamma R \qquad (2.2.59)$$

Interestingly, the pull-off force is independent of the Young's modulus of the material.

The pull-off force is typically measured from a force–distance curve, recorded with the AFM, as the depth of the adhesion well in the retracting branch. In principle, measuring the pull-off force and knowing the radius of the probe can provide the surface energy γ for both the JKR and the DMT models, through eqs. (2.2.59) and (2.2.65).

In the JKR model, the contact radius a does not go smoothly to zero, but at pull-off it is still finite and equal to:

$$a_{PO} = \left(\frac{9\pi\gamma R^2}{4E*}\right)^{\frac{1}{3}} \qquad (2.2.60)$$

that is, $a_{PO} \approx 0.63a_0$.

The Hertzian formula $a = \sqrt{R\delta}$ does not hold for the JKR model and the following equation replaces the Hertzian one:

$$\delta = \frac{a^2}{R} - \frac{2}{3}\sqrt{\frac{9\pi\gamma a}{E*}} \qquad (2.2.61)$$

where the contact radius for a given indentation is larger than for the non-adhesive case because of the surface energy.

In the case of the JKR model, it is not possible to obtain a single equation relating the applied force F_n and the indentation δ, as in the Hertz model. Nevertheless, it is possible to obtain a system of equations, which can be solved recursively to obtain the force–indentation relationship, similar to the case of the nonadhesive Sneddon model for the spherical indenter, eq. (2.2.13). This system of equations consists of eqs. (2.2.57) and (2.2.61). Alternatively, in this system, eq. (2.2.57) can be replaced by eq. (2.2.62), obtained from eqs. (2.2.56) and (2.2.57) after specifying the pull-off force, $F_{JKR,PO}$, using eq. (2.2.59). Equation (2.2.62) has the advantage that it depends on the measurable forces, F_n and $F_{JKR,PO}$, and the contact radius a appears only in the right side:

$$\left(\sqrt{F_n + |F_{JKR, PO}|} + \sqrt{|F_{JKR, PO}|}\right)^2 = \frac{4}{3}\frac{E*}{R}a^3 \qquad (2.2.62)$$

The DMT model is derived from the Derjaguin approximation (Derjaguin, 1934), and assumes that the adhesion only acts outside the contact area, which is negligibly small; this assumption leads to a constant adhesive force, $F_{adh} = 4\pi\gamma R$ (see footnote[10]), which adds to the external applied force F_n, determining a total normal force, which causes an increased radius of contact a, as in the case of the JKR model. In contrast to the JKR case, however, the adhesion force does not increase with indentation and does not depend on the elastic properties of the sample.

Thus, force versus contact radius relationship is still Hertzian, but with a total normal force $F_n + 4\pi\gamma R$:

---

10 According to the Derjaguin approximation, the force between a curved surface of radius R and a flat surface is equal to $F = 2\pi RW$, where W is the interaction energy per unit area. Assuming $W = 2\gamma$ and neglecting the area of the contact, this leads to a constant adhesion force $F_{adh} = 4\pi\gamma R$.

$$F_n + 4\pi\gamma R = \frac{4}{3}\frac{E^*}{R}a^3 \tag{2.2.63}$$

It follows that a force–indentation relationship similar to the Hertz model holds, with a constant force offset:

$$F_n = \frac{4}{3}E^*\sqrt{R}\delta^{\frac{3}{2}} - 4\pi\gamma R \tag{2.2.64}$$

Here, the critical or pull-off negative force, at which the tip detaches from the sample, is

$$F_{DMT,PO} = -4\pi\gamma R \tag{2.2.65}$$

and the radius of contact at pull-off is zero, at odds with the JKR case.

The JKR and DMT models are not commonly used for cell and tissue measurements, where it is typically assumed that nonspecific adhesion is negligible but may help in realizing the effect of adhesion on the measured elastic parameters. This is particularly true for the DMT model, where only adhesion forces outside the contact region, that is, over a distance, are considered. Usually this could be due to van der Waals forces, which act over hundreds of nm and will always be attractive (except some exotic special cases). Since, in cells, long-range van der Waals forces are compensated by long-range polymer forces (generated by the glycocalix) of cells, we expect that this model is not relevant for interactions between cells and probes. The JKR model, on the other hand, assumes wetting-like adhesion between the sample and the probe, characterized by a surface energy γ. A wetting-like behavior has been described for spreading and adhesion of cells on solid supports by Sackmann and Bruinsma (2002). However, this process is slow, since it requires diffusion of adhesion molecules on the cell surface, which requires some time to

Table 2.2.4: Force–indentation function of the JKR and DMT models considering adhesion, where F_n is the applied normal force and F_{PO} is the pull-off force.

Probe geometry	Force–indentation function, $F(\delta)$	Reference	Limits of validity
JKR model Sphere of radius R	$\begin{cases} \left(\sqrt{F_n + |F_{JKR,PO}|} + \sqrt{|F_{JKR,PO}|}\right)^2 = \frac{4}{3}\frac{E^*}{R}a^3 \\ \delta = \frac{a^2}{R} - \frac{2}{3}\sqrt{\frac{9\pi\gamma a}{E^*}} \\ F_{JKR,PO} = -3\pi\gamma R \end{cases}$	Johnson et al. (1971) Other probe shapes Popov et al. (2019b)	Short-range adhesive interactions Deformable interface/ large probe (large contact area)
DMT model Sphere of radius R	$F_n = \frac{4}{3}E^*\sqrt{R}\delta^{\frac{3}{2}} - |F_{DMT,PO}|$ $F_{DMT,PO} = -4\pi\gamma R$	Derjaguin et al. (1975)	Long-ranged adhesive interactions Rigid interface/small probe (small contact area)

establish. Thus, it is safe to assume for most applications where contact time is short (below 1 s) that adhesion forces can be neglected in AFM mechanical data. The force-indentation functions of the JKR and DMT models considering adhesion are listed in table 2.2.4.

2.2.7 Thin Shells

The theoretical framework described so far to determine the mechanical properties of living cells assumes a contact between two solid bodies. This has been shown to be a good description for eukaryotic cells but might not be the case for other types of cells, such as bacteria or plant cells. In that case, unlike the very compliant plasma membrane of the eukaryotic cells, bacteria present a rigid cell wall that may be deformed by bending at very small depths caused by the AFM tip. Thus, the theory described above may not be valid except for a very small indentation depth (Loskill et al., 2014). In the case of larger deformations, the theory of thin shells has been used. The simplest model might be that of a convex spherical cap of thickness h and radius R, loaded by a force F (the AFM tip) at one point. In this context, we can define two characteristic quantities, the bending stiffness:

$$K = \frac{Eh^3}{12(1-v^2)} \qquad (2.2.66)$$

and the extensional stiffness:

$$\eta = \frac{Eh}{(1-v^2)} \qquad (2.2.67)$$

The relationship between force and deformation can be derived assuming that bending prevails over stretching and that the deformation δ and thickness h of the shell are much smaller than the radius of the spherical cap (δ, $h \ll R$), as shown in figure 2.2.7, (Landau and Lifshitz, 1986):

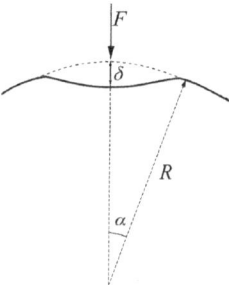

Figure 2.2.7: Schematic drawing of the thin-film model (inspired from Landau and Lifshitz, 1986).

$$F = \frac{4}{\sqrt{3(1-v^2)}} \frac{Eh^2}{R} \delta. \qquad (2.2.68)$$

As can be seen from the equation above, the force-deformation relationship is linear in this case, with a slope related to the geometry (R, h) and the mechanical properties of the shell (E, v).

This theoretical framework has been used, for example, to determine the stiffness of microcapsules or bacterial cell walls (Arnoldi et al., 2000, Dubreuil et al., 2003, Fery and Weinkamer, 2007, Gaboriaud et al., 2005). The force-indentation equation corresponding to the thin shell model is shown in table 2.2.5.

Table 2.2.5: Thin shell model force–deformation relationship.

Probe geometry	Force–indentation function, $F(\delta)$	Reference	Limits of validity
Thin spherical shell stiffer than the material inside with thickness h and radius R	$F = \dfrac{4}{\sqrt{3(1-v^2)}} \dfrac{Eh^2}{R} \delta$	Landau and Lifshitz (1986)	δ, $h \ll R$

2.2.8 Finite Element Modeling

In the previous sections of this chapter, we presented some of the principal analytical solutions of contact problems for configurations commonly used in indenter experiments. As we explained, these analytical solutions can be obtained when the symmetry of the contact between the two bodies allows simplifying the geometry of the contact problem. This is possible when the external force applied is normal to the sample's surface, the probe geometry is axisymmetric, the bodies are isotropic, and the strains are small (even if some analytical solutions are available for tangent contact (Popov et al., 2019)). These important assumptions allow the method of dimensionality reduction (MDR) (Popov and Heß, 2013, 2015) to reduce a three-dimensional contact mechanics problem to a two-dimensional problem.

This reduction, thus, is not always possible with all the probe geometries and the analytical solutions described in the previous sections are accurate only within the limits of validity mentioned for each model. In the cases in which analytical solutions are difficult to achieve, because of the complexity of the contact problem resulting from intricate geometries of contacting bodies or from large deformations and nonlinear materials, the most general Bousinnesq problem of stresses and deformations or strains arising from bodies in contact can often be solved numerically. In these cases, the stress-strain relation cannot be anymore simplified and in the case of a linear elastic material, it is described by the following general equation:

$$\boldsymbol{\sigma} = \boldsymbol{D}{:}\varepsilon, \quad \sigma_{ij} = D_{ijkl}\varepsilon_{kl} \tag{2.2.69}$$

where σ is the rank-2 stress tensor, ε the rank-2 strain tensor, and $\boldsymbol{D}$ is a rank-4 elasticity tensor. At this point, many contact mechanics problems are described by the theory of mechanics of continuum bodies, where the structural problem is formulated as a set of differential equations that are satisfied at every point in the domain. The system of differential equations is obtained considering the principal three fundamental laws of mechanics: conservation of mass, conservation of linear momentum, and conservation of angular momentum.

The principle of the conservation of mass allows a Lagrangian description of the contact problem; the principle of the conservation of the angular momentum guarantees that the stress tensor is symmetric; while the principle of conservation of linear momentum is imposed on an infinitesimal element of a structure to impose the principle of force equilibrium and obtain a system of partial differential equations (PDE) along with boundary conditions that is called the boundary-valued problem (BVP). If the system is conservative, also a variational method based on the principle of minimum potential energy can be developed to solve the problem.

The principle numerical implementations to solve variational equations or BVP problems widely used and validated are the finite element method (FEM) and boundary element method (BEM). While the BEM has been used for general contact problems, the FEM method is specific to solid mechanics. An exhaustive description of the numerical methods, with the derivation of all the equations, would need a specific manual, which goes beyond the aim of this chapter. Accordingly, we briefly illustrate the basic concept of the FEM method, specific to solid mechanics, and we refer to more exhaustive works for further details.

The FEM method allows solving partial differential equations in two or three space variables. FEM analysis is widely used in many fields of physics or mechanical engineering to solve linear or nonlinear problems, which cannot be solved analytically, such as contact problems for the study of the mechanical properties of new materials, climate models for atmospherical predictions, mechanical problems related to seismology and geophysics, and many others. Nowadays, the FEM analysis is used in several fields of industry, such as: biomechanical industry for the design of new prosthesis, cosmetics, aeronautical or automotive industry to study the resistance of new materials to heat and impact, for example, in case of vehicle crash.

The method consists in simplifying a complex problem – dividing a large system into smaller subunits (Figure 2.2.8), for which the solution is simpler. The small subunits are called finite elements.

The subunits are obtained thanks to a discretization of the space in two or three dimensions, implementing a mesh of the object with finite number of points, which is the domain of the unknown function. Indeed, for each finite element, simpler

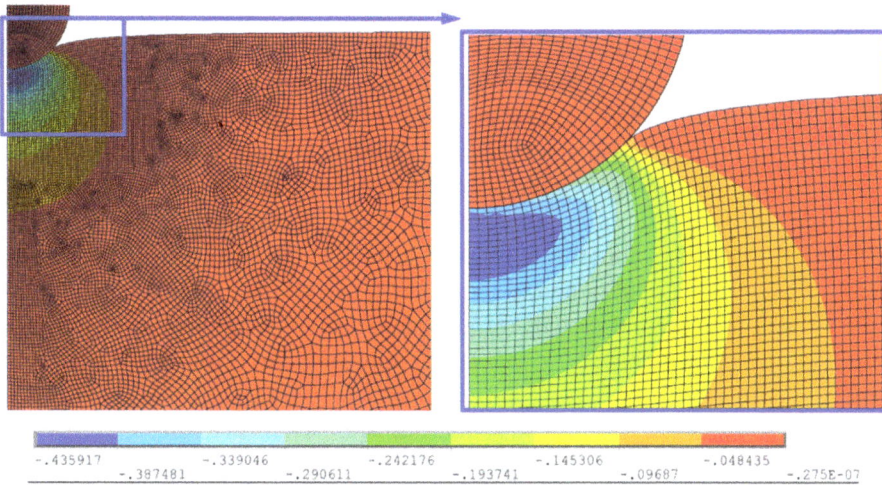

Figure 2.2.8: Example of a finite element simulation for a rigid indenter indenting orthogonally a soft semi-infinite half space in the case of large deformations, using finite element package ANSYS. The figure was reproduced from Wu et al. (2016).

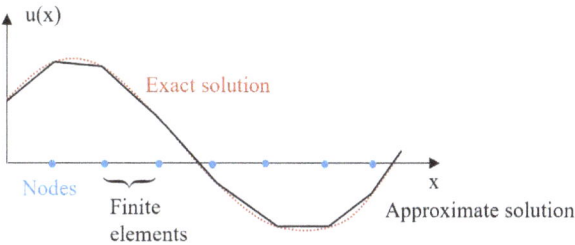

Figure 2.2.9: Graphical representation of the linear approximation for the function u(x) with the finite element method used to solve a one-dimensional problem. Adapted from Wu et al. (2016).

equations are defined, which are then assembled in a large system of algebraic equations over the entire domain.

The finite elements are connected through nodes to the adjacent elements. The variational equation solution of each finite element is then not solved analytically, but it is approximated in a polynomial form for the solution to the entire problem. The approximate solution $u(x)$ is expressed as a sum of a number of functions that are called trial functions, as shown in figure 2.2.9. The FEM analysis applied in contact mechanics can answer the following questions: (1) if two or more bodies are in contact, (2) where is and what is the region of contact, (3) how much force and pressure is distributed in the contact interface, (4) the magnitude and distribution of the strains in the material due to the stresses, and (5) the relative motion in general of the bodies after the contact. Therefore, the FEM method can be very useful to study contact problems that cannot be solved analytically. It can also be used as a supportive study to validate a contact mechanics analytical model. Indeed, FEM analysis can provide some additional information, for example, the lateral displacements due to the probe indentation of a soft material, which are not considered in analytical models due to the assumption of perfect normal penetration of the indenter.

In many recent works, in which contact mechanics models are developed, the results of the analytical model are compared to those obtained, thanks to a FEM analysis, in order to validate the analytical model. For example, in Doss et al. (2019), Efremov et al. (2017), and in Garcia et al. (2018), the FEM simulations were used to compare different models (or just numerical, or numerical and analytical relationships) to then choose the theoretical approach that could better model the experimental data.

The principal steps for the contact analysis with FEM method are:
1. Defining the problem geometry and the contact pairs and types
2. Searching for the contact point
3. Calculating contact force and tangent stiffness

The detailed explanation of these three steps and the relative equations are described in a comprehensive but didactic manner in the book *Introduction to Nonlinear Finite Element Analysis* by Nam-Ho Kim (2015), where a section is also dedicated to the Matlab implementation of FEM analysis for contact mechanics problems and where some Matlab codes are available. Apart from Matlab FEM implementation, several software tools are available nowadays for a user-friendly application of FEM analysis; some commercial ones are, for example, ANSYS, COMSOL software (COMSOL Multiphysics; CMSOL AB, Stockholm, Sweden) and Abaqus CAE (Simulia Corp. Providence, RI), and some open source, FreeFem++ and FEBio.

2.2.9 Concluding Remarks

In this chapter, we have tried to provide the most widely used contact models to describe AFM measurements on cells and tissues. Those included purely elastic bodies, viscoelastic samples finite bodies, and adhesion. There are, of course, a large number of models that were not considered. For example, models including different layers of materials with varying elasticity, models considering surface tension, and nonlinear materials (Bhushan and Peng, 2002). We have also constrained to models used to fit the force–indentation curves. Other approaches, such as the Oliver and Pharr method, widely used in material science, or the early A-Hassan approach, developed for AFM measurements on cells, have been omitted as they can be derived from the formalisms used here (A-Hassan et al., 1998, Lin and Horkay, 2008, Pharr et al., 1992). Given the widespread application of AFM as a nanomechanical tool in biology and given the heterogeneity and nonlinearity of biological samples, we expect the emergence of more sophisticated approaches and derivations, likely involving computer simulations that would move further and further away from the seminal Hertz model.

References

A-Hassan, E., Heinz, W., Antonik, M., D'Costa, N., Nageswaran, S., Schoenenberger, C.-A., and Hoh, J. (1998). "Relative Microelastic Mapping of Living Cells by Atomic Force Microscopy." Biophysical Journal **74**: 1564–1578. https://doi.org/10.1016/S0006-3495(98)77868-3.

Akhremitchev, B.B., and Walker, G.C. (1999). "Finite Sample Thickness Effects on Elasticity Determination Using Atomic Force Microscopy." Langmuir **15**: 5630–5634. https://doi.org/10.1021/la980585z.

Alcaraz, J., Buscemi, L., Grabulosa, M., Trepat, X., Fabry, B., Farré, R., and Navajas, D. (2003). "Microrheology of Human Lung Epithelial Cells Measured by Atomic Force Microscopy." Biophysical Journal **84**: 2071–2079. https://doi.org/10.1016/S0006-3495(03)75014-0.

Aleksandrov, V.M. (1968). "Asymptotic methods in contact problems of elasticity theory: PMM vol.32, N   4, 1968, pp.672–683". Journal of Applied Mathematics and Mechanics **32**: 691–703. https://doi.org/10.1016/0021-8928(68)90099-3.

Aleksandrov, V.M. (1969). "Asymptotic solution of the contact problem for a thin elastic layer: PMM vol. 33, no. 1, 1969, pp. 61–73." Journal of Applied Mathematics and Mechanics **33**: 49–63. https://doi.org/10.1016/0021-8928(69)90113-0.

Arnoldi, M., Fritz, M., Bäuerlein, E., Radmacher, M., Sackmann, E., and Boulbitch, A. (2000). "Bacterial turgor pressure can be measured by atomic force microscopy". Phys. Rev. E **62**: 1034–1044. https://doi.org/10.1103/PhysRevE.62.1034.

Balland, M., Desprat, N., Icard, D., Féréol, S., Asnacios, A., Browaeys, J., Hénon, S., and Gallet, F. (2006). "Power laws in microrheology experiments on living cells: Comparative analysis and modeling". Phys. Rev. E **74**: 021911. https://doi.org/10.1103/PhysRevE.74.021911.

Barber, J.R., and Billings, D.A. (1990). "An approximate solution for the contact area and elastic compliance of a smooth punch of arbitrary shape". International Journal of Mechanical Sciences **32**: 991–997. https://doi.org/10.1016/0020-7403(90)90003-2.

Betti, E. (1872). "Teoria dell'elasticità". Il Nuovo Cimento Serie 2. Volume **VII-VIII**: 69–97.

Bhushan, B., and Peng, W. (2002). "Contact mechanics of multilayered rough surfaces." Appl. Mech. Rev **55**: 435–480. https://doi.org/10.1115/1.1488931.

Bilodeau, G.G. (1992). "Regular Pyramid Punch Problem". Journal of Applied Mechanics **59**: 519–523. https://doi.org/10.1115/1.2893754.

Boussinesq, J. (1885). "Application des potentiels à l'étude de l'équilibre et du mouvement des solides élastiques, principalement au calcul des déformations et des pressions que produisent, dans ces solides, des efforts quelconques exercés sur une petite partie de leur surface ou de leur intérieur : mémoire suivi de notes étendues sur divers points de physique mathématique et d'analyse. Lilliad - Université de Lille - Sciences et technologies.

Briscoe, B.J., Sebastian, K.S., and Adams, M.J. (1994). "The effect of indenter geometry on the elastic response to indentation". J. Phys. D: Appl. Phys. **27**: 1156–1162. https://doi.org/10.1088/0022-3727/27/6/013.

Brückner, B.R., Nöding, H., and Janshoff, A. (2017). "Viscoelastic Properties of Confluent MDCK II Cells Obtained from Force Cycle Experiments." Biophysical Journal **112**: 724–735. https://doi.org/10.1016/j.bpj.2016.12.032.

Chadwick, R.S. (2002). "Axisymmetric Indentation of a Thin Incompressible Elastic Layer". SIAM J. Appl. Math. **62**: 1520–1530. https://doi.org/10.1137/S0036139901388222.

Chen, W.T., and Engel, P.A. (1972). "Impact and contact stress analysis in multilayer media." International Journal of Solids and Structures **8**: 1257–1281. https://doi.org/10.1016/0020-7683(72)90079-0.

Chyasnavichyus, M., Young, S.L., Geryak, R., and Tsukruk, V.V. (2016). "Probing elastic properties of soft materials with AFM: Data analysis for different tip geometries." Polymer **102**: 317–325. https://doi.org/10.1016/j.polymer.2016.02.020.

Darling, E.M., Zauscher, S., Block, J.A., and Guilak, F. (2007). "A Thin-Layer Model for Viscoelastic, Stress-Relaxation Testing of Cells Using Atomic Force Microscopy: Do Cell Properties Reflect Metastatic Potential?." Biophysical Journal **92**: 1784–1791. https://doi.org/10.1529/biophysj.106.083097.

Derjaguin, B. (1934). "Untersuchungen über die Reibung und Adhäsion, IV". Kolloid-Zeitschrift **69**: 155–164. https://doi.org/10.1007/BF01433225.

Derjaguin, B.V., Muller, V.M., and Toporov, Yu.P. (1975). "Effect of contact deformations on the adhesion of particles". Journal of Colloid and Interface Science **53**: 314–326. https://doi.org/10.1016/0021-9797(75)90018-1.

D. Garcia, P., R. Guerrero, C., and Garcia, R. (2020). "Nanorheology of living cells measured by AFM-based force–distance curves". Nanoscale **12**: 9133–9143. https://doi.org/10.1039/C9NR10316C.

Dhaliwal, R.S., and Rau, I.S. (1970). "The axisymmetric boussinesq problem for a thick elastic layer under a punch of arbitrary profile". International Journal of Engineering Science **8**: 843–856. https://doi.org/10.1016/0020-7225(70)90086-8.

Dimitriadis, E.K., Horkay, F., Maresca, J., Kachar, B., and Chadwick, R.S. (2002). "Determination of Elastic Moduli of Thin Layers of Soft Material Using the Atomic Force Microscope." Biophysical Journal **82**: 2798–2810. https://doi.org/10.1016/S0006-3495(02)75620-8.

Doss, B.L., Eliato, K.R., Lin, K., and Ros, R. (2019). "Quantitative mechanical analysis of indentations on layered, soft elastic materials." Soft Matter **15**: 1776–1784. https://doi.org/10.1039/C8SM02121J.

Dubreuil, F., Elsner, N., and Fery, A. (2003). "Elastic properties of polyelectrolyte capsules studied by atomic-force microscopy and RICM". Eur. Phys. J. E **12**: 215–221. https://doi.org/10.1140/epje/i2003-10056-0.

Efremov, Y.M., Wang, W.-H., Hardy, S.D., Geahlen, R.L., and Raman, A. (2017). "Measuring nanoscale viscoelastic parameters of cells directly from AFM force-displacement curves." Sci Rep **7**: 1541. https://doi.org/10.1038/s41598-017-01784-3.

Efremov, Y.M., Kotova, S.L., and Timashev, P.S. (2020). "Viscoelasticity in simple indentation-cycle experiments: a computational study." Sci Rep **10**: 13302. https://doi.org/10.1038/s41598-020-70361-y.

Fabrikant, V.I. (1986). "Flat punch of arbitrary shape on an elastic half-space." International Journal of Engineering Science **24**: 1731–1740. https://doi.org/10.1016/0020-7225(86)90078-9.

Fabry, B., Maksym, G.N., Butler, J.P., Glogauer, M., Navajas, D., and Fredberg, J.J. (2001). "Scaling the Microrheology of Living Cells". Phys. Rev. Lett. **87**: 148102. https://doi.org/10.1103/PhysRevLett.87.148102.

Fery, A., and Weinkamer, R. (2007). Mechanical properties of micro- and nanocapsules: Single-capsule measurements. Polymer **48**: 7221–7235. https://doi.org/10.1016/j.polymer.2007.07.050.

Gaboriaud, F., Bailet, S., Dague, E., and Jorand, F. (2005). "Surface Structure and Nanomechanical Properties of Shewanella putrefaciens Bacteria at Two pH values (4 and 10) Determined by

Atomic Force Microscopy." Journal of Bacteriology **187**: 3864–3868. https://doi.org/10.1128/JB.187.11.3864-3868.2005.

Garcia, R. (2020). "Nanomechanical mapping of soft materials with the atomic force microscope: methods, theory and applications." Chemical Society Reviews **49**: 5850–5884. https://doi.org/10.1039/D0CS00318B.

Garcia, P.D., and Garcia, R. (2018a). "Determination of the Elastic Moduli of a Single Cell Cultured on a Rigid Support by Force Microscopy." Biophysical Journal **114**: 2923–2932. https://doi.org/10.1016/j.bpj.2018.05.012.

Garcia, P.D., and Garcia, R. (2018b). "Determination of the viscoelastic properties of a single cell cultured on a rigid support by force microscopy." Nanoscale **10**: 19799–19809. https://doi.org/10.1039/C8NR05899G.

Gavara, N. (2016). "Combined strategies for optimal detection of the contact point in AFM force-indentation curves obtained on thin samples and adherent cells." Sci Rep **6**: 21267. https://doi.org/10.1038/srep21267.

Gavara, N., and Chadwick, R.S. (2012). "Determination of the elastic moduli of thin samples and adherent cells using conical atomic force microscope tips." Nature Nanotechnology.

Graham, G.A.C. (1965). "The contact problem in the linear theory of viscoelasticity." International Journal of Engineering Science **3**: 27–46. https://doi.org/10.1016/0020-7225(65)90018-2.

Greenwood, J.A., Williamson, J.B.P., and Bowden, F.P. (1966). "Contact of nominally flat surfaces." Proceedings of the Royal Society of London. Series A. Mathematical and Physical Sciences **295**: 300–319. https://doi.org/10.1098/rspa.1966.0242.

Hertz, H. (1881). On the contact of elastic bodies. In Hertz's Miscellaneous Papers (London: Macmillan, New York, Macmillan and co.).

Hertz, H. (1882). "Ueber die Berührung fester elastischer Körper." 1882, 156–171. https://doi.org/10.1515/crll.1882.92.156.

Heuberger, M., Dietler, G., and Schlapbach, L. (1996). "Elastic deformations of tip and sample during atomic force microscope measurements". Journal of Vacuum Science & Technology B: Microelectronics and Nanometer Structures Processing, Measurement, and Phenomena **14**: 1250–1254. https://doi.org/10.1116/1.588525.

Johnson, K.L. (1985). Contact Mechanics (Cambridge: Cambridge University Press).

Johnson, K.L., Kendall, K., Roberts, A.D., and Tabor, D. (1971). "Surface energy and the contact of elastic solids." Proceedings of the Royal Society of London. A. Mathematical and Physical Sciences **324**: 301–313. https://doi.org/10.1098/rspa.1971.0141.

Jorba, I., Uriarte, J.J., Campillo, N., Farré, R., and Navajas, D. (2017). "Probing Micromechanical Properties of the Extracellular Matrix of Soft Tissues by Atomic Force Microscopy." Journal of Cellular Physiology **232**: 19–26. https://doi.org/10.1002/jcp.25420.

Kalcioglu, Z.I., Mahmoodian, R., Hu, Y., Suo, Z., and Van Vliet, K.J. (2012). "From macro-to microscale poroelastic characterization of polymeric hydrogels via indentation." Soft Matter **8**: 3393–3398.

Kim, N.-H. (2015). Finite Element Analysis for Contact Problems. In Introduction to Nonlinear Finite Element Analysis, N.-H. Kim, ed. (New York, NY: Springer US), pp. 367–426.

Kontomaris, S.V., and Malamou, A. (2021). "A novel approximate method to calculate the force applied on an elastic half space by a rigid sphere." Eur. J. Phys. **42**: 025010. https://doi.org/10.1088/1361-6404/abccfb.

Landau, L.D., and Lifshitz, E.M. (1986). Theory of Elasticity - 3rd Edition (Oxford: Pergamon Press).

Lee, E.H., and Radok, J.R.M. (1960). "The Contact Problem for Viscoelastic Bodies." Journal of Applied Mechanics **27**: 438–444. https://doi.org/10.1115/1.3644020.

Lin, D.C., Dimitriadis, E.K., and Horkay, F. (2008). "Robust strategies for automated AFM force curve analysis–I. Non-adhesive indentation of soft, inhomogeneous materials." J Biomech Eng **129**: 430–440. https://doi.org/10.1115/1.2720924.

Long, R., Hall, M.S., Wu, M., and Hui, C.-Y. (2011). "Effects of Gel Thickness on Microscopic Indentation Measurements of Gel Modulus." Biophys J **101**: 643–650. https://doi.org/10.1016/j.bpj.2011.06.049.

Loskill, P., Pereira, P.M., Jung, P., Bischoff, M., Herrmann, M., Pinho, M.G., and Jacobs, K. (2014). "Reduction of the Peptidoglycan Crosslinking Causes a Decrease in Stiffness of the Staphylococcus aureus Cell Envelope." Biophysical Journal **107**: 1082–1089. https://doi.org/10.1016/j.bpj.2014.07.029.

Love, A.E.H. (1929). "IX. The stress produced in a semi-infinite solid by pressure on part of the boundary. Philosophical Transactions of the Royal Society of London." Series A, Containing Papers of a Mathematical or Physical Character **228**: 377–420. https://doi.org/10.1098/rsta.1929.0009.

Love, A.E.H. (1939). "Boussinesq's problem for a rigid cone." The Quarterly Journal of Mathematics **os-10**: 161–175. https://doi.org/10.1093/qmath/os-10.1.161.

Mahaffy, R.E., Shih, C.K., MacKintosh, F.C., and Käs, J. (2000). "Scanning Probe-Based Frequency-Dependent Microrheology of Polymer Gels and Biological Cells". Phys. Rev. Lett. **85**: 880–883. https://doi.org/10.1103/PhysRevLett.85.880.

Managuli, V., and Roy, S. (2018). "Asymptotical Correction to Bottom Substrate Effect Arising in AFM Indentation of Thin Samples and Adherent Cells Using Conical Tips." Exp Mech **58**: 733–741. https://doi.org/10.1007/s11340-018-0373-8.

Maugis, D. (1992). "Adhesion of spheres: The JKR-DMT transition using a dugdale model." Journal of Colloid and Interface Science **150**: 243–269. https://doi.org/10.1016/0021-9797(92)90285-T.

Moeendarbary, E., Valon, L., Fritzsche, M., Harris, A.R., Moulding, D.A., Thrasher, A.J., Stride, E., Mahadevan, L., and Charras, G.T. (2013). "The cytoplasm of living cells behaves as a poroelastic material". Nature Materials **12**: 253–261. https://doi.org/10.1038/nmat3517.

Müller, P., Abuhattum, S., Möllmert, S., Ulbricht, E., Taubenberger, A.V., and Guck, J. (2019). "nanite: using machine learning to assess the quality of atomic force microscopy-enabled nano-indentation data". BMC Bioinformatics **20**: 465. https://doi.org/10.1186/s12859-019-3010-3.

Pharr, G.M., Oliver, W.C., and Brotzen, F.R. (1992). "On the generality of the relationship among contact stiffness, contact area, and elastic modulus during indentation." Journal of Materials Research 7: 613–617. https://doi.org/10.1557/JMR.1992.0613.

Popov, V.L., and Heß, M. (2013). Methode der Dimensionsreduktion in Kontaktmechanik und Reibung (Springer).

Popov, V.L., and Heß, M. (2015). Method of Dimensionality Reduction in Contact Mechanics and Friction.

Popov, V.L., Heß, M., and Willert, E. (2019a). Introduction (Berlin, Heidelberg: Springer Berlin Heidelberg).

Popov, V.L., Heß, M., and Willert, E. (2019b). Tangential Contact. In Handbook of Contact Mechanics: Exact Solutions of Axisymmetric Contact Problems, V.L. Popov, M. Heß, and E. Willert, eds. (Berlin, Heidelberg: Springer), pp. 125–173.

Radmacher, M., Tillmann, R.W., Fritz, M., and Gaub, H.E. (1992). "From Molecules to Cells: Imaging Soft Samples with the Atomic Force Microscope." Science **257**: 1900–1905. https://doi.org/10.1126/science.1411505.

Radmacher, M., Fritz, M., Kacher, C.M., Cleveland, J.P., and Hansma, P.K. (1996). "Measuring the viscoelastic properties of human platelets with the atomic force microscope." Biophys J **70**: 556–567.

Rayleigh, Lord (1873). "Note on the Numerical Calculation of the Roots of Fluctuating Functions." Proceedings of the London Mathematical Society s1-5, 119–124. https://doi.org/10.1112/plms/s1-5.1.119.

Rheinlaender, J., Dimitracopoulos, A., Wallmeyer, B., Kronenberg, N.M., Chalut, K.J., Gather, M.C., Betz, T., Charras, G., and Franze, K. (2020). "Cortical cell stiffness is independent of substrate mechanics." Nat. Mater. **19**: 1019–1025. https://doi.org/10.1038/s41563-020-0684-x.

Rico, F., Roca-Cusachs, P., Gavara, N., Farré, R., Rotger, M., and Navajas, D. (2005). "Probing mechanical properties of living cells by atomic force microscopy with blunted pyramidal cantilever tips." Phys. Rev. E **72**: 021914. https://doi.org/10.1103/PhysRevE.72.021914.

Rico, F., Roca-Cusachs, P., Sunyer, R., Farré, R., and Navajas, D. (2007). "Cell dynamic adhesion and elastic properties probed with cylindrical atomic force microscopy cantilever tips." J Mol Recognit **20**: 459–466. https://doi.org/10.1002/jmr.829.

Rigato, A., Miyagi, A., Scheuring, S., and Rico, F. (2017). "High-frequency microrheology reveals cytoskeleton dynamics in living cells." Nature Phys **13**: 771–775. https://doi.org/10.1038/nphys4104.

Sackmann, E., and Bruinsma, R. (2002). Cell Adhesion as Wetting Transition? In Physics of Bio-Molecules and Cells. Physique Des Biomolécules et Des Cellules, F. Flyvbjerg, F. Jülicher, P. Ormos, and F. David, eds. (Berlin, Heidelberg: Springer), pp. 285–309.

Shield, R.T. (1967). "Load-displacement relations for elastic bodies." Journal of Applied Mathematics and Physics (ZAMP) **18**: 682–693. https://doi.org/10.1007/BF01602041.

Slaughter, W.S. (2002). Linearized Elasticity Problems. In The Linearized Theory of Elasticity, W.S. Slaughter, ed. (Boston, MA: Birkhäuser), pp. 221–254.

Sneddon, I.N. (1965). "The relation between load and penetration in the axisymmetric boussinesq problem for a punch of arbitrary profile." International Journal of Engineering Science **3**: 47–57. https://doi.org/10.1016/0020-7225(65)90019-4.

Tabor, D. (1977). "Surface forces and surface interactions." Journal of Colloid and Interface Science **58**: 2–13. https://doi.org/10.1016/0021-9797(77)90366-6.

Ting, T.C.T. (1966). "The Contact Stresses Between a Rigid Indenter and a Viscoelastic Half-Space." Journal of Applied Mechanics **33**: 845–854. https://doi.org/10.1115/1.3625192.

Tu, Y.-O., and Gazis, D.C. (1964). "The Contact Problem of a Plate Pressed Between Two Spheres". Journal of Applied Mechanics **31**: 659–666. https://doi.org/10.1115/1.3629728.

Wu, C.-E., Lin, K.-H., and Juang, J.-Y. (2016). "Hertzian load–displacement relation holds for spherical indentation on soft elastic solids undergoing large deformations." Tribology International **97**: 71–76. https://doi.org/10.1016/j.triboint.2015.12.034.

Yang, F. (1998). "Indentation of an incompressible elastic film" (1 This work was finished when the author was at the University of Rochester, Rochester, NY, USA.1). Mechanics of Materials **30**: 275–286. https://doi.org/10.1016/S0167-6636(98)00035-0.

Instruments and Methods

Ignacio Casuso, Matteo Chighizola, Hatice Holuigue,
Ewelina Lorenc, Alessandro Podestà, Manfred Radmacher,
Felix Rico, Jorge Rodriguez-Ramos

3.1 Atomic Force Microscope AFM

Atomic force microscopy (AFM) was introduced in 1986 (Binnig et al., 1986). Based on the scanning tunneling microscope (STM) (Binnig and Rohrer, 1982, Binnig et al., 1982), AFM appeared as a solution to characterize nonconductive surfaces. Unlike the STM, where the probe measures the tunneling current between the tip and the sample, AFM detects mechanical forces between the tip and the atoms on the surface. Both STM and AFM can be operated in liquid conditions, which opened the door to work on biological material (Hansma et al., 1988, Marti et al., 1987). However, given the ability of AFM to scan the surface of insulating materials, AFM was more adapted than the STM to characterize biological samples which are poorly conductive (Radmacher et al., 1992). Additionally, the application of AFM as a force probe to characterize the mechanical properties of samples occurred soon after its invention (Maivald et al., 1991, Weisenhorn et al., 1989, 1992).

References

Binnig, G. and H. Rohrer (1982). "Scanning tunneling microscopy." Helvetica Physica Acta **55**: 726–735.
Binnig, G., C. F. Quate and C. Gerber (1986). "Atomic force microscope." Physical Review Letters **56**: 930–933.
Hansma, P. K., V. B. Elings, O. Marti and C. E. Bracker (1988). "Scanning tunneling microscopy and atomic force microscopy: Application to biology and technology." Science **242**: 209–216.
Marti, O., B. Drake and P. K. Hansma (1987). "Atomic force microscopy of liquid-covered surfaces: Atomic resolution images." Applied Physics Letters **51**: 484–486.
Radmacher, M., R. W. Tillmann, M. Fritz and H. E. Gaub (1992). "From molecules to cells: Imaging soft samples with the atomic force microscope." Science **257**: 1900–1905.
Maivald, P., H. J. Butt, S. A. C. Gould, C. B. Prater, B. Drake, J. A. Gurley, V. B. Elings and P. K. Hansma (1991). "Using force modulation to image surface elasticities with the atomic force microscope." Nanotechnology **2**: 103–106.

Ignacio Casuso, Aix-Marseille Univ, CNRS, INSERM, LAI, Turing centre for biological systems, Marseille, France
Matteo Chighizola, Hatice Holuigue, Ewelina Lorenc, Alessandro Podestà, Dipartimento di Fisica "Aldo Pontremoli" and CIMaINa, Università degli Studi di Milano, Milano, Italy
Manfred Radmacher, Institute of Biophysics, University Bremen, Bremen, Germany
Felix Rico, Aix-Marseille Univ, CNRS, INSERM, LAI, Turing centre for biological systems, Marseille, France
Jorge Rodriguez-Ramos, Aix-Marseille Univ, CNRS, INSERM, LAI, Turing centre for living systems

https://doi.org/10.1515/9783110640632-004

Weisenhorn, A. L., P. K. Hansma, T. R. Albrecht and C. F. Quate (1989). "Forces in atomic force microscopy in air and water." Applied Physics Letters **54**: 2651–2653.

Weisenhorn, A. L., P. Maivald, H. J. Butt and P. K. Hansma (1992). "Measuring Adhesion, Attraction, and Repulsion Between Surfaces In Liquids with an Atomic-Force Microscope." Physical Review B **45**: 11226–11232.

Felix Rico, Ignacio Casuso, Manfred Radmacher
3.1.1 AFM Instrumentation

3.1.1.1 Introduction

In this chapter, we describe the main characteristics and principles of operation of the instrumentation involved in AFM. We place special attention to the AFM components required to characterize biological material and, in particular, for mechanical probing of cells and tissues. We also introduce the coupling of AFM with other microscopy techniques.

3.1.1.2 AFM Components

An atomic force microscope is constituted of four main elements: (1) the probe, consisting of a tip attached at the free end of a flexible cantilever, (2) the cantilever deflection detection system, (3) the piezoelectric elements that allow positioning and displacing the tip relative to the sample, and (4) the control electronics.

3.1.1.2.1 Cantilever and Tip

The AFM probe comprises a tip mounted or attached at the end of a flexible cantilever. Users tend to use the terms tip and cantilever interchangeable even though this is not appropriate. The cantilever is commonly made of silicon, silicon nitride, or silicon dioxide, often with a reflective coating at its back surface. Cantilevers come in different geometries (mainly rectangular or V-shaped) and various dimensions that determine the force sensitivity and dynamic response. The tip can be fabricated during the manufacturing process of the cantilever or attached or grown after fabrication using glue or other methods and also come in different geometries and dimensions depending on the application (Tortonese, 1997). Sharp tips are commonly

Acknowledgments: We acknowledge the support of the European Union's Horizon 2020 research and innovation program under the Marie Skłodowska-Curie grant agreement no. 812772, Project Phys2-BioMed. This project has received funding from the European Research Council (ERC, grant agreement no. 772257), from the Agence Nationale de la Recherche (ANR, grant no. ANR-19-CE11-0024), and from the Excellence Initiative of Aix-Marseille University – A*MIDEX, a French "Investissement d'avenir" programme managed by the ANR (no. ANR-11-IDEX-0001-02).

Felix Rico, Ignacio Casuso, Aix-Marseille University, CNRS, INSERM, LAI, Turing Centre for Biological Systems, Marseille, France
Manfred Radmacher, Institute of Biophysics, University Bremen, Bremen, Germany

https://doi.org/10.1515/9783110640632-005

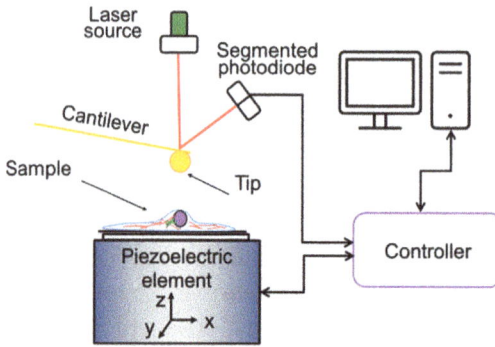

Figure 3.1.1.1: Schematic drawing of an atomic force microscope featuring the optical lever deflection detection system (laser beam and photodiode), the piezoelectric elements for tip positioning, the cantilever, and the control electronics.

used for imaging at high resolution, while blunt or even spherical tips are often applied for mechanical probing soft materials.

In the range of deformations used during AFM operation, the AFM cantilever can be modeled as a linear spring with force constant, k, whose deflection (d) is proportional to the applied force (F) invoking the Hooke's law:

$$F = kd \tag{3.1.1.1}$$

where the force constant k is determined by the geometry and dimensions of the cantilever. The deflection is typically monitored using optical detection systems (see next section).

For a more detailed description of AFM cantilevers and tips refer to Chapter 3.1.2.

3.1.1.2.2 Cantilever Deflection Detection System

The deflection of the cantilever is the main measured quantity in AFM as it provides a measure of the interaction force between the tip and the sample. While different approaches were introduced for detection of the cantilever deflection, like an STM tip (Binnig et al., 1986), interferometry (Rugar et al., 1988, 1989), or piezoresistive cantilevers (Tortonese et al., 1993), the most widely used is the optical lever or optical beam deflection method (OLDM) (Alcaraz, 2001, Alexander et al., 1989, D'Costa and Hoh, 1995, Hansma et al., 1994, Meyer and Amer, 1988, Putman et al., 1992a). In the optical lever method, a collimated laser beam of diameter ~10 µm is reflected on the back side of the AFM cantilever (opposite to the tip side) that is commonly coated with gold. The laser beam is reflected back to a segmented photodiode (Figure 3.1.1.1), resulting in a photocurrent generated in each of the photodiode segments. A four-segmented photodiode allows the detection of the vertical and lateral deflection of a cantilever. The

vertical deflection of the cantilever is monitored using the difference in the signal between the two upper and the two lower segments, while the lateral deflection is similarly calculated from the horizontal segments (see Figure 3.1.1.2). The photocurrent at
each segment (A to D) is transformed into a voltage using a first transimpedance operational amplifier. Differential circuits are then used to perform the operations to calculate the vertical (A+B−C−D) and lateral deflection (A+C−B−D). A circuit to sum the
signal from all segments (A+B+C+D) is also commonly used to normalize the differential signals. The rising time of the photodiode together with the response
time of the electronic board determine the bandwidth (BW) of the optical lever
readout that easily reaches several MHz, enough from most of today's AFM applications. In general terms, for optimal AFM operation, the noise level in the deflection detection optics and circuitry should be sufficiently low to enable the
detection of the thermal-noise vibration of the cantilever in a range of frequencies several times larger than the resonance frequency of the cantilever: a condition that changes depending on the imaging media (vacuum, air, or liquid) and
the dimensions, and thus speed of response, of the cantilevers. Typically, values
of ~50 fm/√Hz are sufficient for cantilevers of conventional AFM, whereas sensitivities of ~10 fm/√Hz are required when faster high-speed AFM cantilevers are
used (Ando et al., 2012).

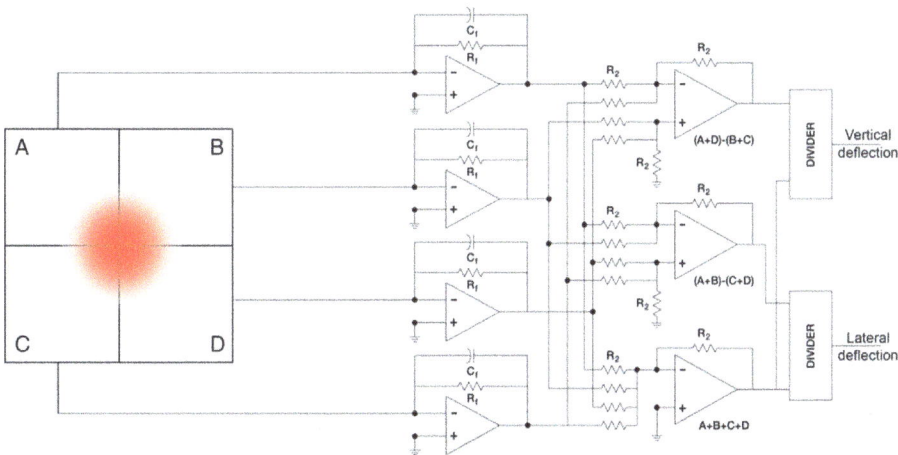

Figure 3.1.1.2: Four quadrant photodiode (left) with a centered laser spot (red) and example of
electronic circuit (right) to determine the vertical and lateral deflection signals.

The OLDM works as follows: when the cantilever deflects, the laser beam changes the
reflection angle, which translates into a change in the position of the laser spot on the
photodiode (Figure 3.1.1.3), and, thus, on the output voltage out of the circuitry. Conversion factors are required to properly translate the photodiode signal in volts into a
cantilever deflection in nanometers (see Chapter 3.1.3): the change in position of the

laser spot (ΔS) on the photodiode (the output in volts is linear to the laser position) depends on the change in the angle of the reflected beam ($\Delta \varphi$) and the distance from the cantilever to the photodiode (D_{CP}) and, thus, on the change in the cantilever deflection (Δd) and the length of the cantilever (L) itself (Figure 3.1.1.3):

$$\Delta S(V) = 2D_{CP}\tan\Delta \varphi \approx 3D_{CP}\frac{\Delta d}{L} \qquad (3.1.1.2)$$

As, for small angles, the change in the angle is proportional to the deflection of the cantilever Δd, consequently, the change in the detector voltage $\Delta S(V)$ is also proportional to Δd. The linear conversion factor is termed optical lever sensitivity (OLS). Given a fixed geometry of the AFM system, that is, the distance D_{CP}, the shorter the cantilever L, the more sensitive the detection method is, that is, the larger the output voltage is for the same cantilever deflection.

The OLS requires calibration. By deflecting the lever by a known amount (Δd) and measuring the resulting voltage change (ΔV), the ratio $\Delta d/\Delta V$ in units of nanometer/volt (nm/V) is obtained, known as the inverse of the OLS (invOLS, sometimes also referred to as photodiode or cantilever or deflection sensitivity). Proper determination of the invOLS is essential for accurate force measurements (see Chapter 3.1.3).

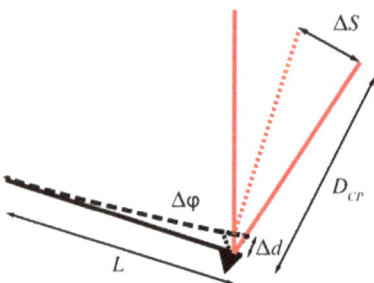

Figure 3.1.1.3: Sketch diagram of the optical lever deflection method.

The inherent mechanism of the OLDM leads to possible artifacts that should be noted here. For large deflections, thus, large angles, the above equation will not hold anymore because the linear approximation between angle and deflection is not valid. In addition, the laser beam usually has a Gaussian intensity profile, which also leads to nonlinearities of the photodiode signal for large spot displacements (ΔS) (Proksch et al., 2004). These nonlinearities are considered and prevented in current commercial AFM systems and could be mainly ignored by the user. Other possible artifacts are nonetheless commonly observed in AFM measurements, mainly reflected in force–distance curves, and will be discussed below.

A possible artifact observed in systems using the OLDM can be observed in systems where the tip is moved with respect the position of the laser beam in z-direction (so-called tip scanning systems) and is commonly referred to as the virtual deflection effect. As shown in Figure 3.1.1.4A, given the geometry of the OLDM, if the tip is moved in the vertical direction while the laser is fixed, a displacement of the spot on the photodiode is detected. For a cantilever moving a distance (Δz) and mounted at an angle (θ), the theoretical displacement of the spot on the photodiode is

$$\Delta S = \Delta z \sin 2\theta \qquad (3.1.1.3)$$

This leads to a "virtual" deflection reflected, visible as a tilt in the noncontact part of the force curve. This tilt is clearly noticed at large cantilever travels (several microns, as reflected in Figure 3.1.1.4A) while it may pass unnoticed at shorter ones. This artifact does not appear, obviously, on systems in which the laser beam moves concomitantly with the cantilever or on sample scanning systems. Most AFM data processing software packages provide an option to correct this tilt. However, the proper method to correct for this artifact is still a matter of debate.

Another common problem of the OLDM is the appearance of interference fringes leading to an undulating signal in the z-direction superimposed on the deflection signal (Figure 3.1.1.4B). This common effect, sometimes referred to as "optical interference," mostly appears on highly reflective surfaces such as gold, but glass and mica can also lead to them and was reported in the literature earlier (Weisenhorn et al., 1992). The interference arises from laser light being reflected from the back side of the cantilever and from the sample. These two parallel beams of highly coherent light generate an interference pattern that depends on the difference in optical path between the two beams. In the OLDM geometry, the change in z of the cantilever generates a change in the path difference leading to interference fringes detected in the photodiode superimposed to the deflection signal (Figure 3.1.1.4B). The sinusoidal wave superimposed in the noncontact part of the curve has a periodicity (λ_i) that depends on the wavelength of the laser source (λ), the refractive index of the surrounding medium, and on the incidence angle (θ) between the cantilever and the laser (Cappella and Dietler, 1999):

$$\lambda_i = \frac{\lambda}{n} \frac{\cos\theta}{(1 + \cos\theta)} \qquad (3.1.1.4)$$

Using this or similar relationship,[1] the interference pattern has been previously used to calibrate the z-piezo of the AFM system (Jaschke and Butt, 1995). However, the interference artifact is most of the time undesirable, and different approaches have been implemented to minimize it. Since light with low coherence length does not generate detectable interferences, much effort has been made in reducing the

---

1 Various relationships have been reported, leading to similar results (Burnham et al., 2003, Jaschke and Butt, 1995).

coherence of the light source. For example, by superimposing a modulation of the power supply of the laser current at high frequencies (hundreds of MHz) that destabilizes the coherency at the cavity of resonance of the laser. This induces a decrease in the coherence length at every modulating cycle that reduces the interference phenomena (Kassies et al., 2004). Another more common option is replacing the laser source with a superluminescence diode, which has an inherently small coherence length. Most commercial AFM systems currently feature a superluminescence diode instead of a laser to minimize interferences. Nevertheless, interference patterns do not disappear completely, and the optical interference is still an issue to be solved in AFM.

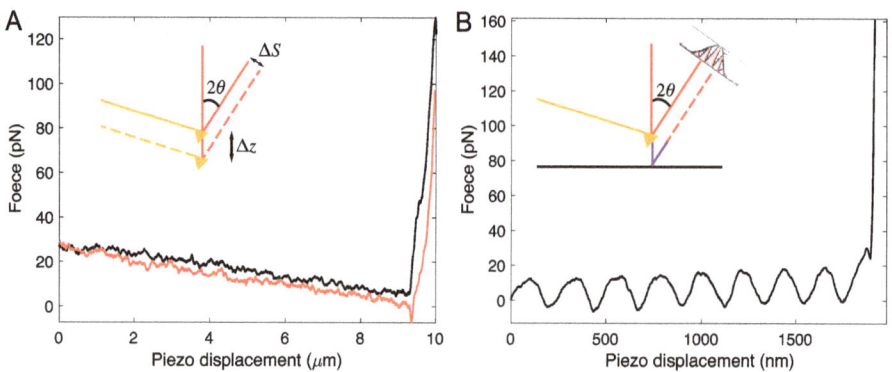

Figure 3.1.1.4: Typical artifacts using the optical lever deflection method. (A) Force curve showing a "virtual" tilt in the noncontact region (left) due to translation of the cantilever relative to the laser beam leading to the virtual deflection artifact (black is approach and red is retract). (B) Force curve showing an interference pattern in the noncontact region (left) caused by optical interferences between the beams reflected by the cantilever and by the sample surface (inset). The periodicity depends on the difference in the optical paths (purple line in the inset). The periodicity of ~220 nm is in relatively good agreement with the prediction from eq. (3.1.1.4) in the text (253 nm) for a laser of wavelength 680 nm, the refractive index of water, and an angle of 12°.

3.1.1.2.3 Piezoelectric Elements

Whether AFM is used for topography imaging, force measurements, or mechanical mapping, the tip has to be moved relative to the sample. This is accomplished using piezoelectric actuators. The piezoelectric effect was discovered by Jacques Curie and Pierre Curie (1880) and describes the capability of a material to generate a gradient of electrical (surface) charge when being compressed and vice versa. For the piezoelectric effect to take place, the temperature of the material must be below its Curie temperature, which is specific to each piezoelectric material. In piezoelectric elements commonly used in AFM, an applied voltage of tens or hundreds of volts results in a displacement in the nano- to micrometer scale, the applied voltage creates an electric field inside the piezoelectric minerals that tends to align the polarization dipoles inside the material. In a piezoelectric material, for each unit cell, due to the applied electric field, the displacement of the electron clouds induces deformation, creating an overall extension or compression of the material. Piezoelectricity is weak in natural piezoelectric materials like quartz or tourmaline. Instead, in polycrystalline ferroelectric ceramics (such as barium titanate and lead zirconate titanate (PZT)) used for the fabrication of piezoelectric actuators in AFM, the boundaries between ferroelectric electric polarization domains (domain walls) easily move in response to an external applied field which creates an internal reorganization self-enhancement process that results in a much higher efficiency over natural piezoelectrics. The PZT ceramics, developed in 1952 by the Tokyo Institute of Technology, are the mostly used piezoelectric materials. They are commonly shaped as thin piezoelectric layers (small thickness t) that can be easily fabricated and reach high electric fields E because $E = V/t$. For the generation of large displacements, piezoelectric stack actuators are generally used. Please note that the maximal piezoelectric elongation is around 0.2% of the distance between electrodes.

AFM systems are equipped with piezoelectric elements to move the tip relative to the sample in the x- and y-directions (parallel to the sample plane) and z-direction (normal to the sample plane). The most common configurations are tip-scanning, in which the piezo elements position the tip relative to the sample, and sample-scanning, in which the sample is positioned relative to the tip. In large scan systems, often a hybrid approach is used: tip positioning in the z-direction and sample positioning in x and y.

Piezoelectric elements also come in different configurations like tubes, longitudinal stack actuators, and shear actuators. The use of one or other configuration depends on the application and the desired range of displacement. In the 1990s and 2000s, most AFM systems used a piezotube to move the tip or sample in the three dimensions. This, however, resulted in a large crosstalk between the different axes, leading to inaccurate positioning. Most current systems use longitudinal stack actuators combined with flexure guide stages that allow amplification of the travel.

The range of movement and the dynamic response accessible in a given piezo element depends on the dimensions and material properties of the elements. Most AFM systems dedicated to work on living cell samples feature large xy scanning areas of $\sim100 \times 100$ μm^2 and a z-range of ~15 μm. To reach these displacements, voltages in the order of ~150 V are required. Therefore, the relatively low sensitivity of piezo elements demand dedicated high-voltage amplifiers to drive the piezo. The dynamic range of the movement depends again on the dimensions and material properties of the piezo elements and is generally limited by resonances within the piezo element, for example, caused by the reflection of the ultrasound waves generated during operation. As a rule of thumb, the maximal speed of operation of a piezoelectric actuator is defined by one-third of its resonant frequency at nominal displacement. In piezoelectric stacks, the maximal elongation ΔL_0 is $\sim1/500$ of its length L_0. The first resonance frequency

f_0 and the maximal operational frequency of the actuator $f_0/3$ will be proportional to $1/\sqrt{\Delta L_0}$. Thus, the speed of response of the piezoelectric element in AFM gets smaller, the larger the maximal elongation. Consequently, systems used for cell imaging asking for scan sizes of up to 100 µm will be slower that smaller scan-size systems, which can be used for imaging molecular size samples of some tens of nanometers. Another factor that slows down the maximal operational frequency is the weight of any attached mass m, as the sample holder, by a factor $1/\sqrt{m}$.

A strategy commonly used to extend the maximal displacement and maximal speed of a piezoelectric actuator is to couple a flexure-guided mechanical amplifier; the trade-off is a reduction in the reaction speed by $(\Delta(\Delta L_0)/\Delta L_0)^2$, where $\Delta(\Delta L_0)$ is the extension in the maximal displacement. Flexure-guided mechanical amplifiers are more and more commonly used in commercial AFM systems, thanks, in part, to the development of high-speed AFM systems (Ando et al., 2001, Viani et al., 1999, Watanabe et al., 2013).

Piezoelectric elements present nonlinearities, hysteresis, creep, temperature dependence, aging, and damages derived from excessive tensile peak forces, which will introduce uncertainties in positioning (Figure 3.1.1.5). The displacement of a piezo element is not directly proportional to the applied voltage; the nonlinearity of the response of polycrystalline ferroelectric ceramics that occurs at high applied voltages divides by a factor ~2 to 4 the ratio "displacement over applied voltage" with respect to low applied voltages; while the hysteresis creates a maximal error in the piezo displacement between compression and extension of ~10%; and creep changes the displacement over time by ~1% for a few seconds at a constant driven voltage (Figure 3.1.1.5A).

To achieve a more accurate control of the piezo movement, different approaches can be applied such as controlling the applied charge instead of voltage, applying preconditioned reverse signals on previously calibrated piezo elements, and most commonly, using position sensors. There are different types of position sensors such as strain gauge, capacitive, and linear variable differential transformer sensors (Fleming, 2013). The use of position sensors allows for a direct measurement of the applied movement, irrespective of the nonlinearities always present in piezo elements. In closed-loop mode, a feedback loop circuit continuously corrects the voltage applied to the piezo to follow the desired movement, monitored by the position sensor. In contrast, open-loop operation applies a voltage to the piezo while the position is constantly monitored, but not corrected. The main drawbacks of closed-loop mode are the limited response time, which now mainly depends on the feedback loop electronics and its adjustments, and the decrease of the small displacement resolution because the sensor and electronics introduce extra noise. Figure 3.1.1.5B shows examples of the z-displacement of a piezo element on a commercial AFM system in open- and closed-loop modes.

AFM systems designed to work with living cell samples feature a relatively large z-range of 10–20 µm (Lehenkari et al., 2000). This is reasonable for most applications since the thickness of most cells is in the range of 5–10 μm. However, some applications require even longer ranges. For example, in cell adhesion studies, in which the cantilever tip is coated with ligands to target receptors on the cell surface or in which a whole cell is attached to the cantilever to perform force–distance curves against a ligand-coated surface. Cell adhesion measurements lead often to the formation of long membrane tethers that can be several tens of micrometers long. In that case, longer z-piezos are required, and some home-made and commercial systems have been developed for this purpose (Chu et al., 2013, Puech et al., 2005, 2006). In contrast, AFM systems for molecular studies feature relatively short z-piezo travels (1–5 μm).

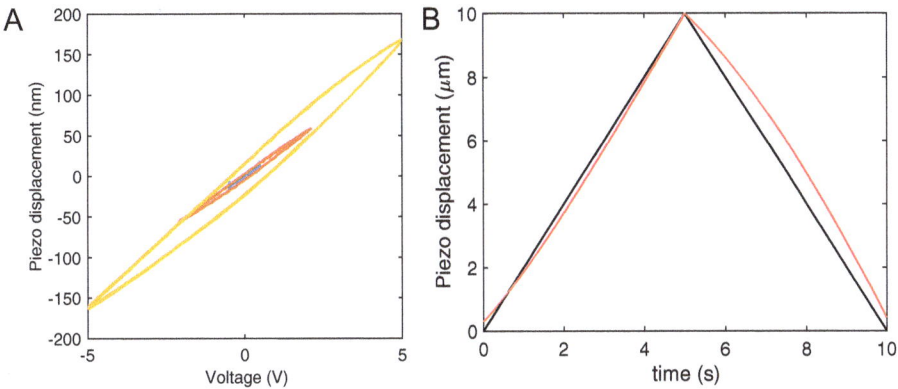

Figure 3.1.1.5: (A) Actual piezo displacement versus applied voltage showing nonlinear behavior and hysteresis for triangular excitations of different voltage amplitude. (B) Displacement in open- (red line) and close (black line)-loop modes of operation of a piezo element with position sensor.

3.1.1.2.4 Control Hardware and Software

AFM systems are controlled using dedicated electronics and control software that allow setting the imaging or force curve parameters and data acquisition. The tasks required during AFM measurements involve digital to analogue (D/A) conversion of the settings (applied piezo voltage), a feedback mechanism for imaging and force control during measurements, analogue to digital (A/D) conversion of the signals (deflection and piezo movement detection), and real-time visualization of the acquired data. Most commercial systems provide complete software packages to perform these tasks.

A/D and D/A conversions are performed by electronic boards usually included in the so-called AFM controller. The controller generally also includes the high-voltage amplifiers to drive the piezo elements, the feedback electronics for closed-loop operation of the piezo elements (if position sensors are present), and the control electronics required for imaging and force measurements. Indeed, an important part of control electronics involves the feedback circuitry required to keep the force constant during imaging. While this can be currently performed using field programmable gated arrays (FPGA) combined or not with digital signal processing modules, the feedback principle remains the same and is commonly based on a proportional–integral–derivative (PID) control circuit. The PID controller regulates the output voltage such that there is zero error between process variable; in this case, the readout of the AFM probe or the position sensor and the desired output (setpoint or piezo position). The speed of reaction of the PID controller must be adjusted by the user to suit the dynamics of the imaging process of the AFM. If values for P, I, and D gains are not correctly tuned, either instabilities or slow control performances occur. During AFM imaging operation, the PID values are fine-adjusted using trial and error, as the reaction speed of the AFM feedback loop depends on uncontrollable factors such as the cantilever response time, the tip contamination, and the type of sample. Out of the three PID parameters, the I component (the integral gain) is more useful for maintaining the setpoint accurate, as it provides an integrated averaged signal on the deviation from the setpoint over a small period of time. The P parameter, proportional to the error between the setpoint and the value, modulates the fast response to the smallest features of the sample. Finally, the D parameter responds quickly to large deviations from the setpoint, and it is normally unnecessary, unless the AFM feedback loop has an accused tendency to erratic behavior.

Force measurements require a control software very different from that used for imaging. While this was an important limitation in early AFM systems, mechanical characterization has become a usual demand, and most systems provide now dedicated environments for advanced force measurements. Indeed, current control software provides tools for custom design of the force ramp routine. For example, microrheology measurements require particular force curve protocols including approach to the sample until a desired applied force is achieved, followed by tip oscillation at different frequencies around a constant indentation and finally return to the initial position out of contact (Alcaraz et al., 2003, Mahaffy et al., 2000, Radmacher et al., 1993, Rigato et al., 2017, Schächtele et al., 2018, Takahashi and Okajima, 2015). Moreover, given the structural and mechanical heterogeneity of living cells and tissues, this typical protocol may be carried out at different positions along the cell surface, which in turn requires mapping over a scanning area applying the designed force protocol (A-Hassan et al., 1998, Hecht et al., 2015, Hiratsuka et al., 2009, Rotsch et al., 1997). The current computational power of desktop computers and the incorporation of FPGA or other types of Programmable Logic Devices allow even the extraction of mechanical parameters, such as elasticity and adhesion, in real time during scanning. While the early force volume mode required tens of minutes of acquisition for a single map, faster modes have been recently developed that allow force maps at high resolution within minutes (Rico et al., 2011, Picas et al., 2013, Medalsy and Müller, 2013, Rigato et al., 2015).

Given the required complexity of AFM control software, home-made systems are commonly adapted to the requirements in the development laboratory. However, open software packages to control home-made or commercial systems are becoming available, and we expect it to be the trend in the following decades (Pawlak and Strzelecki, 2016).

3.1.1.3 Coupling AFM with Optical Microscopy

AFM is a versatile technique that allows visualization and characterization of a number of samples, including living cells. However, the scanning area of most AFM systems is limited to ~100 × 100 µm^2 in the sample plane. Adding the slow imaging rate (an image may take several minutes to be acquired) makes it difficult to work on living cells, which are sometimes as large as the size of the scanning area, and positioning of the cantilever on the desired cell region is tedious and slow. More importantly, most of the cell biology research is carried out using optical microscopy such as bright field, phase contrast, and fluorescence. Thus, combination of AFM with optical microscopy has been an essential demand for biological applications (Lehenkari et al., 2000). Since the introduction of AFM, a number of studies have coupled AFM systems with optical microscopy (Chaudhuri et al., 2009, Horton et al., 2000, Lehenkari et al., 2000, Madl et al., 2006, Putman et al., 1992b). This is now a common feature in commercial systems, allowing concomitant visualization of the sample and the AFM cantilever using the optical microscope and accurate positioning of the tip on the desired cell or cell region using the stage of the optical microscope. Precise positioning is then assured using the AFM's piezoelectric elements.

The most common approach for cell studies comprises mounting the AFM system on the stage of an inverted optical microscope, although other approaches have been proposed, even for high-speed AFM setups and with lateral view (Chaudhuri et al., 2009, Colom et al., 2013, Fukuda et al., 2013). An example of a homebuilt AFM coupled to an optical microscope is shown in Figure 3.1.1.6. Earlier work combining fluorescence microscopy with AFM force measurements showed that both the mechanical maps and cytoskeleton visualization were possible on living cells revealing softening coupled to actin depolymerization on fibroblasts or furrow stiffening during cell division (Matzke et al., 2001, Rotsch and Radmacher, 2000). Other works combined intracellular calcium detection with mechanical perturbation of cells with the AFM tip to reveal calcium influx and modulation of mechanostransduction (Charras and Horton, 2002). Spinning disk confocal microscopes allow fast optical imaging rates and have also been combined with force measurements to study mechanotransduction and to determine the deformation applied by the AFM tip on living cells (Efremov et al., 2019, Melzak and Toca-Herrera, 2015, Trache and Lim, 2009).

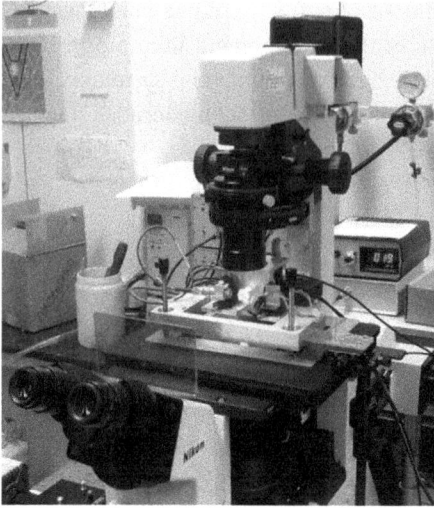

Figure 3.1.1.6: Home-built AFM coupled to an optical microscope (Rico, 2006).

The current success of the so-called super-resolution optical microscopic techniques, such as stochastic optical reconstruction microscopy (Rust et al., 2006), photo-activated localization microscopy (Hess et al., 2006), or stimulated emission depletion (Hell and Wichmann, 1994), in which structural illumination and/or image processing result in fluorescence images with subdiffraction-limited resolution (~10–20 nm) have also led to couple AFM with these novel optical modes (Chacko et al., 2013). For example, Curry et al. (2017) used STED in combination with AFM to reveal the active remodeling of actin and microtubule cytoskeleton of living astrocytes during migration. Lafont and coworkers coupled AFM force measurements with STED microscopy to determine the mechanical properties of intracellular organoids and for revealing early binding of bacteria to the cell surface (Ciczora et al., 2019, Janel et al., 2019). TIRF has also been coupled to AFM allowing fluorescence detection of single molecules (Fukuda et al., 2013, Trache and Lim, 2010). Fluorescence microscopy is not only useful to determine the structure of cellular constituents, and it can also be used to discriminate different phenotypes of the populations of cells. For example, cells that have been transfected with a particular protein of interest may appear fluorescent, while those without it, not. Thus, the combination of AFM to target only transfected cells is another of the many possible applications (Alsteens et al., 2016, Colom et al., 2013). Going a step further, correlation of atomic force, light, and electron microscopy has also been possible shedding light on cell structures and mechanics at different length scales (Janel et al., 2017). The combination of various microscopy techniques is one of the emerging fields, and it has been coined "correlation imaging" (Fantner and Lafont, 2019).

While using optical microscopy, in particular, fluorescence microscopy, is essential for AFM studies on cells, the coupling may have some drawbacks that should be taken into account as well. For example, the sample stage of an optical microscope is less stable than the stage of a common stand-alone AFM system. While this is less important for mechanical measurements of cells, high-resolution images at subnanometer scales are likely compromised. An important source of noise is the cooling fans from sensitive CCD or CMOS cameras and/or the illumination sources for optical microscopy, which transmit mechanical vibrations to the microscope frame. This mechanical noise often interferes with the cantilever, which works as a miniature microphone and to the sample support leading to undesired signals in the recorded force curves or topographic images. To minimize this effect, some solutions are possible. For example, physical uncoupling of the camera and illumination sources from the actual frame of the optical microscope may help. Also, cooling of the camera using a fluid cooling system completely abolishes the need of fans, importantly reducing mechanical noise. Apart from mechanical noise, other interference may occur when using fluorescence illumination. Fluorescence sources use relatively high-power light which may have an effect on the temperature of the sample or may even directly perturb the deflection of the cantilever or even be directly detected by the photodiode. Direct detection by the photodiode is generally not an issue since most AFM systems use filters that only allow the AFM laser source to be detected and spurious light is effectively filtered. Actual perturbation of the cantilever deflection is more crucial. Most cantilevers present a reflective coating generally made of gold. This generates stress in the plane of the cantilever because the thermal conductivity and expansion of reflective coating and that of the cantilever itself (silicon or silicon nitride) are different. This may lead to bending or drift, even without external illumination sources, but can be much more pronounced when using high-intensity light sources. The use of cantilevers without gold coating or with coating limited to the very end of the free end importantly minimizes this effect (Radmacher et al., 1995, Edwards and Perkins, 2016, Churnside et al., 2012). The advantage is so obvious that currently commercial cantilevers are available presenting this option. The effect of the illumination has been well-characterized recently using a fluorescence source with different wavelengths showing that accurate synchronization of light and force signals is required to precisely determine immune cell activation (Cazaux et al., 2016).

AFM was also combined with different cell culture methods. For example, force measurements on cells grown on soft substrates have shown the relevance between the elasticity of the extracellular environment and the biological and mechanical response of the cell (Discher et al., 2005, Domke et al., 2000, Engler et al., 2004, Solon et al., 2007). The use of micropatterns to control the size and shape of grown cells has also provided improved robustness and quantification on the mechanical heterogeneity of living cells, allowing averaging AFM mechanical maps of tens of cells (Rigato et al., 2015). Other approaches requiring the combination of optical and atomic force microscopy include, for example, the application of traction force microscopy.

In this approach, an elastic substrate embedded with fluorescent beads is used to grow cells on. Cells deform the gel and, knowing the modulus of elasticity of the substrate and measuring the relative displacement of the fluorescent beads, the applied force and strain energy can be calculated (Butler et al., 2002, Dembo and Wang, 1999, Dembo et al., 1996). The group of T. Schaeffer has recently combined AFM force mapping and traction microscopy on living cells observing interesting relationships between the generated forces and viscoelastic state of the cells (Schierbaum et al., 2019). For a more in-depth description of the integration of AFM with other techniques, see the dedicated Chapter 3.8.

References

A-Hassan, E., W. Heinz, M. D. Antonik, N. P. D'Costa, S. Nageswaran, C. A. Schoenenberger and J. H. Hoh (1998). "Relative microelastic mapping of living cells by atomic force microscopy." Biophysical Journal **74**: 1564–1578.

Alcaraz, J. (2001). Micromechanics of cultured human bronchial epithelial cells measured with atomic force microscopy. Universitat de Barcelona.

Alcaraz, J., L. Buscemi, M. Grabulosa, X. Trepat, B. Fabry, R. Farre and D. Navajas (2003). "Microrheology of human lung epithelial cells measured by atomic force microscopy." Biophysical Journal **84**: 2071–2079.

Alexander, S., L. Hellemans, O. Marti, J. Schneir, V. Elings, P. K. Hansma, M. Longmire and J. Gurley (1989). "An Atomic-Resolution Atomic-Force Microscope Implemented Using an Optical-Lever." Journal of Applied Physics **65**: 164–167.

Alsteens, D., R. Newton, R. Schubert, D. Martinez-Martin, M. Delguste, B. Roska and D. J. Müller (2016). "Nanomechanical mapping of first binding steps of a virus to animal cells." Nature Nanotechnology **12**: 177–183.

Ando, T., N. Kodera, E. Takai, D. Maruyama, K. Saito and A. Toda (2001). "A high-speed atomic force microscope for studying biological macromolecules." Proceedings of the National Academy of Sciences **98**: 12468–12472.

Ando, T., U. Takayuki and K. Noriyuki (2012). "High-speed atomic force microscopy." Japanese Journal of Applied Physics **51**: 08KA02.

Binnig, G. and H. Rohrer (1982). "Scanning tunneling microscopy." Helvetica Physica Acta **55**: 726–735.

Binnig, G., H. Rohrer, C. Gerber and E. Weibel (1982). "Surface studies by scanning tunneling microscopy." Physical Review Letters **49**: 57.

Binnig, G., C. F. Quate and C. Gerber (1986). "Atomic force microscope." Physical Review Letters **56**: 930–933.

Burnham, N. A., X. Chen, C. S. Hodges, G. A. Matei, E. J. Thoreson, C. J. Roberts, M. C. Davis and S. J. B. Tendler (2003). "Comparison of calibration methods for atomic-force microscopy cantilevers." Nanotechnology **14**: 1–6.

Butler, J. P., I. M. Tolic-Norrelykke, B. Fabry and J. J. Fredberg (2002). "Traction fields, moments, and strain energy that cells exert on their surroundings." AJP – Cell Physiology **282**: C595–C605.

Cappella, B. and G. Dietler (1999). "Force-distance curves by atomic force microscopy." Surface Science Reports **34**: 1-+.

Cazaux, S., A. Sadoun, M. Biarnes-Pelicot, M. Martinez, S. Obeid, P. Bongrand, L. Limozin and P.-H. Puech (2016). "Synchronizing atomic force microscopy force mode and fluorescence microscopy in real time for immune cell stimulation and activation studies." Ultramicroscopy **160**: 168–181.

Chacko, J. V., F. C. Zanacchi and A. Diaspro (2013). "Probing cytoskeletal structures by coupling optical superresolution and AFM techniques for a correlative approach." Cytoskeleton **70**: 729–740.

Charras, G. T. and M. A. Horton (2002). "Single cell mechanotransduction and its modulation analyzed by atomic force microscope indentation." Biophysical Journal **82**: 2970–2981.

Chaudhuri, O., S. H. Parekh, W. A. Lam and D. A. Fletcher (2009). "Combined atomic force microscopy and side-view optical imaging for mechanical studies of cells." Nature Methods **6**: 383–387.

Chu, C., E. Celik, F. Rico and V. T. Moy (2013). "Elongated Membrane Tethers, Individually Anchored by High Affinity α4β1/VCAM-1 Complexes, Are the Quantal Units of Monocyte Arrests." PloS One **8**: e64187.

Churnside, A. B., R. M. A. Sullan, D. M. Nguyen, S. O. Case, M. S. Bull, G. M. King and T. T. Perkins (2012). "Routine and timely sub-piconewton force stability and precision for biological applications of atomic force microscopy." Nano Letters **12**: 3557–3561.

Ciczora, Y., S. Janel, M. Soyer, M. Popoff, E. Werkmeister and F. Lafont (2019). "Blocking bacterial entry at the adhesion step reveals dynamic recruitment of membrane and cytosolic probes." Biology of the Cell **111**: 67–77.

Colom, A., I. Casuso, F. Rico and S. Scheuring (2013). "A hybrid high-speed atomic force–optical microscope for visualizing single membrane proteins on eukaryotic cells." Nature Communications **4**: 8.

Curie, J. and P. Curie (1880). "Développement par compression de l'électricité polaire dans les cristaux hémièdres à faces inclinées." Bulletin de Minéralogie **3**: 90–93.

Curry, N., G. Ghézali, G. S. Kaminski Schierle, N. Rouach and C. F. Kaminski (2017). "Correlative STED and atomic force microscopy on live astrocytes reveals plasticity of cytoskeletal structure and membrane physical properties during polarized migration." Frontiers in Cellular Neuroscience **11**: 104.

D'Costa, N. P. and J. H. Hoh (1995). "Calibration of optical lever sensitivity for atomic force microscopy." Review of Scientific Instruments **66**: 5096–5097.

Dembo, M. and Y.-L. Wang (1999). "Stresses at the cell-to-substrate interface during locomotion of fibroblasts." Biophysical Journal **76**: 2307–2316.

Dembo, M., T. Oliver, A. Ishihara and K. Jacobson (1996). "Imaging the traction stresses exerted by locomoting cells with the elastic substratum method." Biophysical Journal **70**: 2008–2022.

Discher, D. E., P. Janmey and Y. L. Wang (2005). "Tissue cells feel and respond to the stiffness of their substrate." Science **310**: 1139–1143.

Domke, J., S. Dannohl, W. J. Parak, O. Muller, W. K. Aicher and M. Radmacher (2000). "Substrate dependent differences in morphology and elasticity of living osteoblasts investigated by atomic force microscopy." Colloids and Surfaces. B, Biointerfaces **19**: 367–379.

Edwards, D. T. and T. T. Perkins (2016). "Optimizing force spectroscopy by modifying commercial cantilevers: Improved stability, precision, and temporal resolution." Journal of Structural Biology **197** (1): 13–25.

Efremov, Y. M., M. Velay-Lizancos, C. J. Weaver, A. I. Athamneh, P. D. Zavattieri, D. M. Suter and A. Raman (2019). "Anisotropy vs isotropy in living cell indentation with AFM." Scientific Reports **9**: 5757.

Engler, A. J., M. A. Griffin, S. Sen, C. G. BVönnemann, H. L. Sweeney and D. E. Discher (2004). "Myotubes differentiate optimally on substrates with tissue-like stiffness." The Journal of Cell Biology **166**: 877–887.

Fantner, G. and F. Lafont (2019). Correlative microscopy using scanning probe microscopes. Correlative Imaging, Focusing on the Future, 99–118.

Fleming, A. J. (2013). "A review of nanometer resolution position sensors: Operation and performance." Sensors and Actuators. A, Physical **190**: 106–126.

Fukuda, S., T. Uchihashi, R. Iino, Y. Okazaki, M. Yoshida, K. Igarashi and T. Ando (2013). "High-speed atomic force microscope combined with single-molecule fluorescence microscopy." Review of Scientific Instruments **84**: 073706.

Hansma, P. K., V. B. Elings, O. Marti and C. E. Bracker (1988). "Scanning tunneling microscopy and atomic force microscopy: Application to biology and technology." Science **242**: 209–216.

Hansma, P. K., B. Drake, D. Grigg, C. B. Prater, F. Yashar, G. Gurley, S. V. Eling, S. Feinstein and R. Lal (1994). "A New, Optical-Lever Based Atomic-Force Microscope." Journal of Applied Physics **76**: 796–799.

Hecht, F. M., J. Rheinlaender, N. Schierbaum, W. H. Goldmann, B. Fabry and T. E. Schaffer (2015). "Imaging viscoelastic properties of live cells by AFM: Power-law rheology on the nanoscale." Soft Matter **11**: 4584–4591.

Hell, S. W. and J. Wichmann (1994). "Breaking the diffraction resolution limit by stimulated emission: Stimulated-emission-depletion fluorescence microscopy." Optics Letters **19**: 780–782.

Hess, S. T., T. P. Girirajan and M. D. Mason (2006). "Ultra-high resolution imaging by fluorescence photoactivation localization microscopy." Biophysical Journal **91**: 4258–4272.

Hiratsuka, S., Y. Mizutani, M. Tsuchiya, K. Kawahara, H. Tokumoto and T. Okajima (2009). "The number distribution of complex shear modulus of single cells measured by atomic force microscopy." Ultramicroscopy **109**: 937–941.

Horton, M. A., G. T. Charras, G. Ballestrem and P. P. Lehenkari (2000). "Integration of Atomic Force and Confocal Microscopy." Single Molecules **1**: 135–137.

Janel, S., E. Werkmeister, A. Bongiovanni, F. Lafont and N. Barois (2017). CLAFEM: Correlative light atomic force electron microscopy. In Methods in Cell Biology, (Elsevier), pp. 165–185.

Janel, S., M. Popoff, N. Barois, E. Werkmeister, S. Divoux, F. Perez and F. Lafont (2019). "Stiffness tomography of eukaryotic intracellular compartments by atomic force microscopy." Nanoscale **11**: 10320–10328.

Jaschke, M. and H. Butt (1995). "Height calibration of optical lever atomic force microscopes by simple laser interferometry." Review of Scientific Instruments **66**: 1258–1259.

Kassies, R., K. O. van der Werf, M. L. Bennink and C. Otto (2004). "Removing interference and optical feedback artifacts in atomic force microscopy measurements by application of high frequency laser current modulation." Review of Scientific Instruments **75**: 689–693.

Lehenkari, P. P., G. T. Charras, A. Nykanen and M. A. Horton (2000). "Adapting atomic force microscopy for cell biology." Ultramicroscopy **82**: 289–295.

Madl, J., S. Rhode, H. Stangl, H. Stockinger, P. Hinterdorfer, G. J. Schutz and G. Kada (2006). "A combined optical and atomic force microscope for live cell investigations." Ultramicroscopy **106**: 645–651.

Mahaffy, R. E., C. K. Shih, F. C. MacKintosh and J. Kas (2000). "Scanning Probe-Based Frequency-Dependent Microrheology of Polymer Gels and Biological Cells." Physical Review Letters **85**: 880–883.

Maivald, P., H. J. Butt, S. A. C. Gould, C. B. Prater, B. Drake, J. A. Gurley, V. B. Elings and P. K. Hansma (1991). "Using force modulation to image surface elasticities with the atomic force microscope." Nanotechnology **2**: 103–106.

Marti, O., B. Drake and P. K. Hansma (1987). "Atomic force microscopy of liquid-covered surfaces: Atomic resolution images." Applied Physics Letters **51**: 484–486.

Matzke, R., K. Jacobson and M. Radmacher (2001). "Direct, high-resolution measurement of furrow stiffening during division of adherent cells." Nature Cell Biology **3**: 607–610.

Medalsy, I. D. and D. J. Müller (2013). "Nanomechanical properties of proteins and membranes depend on loading rate and electrostatic interactions." ACS Nano **7**: 2642–2650.

Melzak, K. A. and J. L. Toca-Herrera (2015). "Atomic force microscopy and cells: Indentation profiles around the AFM tip, cell shape changes, and other examples of experimental factors affecting modeling." Microscopy Research and Technique **78**: 626–632.

Meyer, G. and N. M. Amer (1988). "Novel optical approach to atomic force microscopy." Applied Physics Letters **53**: 1045–1047.

Pawlak, K. and J. Strzelecki (2016). "Nanopuller-open data acquisition platform for AFM force spectroscopy experiments." Ultramicroscopy **164**: 17–23.

Picas, L., F. Rico, M. Deforet and S. Scheuring (2013). "Structural and Mechanical Heterogeneity of the Erythrocyte Membrane Reveals Hallmarks of Membrane Stability." ACS Nano **7**: 1054–1063.

Proksch, R., T. E. Schaffer, J. P. Cleveland, R. C. Callahan and M. B. Viani (2004). "Finite optical spot size and position corrections in thermal spring constant calibration." Nanotechnology **15**: 1344–1350.

Puech, P. H., A. Taubenberger, F. Ulrich, M. Krieg, D. J. Muller and C. P. Heisenberg (2005). "Measuring cell adhesion forces of primary gastrulating cells from zebrafish using atomic force microscopy." Journal of Cell Science **118**: 4199–4206.

Puech, P. H., K. Poole, D. Knebel and D. J. Muller (2006). "A new technical approach to quantify cell-cell adhesion forces by AFM." Ultramicroscopy **106**: 637–644.

Putman, C. A. J., B. G. Degrooth, N. F. Vanhulst and J. Greve (1992a). "A Detailed Analysis of the Optical Beam Deflection Technique for Use in Atomic Force Microscopy." Journal of Applied Physics **72**: 6–12.

Putman, C. A. J., K. O. Vanderwerf, B. G. Degrooth, N. F. Vanhulst, F. B. Segerink and J. Greve (1992b). "Atomic Force Microscope with Integrated Optical Microscope for Biological Applications." Review of Scientific Instruments **63**: 1914–1917.

Radmacher, M., R. W. Tillmann, M. Fritz and H. E. Gaub (1992). "From molecules to cells: Imaging soft samples with the atomic force microscope." Science **257**: 1900–1905.

Radmacher, M., R. W. Tilmann and H. E. Gaub (1993). "Imaging Viscoelasticity by Force Modulation with the Atomic Force Microscope." Biophysical Journal **64**: 735–742.

Radmacher, M., J. P. Cleveland and P. K. Hansma (1995). "Improvement of thermally induced bending of cantilevers used for atomic force microscopy." Scanning **17**: 117–121.

Rico, F. (2006). Study of viscoelasticity and adhesion of human alveolar epithelial cells by atomic force microscopy. The importance of probe geometry. University of Barcelona, Department of Physiological Sciences I.

Rico, F., C. Su and S. Scheuring (2011). "Mechanical mapping of single membrane proteins at submolecular resolution." Nano Letters **11**: 3983–3986.

Rigato, A., F. Rico, F. Eghiaian, M. Piel and S. Scheuring (2015). "Atomic force microscopy mechanical mapping of micropatterned cells shows adhesion geometry-dependent mechanical response on local and global scales." ACS Nano **9**: 5846–5856.

Rigato, A., A. Miyagi, S. Scheuring and F. Rico (2017). "High-frequency microrheology reveals cytoskeleton dynamics in living cells." Nature Physics **13**: 771–775.

Rotsch, C. and M. Radmacher (2000). "Drug-Induced changes of cytoskeletal structure and mechanics in fibroblasts: An atomic force microscopy study." Biophysical Journal **78**: 520–535.

Rotsch, C., F. Braet, E. Wisse and M. Radmacher (1997). "AFM imaging and elasticity measurements on living rat liver macrophages." Cell Biology International **21**: 685–696.

Rugar, D., H. J. Mamin, R. Erlandsson, J. E. Stern and B. D. Terris (1988). "Force microscope using a fiber-optic displacement sensor." Review of Scientific Instruments **59**: 2337–2340.

Rugar, D., H. J. Mamin and P. Guethner (1989). "Improved fiber-optic interferometer for atomic force microscopy." Applied Physics Letters **55**: 2588–2590.

Rust, M. J., M. Bates and X. Zhuang (2006). "Sub-diffraction-limit imaging by stochastic optical reconstruction microscopy (STORM)." Nature Methods **3**: 793.

Schächtele, M., E. Hänel and T. E. Schäffer (2018). "Resonance compensating chirp mode for mapping the rheology of live cells by high-speed atomic force microscopy." Applied Physics Letters **113**: 093701.

Schierbaum, N., J. Rheinlaender and T. E. Schäffer (2019). "Combined atomic force microscopy (AFM) and traction force microscopy (TFM) reveals a correlation between viscoelastic material properties and contractile prestress of living cells." Soft Matter **15**: 1721–1729.

Solon, J., I. Levental, K. Sengupta, P. C. Georges and P. A. Janmey (2007). "Fibroblast adaptation and stiffness matching to soft elastic substrates." Biophysical Journal **93**: 4453–4461.

Takahashi, R. and T. Okajima (2015). "Mapping power-law rheology of living cells using multi-frequency force modulation atomic force microscopy." Applied Physics Letters **107**: 173702.

Tortonese, M. (1997). "Cantilevers and tips for atomic force microscopy." IEEE Engineering in Medicine and Biology Magazine **16**: 28–33.

Tortonese, M., R. C. Barrett and C. F. Quate (1993). "Atomic resolution with an atomic force microscope using piezoresistive detection." Applied Physics Letters **62**: 834–836.

Trache, A. and S.-M. Lim (2009). "Integrated microscopy for real-time imaging of mechanotransduction studies in live cells." Journal of Biomedical Optics **14**: 034024.

Trache, A. and S.-M. Lim (2010). "Live cell response to mechanical stimulation studied by integrated optical and atomic force microscopy." JoVE (Journal of Visualized Experiments). (44): e2072.

Viani, M. B., T. E. Schäffer, G. T. Paloczi, L. I. Pietrasanta, B. L. Smith, J. B. Thompson, M. Richter, M. Rief, H. E. Gaub, K. W. Plaxco, et al.. (1999). "Fast imaging and fast force spectroscopy of single biopolymers with a new atomic force microscope designed for small cantilevers." Review of Scientific Instruments **70**: 4300–4303.

Watanabe, H., T. Uchihashi, T. Kobashi, M. Shibata, J. Nishiyama, R. Yasuda and T. Ando (2013). "Wide-area scanner for high-speed atomic force microscopy." Review of Scientific Instruments **84**: 053702.

Weisenhorn, A. L., P. K. Hansma, T. R. Albrecht and C. F. Quate (1989). "Forces in atomic force microscopy in air and water." Applied Physics Letters **54**: 2651–2653.

Weisenhorn, A. L., P. Maivald, H. J. Butt and P. K. Hansma (1992). "Measuring Adhesion, Attraction, and Repulsion Between Surfaces In Liquids with an Atomic-Force Microscope." Physical Review B **45**: 11226–11232.

Ewelina Lorenc, Hatice Holuigue, Felix Rico, Alessandro Podestà
3.1.2 AFM Cantilevers and Tips

As a member of the family of scanning probe microscopies, atomic force microscopy (AFM) derives its remarkable versatility and power from the availability of a variety of probes, which serve specific applications (Alessandrini and Facci, 2005, Trache and Meininger, 2008, Eaton and West, 2010, Laat et al., 2016). This chapter provides an overview of the main features of AFM probes, in relation to their use and applications; the subject has been extensively reviewed and discussed elsewhere (Gahan, 2004, Voigtländer, 2015, Gavara, 2017). For a complete overview of the available probes for AFM, the reader is invited to browse the online catalogues of the companies that sell AFM probes in the market. However, caution should be practiced on the information from the companies, and a self-analysis of the probe geometry and dimensions is encouraged.

An AFM probe typically consists of i) a cantilever beam clamped at one end, typically made of silicon or silicon nitride, which, together with the optical beam deflection apparatus (see Chapters 3.1.1 and 3.1.3), acts as both a force sensor and transducer and ii) a tip attached/integrated into the cantilever, typically at its free end (Figure 3.1.2.1).

Specific features of the probe provide specific functionalities. A small radius of curvature, down to 1–2 nm, is necessary for a high spatial resolution in both force and imaging modes. A small spring constant k, down to 0.01 N/m, provides high force sensitivity, meaning that a small force F produces a large, easily measurable, deflection z; indeed, according to Hook's law:

$$z = F/k \tag{3.1.2.1}$$

A lower limit to the measurable deflection (and therefore to the measurable force) is set by the thermal noise z_{th} of the cantilever, which can be estimated from the equipartition theorem as $z_{th} = \sqrt{k_B T/k}$, k_B, and T being the Boltzmann constant and the absolute temperature, respectively (see Chapter 3.1.3).

The minimum thermal noise-limited detectable force that can be measured dynamically with an instrumental bandwidth BW is

$$F_{th,min} = \sqrt{4k_B T b \text{BW}} \tag{3.1.2.2}$$

or, equivalently (since $b = k/(2\pi f_R Q)$),

Ewelina Lorenc, Hatice Holuigue, Alessandro Podestà, Dipartimento di Fisica "Aldo Pontremoli" and CIMaINa, Università degli Studi di Milano, Milano, Italy
Felix Rico, Aix-Marseille Univ, CNRS, INSERM, LAI, Turing centre for biological systems, Marseille, France

https://doi.org/10.1515/9783110640632-006

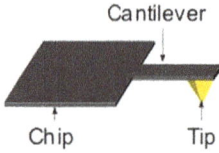

Figure 3.1.2.1: Schematics (not in scale) of an AFM probe, composed of a millimetric chip, a cantilever, and an integrated tip. The typical dimensions of cantilevers are: 100–400 μm (length) x 20–50 μm (width) x 0.5–2 μm (thickness); tip height is typically 5–15 μm. Different tip geometries are available (see main text), including conical, pyramidal, and spherical.

$$F_{\text{th,min}} = \sqrt{2k_{\text{B}}TkBW/(\pi f_{\text{R}}Q)} \tag{3.1.2.3}$$

where b is the damping coefficient (the proportionality factor between the tip veloc-ity and the viscous force), Q, and f_{R} are the quality factor and the resonance fre-quency of the cantilever, respectively (Smith, 1995, Viani et al., 1999).

Soft cantilevers also provide less invasive imaging conditions (i.e., they allow imaging at lower applied force in conditions where other forces, such as capillary adhesion, can be minimized). High resonance frequency and quality factor are nec-essary for dynamic imaging modes, where the cantilever oscillates near resonance, although for specific applications, for example, for high-speed scanning, a small quality factor is preferable since it allows for reducing the duration of the transients and ringing (Adams et al., 2016, Hosseini et al., 2019).

The above-mentioned probe characteristics are usually interconnected, and several constraints exist, which must be considered when a probe for a specific ap-plication is chosen or designed. For example, the spring constant k and the reso-nance frequency f_{R} are related by the equation:

$$2\pi f_{\text{R}} = \sqrt{k/m} \tag{3.1.2.4}$$

where m is the effective mass of the cantilever, and the equation for the quality factor Q is as follows:

$$Q = 2\pi f_{\text{R}}m/b \tag{3.1.2.5}$$

where b is the damping coefficient (in units of force/velocity).

As a consequence of the above-mentioned constraints, a small spring constant, necessary for high force sensitivity, usually comes with a low resonance frequency, which makes a soft probe typically unsuited for dynamic applications. On the other hand, high frequency, high Q cantilevers are typically stiff, which keeps them from imaging in contact mode. When the probe requirements for a specific application are apparently mutually exclusive, specific design strategies can be adopted. For example, vertical approach modes (such as peak force tapping and fast force vol-ume) require the cantilever to be inertially ramped along the vertical direction at a

relatively high frequency (up to 8–10 kHz), avoiding driving it into dynamic oscillation. To this purpose, the ramping frequency must be kept well below the first resonance frequency ($\leq f_R/3$). High-frequency cantilevers are therefore required, but with
small, contact mode-like, spring constant, for high force sensitivity and low invasiveness. The solution to the problem is to make the cantilever extremely thin to obtain a
small spring constant, which points to the use of silicon nitride, rather than silicon
(Eaton and West, 2010), but short, to obtain a high resonance frequency. When a small
tip radius is also required to boost the spatial resolution, a crystalline silicon tip can be
integrated onto the silicon nitride cantilever; in this case, the material of the cantilever
and of the tip is not the same. For fast vertical approach modes, minimizing the increase in the effective viscous drag coefficient when approaching the surface at high
velocity, long tips (several microns) are also useful.

The shape and dimensions of AFM probes are continuously evolving and changing, along with the increasing number of AFM applications. Contact and tapping
mode cantilevers may have rectangular planar and transversal sections, and are
made of both silicon and silicon nitride; this is the simplest and easy-to-model geometry (it is, in fact, an ideal geometry, since several deviations from the rectangular
geometry are typically present). Contact mode cantilevers are mostly made of silicon
nitride since it allows making them thinner and, therefore, softer, despite silicon nitride usually having higher residual stress than silicon, which causes a bending of
the cantilever along its main axis (Eaton and West, 2010). Silicon nitride cantilevers
are often triangular (V-shaped) as it was believed that these cantilevers could provide
low resistance to vertical deflection whilst resisting lateral torsion, although this assumption was then demonstrated to be erroneous (Sader, 2003, Sader and Sader
2003). New geometries, such as paddle-shaped, are now emerging, mainly to reduce
the viscous drag and increase the stability and force precision (Edwards et al., 2017).

The tips are made of either silicon nitride, which offers higher resistance to
wear and is good for contact mode, or crystalline silicon, which provides smaller
radii of curvature and is good for high-resolution dynamic modes, despite being
more brittle. Mass and dimensions of the tip are typically negligible compared to
those of the cantilever.

With the advent of high-speed (i.e., high-frequency) modes, cantilevers have
become shorter, and their shape often deviates from the standard rectangular or
V-shaped ones, including arrow-shaped ends, irregular rectangular geometries,
small aspect ratios, nonideal trapezoidal cross-sections, paddles (for ease of laser
alignment) (Slattery et al., 2014, 2019). The mass distribution along the cantilever
axis is often irregular, with massive, tall tips (the taller the tip, the weaker the hydrodynamic interaction between the cantilever and the sample), and nonnegligible
back-offset of the tip position (the loading point) with respect to the cantilever end.
All these deviations from the ideal geometry may impact the accuracy of the calibration of AFM probes, as discussed in Chapter 3.1.3.

Table 3.1.2.1 presents a rough classification of the AFM probes according to their use. Table 3.1.2.1 is intentionally limited to the more standard probe geometries and applications; nevertheless, the AFM tip catalogue is ample, as required by the increasing number of applications of the AFM methods. The reader is invited to browse the online catalogues of the tip manufacturers for a complete overview of the available probes.

3.1.2.1 Production of Standard AFM Probes

AFM probes are mostly made of silicon, silicon nitride, or a combination of the two materials (for example, silicon tips on a silicon nitride cantilever), and produced by micro-lithographic processes; a typical process would include a multilayer film deposition (for example silicon/silicon oxide) – a deposition of suitable photoresist, followed by activation through masks (photolithography), and chemical etching – to obtain monolithic microcantilevers with integrated tips (Albrecht et al., 1990, Gupta et al., 2003, Yu et al., 2006, Krause and Russell, 2008).

While silicon and silicon nitride are the materials typically used to produce AFM probes, SiO2 (glass) (Tang et al., 2004) can also be used, as well as metal cantilevers with integrated silicon tips can be produced (Chand et al., 2000). All-metal and all-diamond probes will be discussed in a specific section.

Exploiting the crystalline planes of silicon, it is possible to produce tips with a very sharp apical region, down to a couple of nm (Moldovan et al., 2012). Crystalline silicon is however very brittle, and silicon tips are prone to get blunt when used with relatively high forces. Silicon nitride tips are more wear-resistant, yet their obtained radius of curvature is typically larger (20–60 nm). Also, silicon nitride is usually used to produce extremely thin cantilevers. As mentioned already, producing short, silicon nitride cantilevers with integrated sharp silicon tip is a good strategy to obtain soft high-frequency, high-quality factor cantilevers, which provide high force sensitivity, and can be driven inertially at kHz frequencies, far from the natural resonances. Moreover, a tall tip (like Si tips typically are) will reduce the unwanted effects related to the squeezing of liquid between the cantilever and the sample, while ramping up the tip vertically.

The cantilever backside can be metallized by depositing a thin (<50 nm) layer of aluminum or gold to increase the reflectivity and minimize the interference effect on the detector (the laser beam that reaches the detector should primarily be reflected by the cantilever, rather than by the sample). Titanium or chromium layers are usually deposited on the cantilever to improve the adhesion of the reflective coating.

Table 3.1.2.1: Classification of standard AFM probes according to their main application, geometry and composition, typical parameters such as spring constant, resonance frequency, and radius.

Mode	Application	Geometry	Material for cantilever/tip	Typical k (N/m)	Typical res. freq. (kHz)	Typical tip radius (nm)/height (μm)
Contact mode Force curve acquisition	Imaging in air and liquid Adhesion and force spectroscopy (pulling) Nanomechanics of cells and tissues	V-shaped and rectangular	SiNx/Si	0.01–0.5 N/m	11–75 kHz	2-30/3-20 Colloidal probes: radius 2–50 μm
Dynamic modes (tapping or noncontact)	Imaging in air and liquid	Rectangular (air) or V-shaped (liquid)	Si/Si (air) or SiNx/SiNx and SiNx/Si (liquid)	40–80 N/m (air) or 0.1–1 N/m (liquid)	250–350 kHz (air) or 10–75 kHz (liquid)	2-30/3-20
Vertical approach modes (PeakForce tapping, QI-CPI, etc.)	(High-resolution) imaging (Quantitative) nanomechanical mapping on soft samples	Rectangular/V-shaped	SiNx/Si	<1 N/m	20–150 kHz (liquid)	<100/5-20

Typical values/conditions are reported.

3.1.2.2 Production of Custom Probes

3.1.2.2.1 Colloidal Probes

Colloidal (or spherical) probes (CPs) are obtained by attaching a spherical bead to a tipless cantilever (Figure 3.1.2.2) (Ducker et al., 1991, 1992).

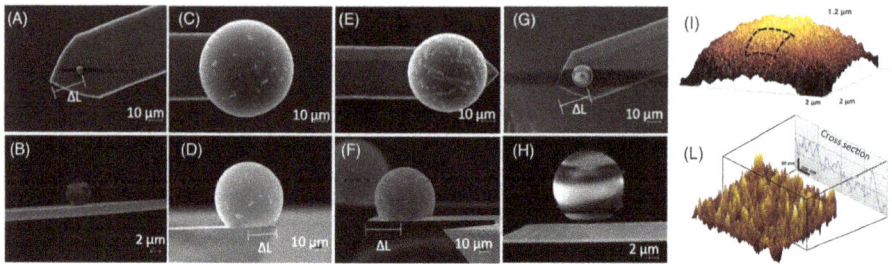

Figure 3.1.2.2: Representative SEM images of CPs. (A, B) A borosilicate glass CP from a contact mode tipless cantilever. (C, D) A soda-lime glass CP from a contact mode tipless cantilever. (E, F) A soda-lime glass CP from a tapping mode tipless cantilever. (G, H) A borosilicate glass CP from a tapping mode tipless cantilever. All probes shown at increasing magnification. (I, L) AFM image obtained in tapping mode of the surface morphology of a borosilicate glass CP coated with a thin film of biocompatible nanostructured zirconia (A–H, adapted with permission from Chighizola et al. (2021b); I, L, adapted with permission from Chighizola et al. (2020)).

The diameter of the spherical bead goes from 1 to 20 µm, typically, although for special applications, diameters up to 100 µm can be used (Chighizola et al., 2021b). Even though sharp conical or pyramidal tips have been mainly used so far and also on biological specimens, the mechanical and adhesive properties of heterogeneous and non-uniform samples such as tissues or extracellular matrices (ECM) can be effectively studied with CPs (Carl and Schillers, 2008, Waters et al., 2012, Puricelli et al., 2015, Rianna and Radmacher, 2016, Nebuloni et al., 2016, Rigato et al., 2017, Schillers et al., 2017, Jorba et al., 2017, 2019, Marsal et al., 2017, Alcaraz et al., 2018, Chighizola et al., 2020, Kubiak et al., 2021). CPs can be advantageous over standard sharp AFM tips because they provide a well-defined interaction (indentation) geometry (sphere on flat surface), which allows a more accurate application of analytical models. Moreover, CPs allow averaging across a larger interaction area or volume in force spectroscopy and nanomechanical measurements, and provide a higher signal-to-noise ratio in spectroscopic experiments; they also present the advantage of being easily functionalized (Noy et al., 1997, Ebner et al., 2019), thanks to the larger area and well-defined surface chemistry; this is useful, for example, to study specific interactions between ligands and receptors at the surface of the cells using force spectroscopy techniques (Carl and Schillers, 2008, Friedrichs et al., 2010, Chaudhuri et al., 2014, Becerra et al., 2015, Puricelli et al., 2015, Nebuloni et al., 2016, Chighizola et al., 2020, Harjumäki et al., 2020).

With regard to the specific advantages for mechanical measurements, due to the larger contact area CPs induce less strain and stress on the sample, which makes the contact mechanics models more applicable. While working on biological samples, such as cells, this configuration can be particularly appreciable. Sharp tips, with their small contact area, can induce high stresses upon contact, are difficult to be accounted for and controlled, and cause strong modulation of the contact area upon interaction with nanometer-sized features. The dimension and geometry of the tip are not very well defined nor can they be easily characterized, while the spherical geometry of CPs can be accurately characterized (Neto and Craig, 2001, Indrieri et al., 2011). Sharp tips are necessary whenever a high spatial resolution is required (Rotsch and Radmacher, 2000, Braunsmann et al., 2014). Despite the reduced radius, they may represent a source of noise, poor reproducibility, and reduced accuracy in the measurements. The advantages of CPs mentioned above come at a price. The first and the obvious drawback is the marked reduction of lateral resolution. Nevertheless, it was shown that CPs with a radius as large as 5 µm still provide enough lateral resolution to distinguish the relevant cellular regions in the AFM-based topographic and mechanical imaging of cells (Puricelli et al., 2015). Another drawback of using CPs is the increased importance of the bottom effect, that is, the fact that on finite-thickness samples, the probe may feel the stiffness of the underlying rigid substrate (see Chapter 2.2) (Dimitriadis et al., 2002, Gavara and Chadwick, 2012, Garcia and Garcia, 2018). Indeed, the bottom-effect correction depends on the ratio of the contact radius a to the sample thickness h; for Hertzian contacts. This ratio is equal to $\sqrt{R\delta}/h$, where d is the indentation; therefore the larger the R, stronger is the bottom effect.

Another drawback is that the added mass of large spherical beads can affect the cantilever dynamics, and therefore have an impact on the accuracy of the thermal noise calibration of the cantilever spring constant (Chighizola et al., 2021b). Moreover, with large colloidal probes, the accuracy of the determination of the deflection sensitivity of the optical beam deflection apparatus can be reduced, primarily because of the stronger friction and irregular interference effects (see Chapter 3.1.3).

The production of CPs is mainly done by attaching a glass microbead on a tipless cantilever, but other materials are also used. CPs can be homemade and customized according to the needs, but they are also produced by companies and sold in the market. Most CPs use glue and adhesives to attach the bead to a tipless cantilever. Monolithic CPs, where no glue is present and the bead is sintered at the cantilever (realizing a unique solid entity), are advantageous since they can be aggressively cleaned in acidic or other solutions to fully restore their surface chemical state (e.g., full hydroxylation in the case of SiO_2 CPs), and reused many times.

For a review on the production of CPs, the user is referred to Yuan et al. (2017). Reports can be found about the production of monolithic CPs (Yapici and Zou, 2009, Indrieri et al., 2011, Kuznetsov and Papastavrou, 2012, Ditscherlein and Peuker, 2017); about the use of glue and micromanipulator or micropipettes (Mak et al., 2006, Plodinec et al., 2010); about the attachment of polystyrene beads by low-temperature sintering (Vinogradova et al., 2001) or about glueing alumina (Pedersen, 1999), zirconia (Hook et al., 1999), titanium (Mak et al., 2006), or even cellulose (quasi)spherical particles (Lai et al., 2019).

As an alternative to the use of 3D nano-micro-manipulators, it is possible to use the XYZ stage of the AFM to locate the beads and bring the cantilever in contact with them (Indrieri et al., 2011, D'Sa et al., 2014). This approach is also compatible with the use of glue or other suitable adhesives (e.g., vaseline or glycerol that dissolve at high temperature during the sintering process); the cantilever can be first dipped into a small droplet of liquid adhesive and then moved to capture a bead.

The radius of CPs, with radius below 5 µm, can be calibrated by reverse AFM imaging (Neto and Craig, 2001, Indrieri et al., 2011), which consists in scanning the CP across a spiked grating and analyzing the obtained replicas of the apical probe region, for example, fitting a spherical cap model to the topographic AFM images. The larger the probe radius, the smaller is the portion of the spherical bead that penetrates within the spikes; therefore, the smaller the probe region that is faithfully characterized. While this is usually not a problem for force spectroscopy applications, for indentation experiments, it would be advisable to characterize a portion of the sphere that is comparable to the maximum indentation; nonstandard gratings with larger spacing and taller spikes should be used. Alternatively, by giving up the three-dimensional, more accurate characterization of the probe shape, one can scan the spherical probe across a stepped grating, obtaining one-dimensional profiles of the spherical bead. Very large CPs can be directly characterized by optical microscopy, with good precision and accuracy (Chighizola et al., 2021b).

3.1.2.2.2 Cellular or Bacterial Probes

Tipless AFM cantilevers can be functionalized by attaching biological samples such as eukaryotic or bacterial cells (even viruses (Rankl et al., 2008)); these probes are mainly used for single-cell force spectroscopy (SCFS) or cell-cell adhesion spectroscopy experiments as presented in several publications (Benoit et al., 2000, Wojcikiewicz et al., 2006, Müller and Dufrêne, 2008, Helenius et al., 2008, Qu et al., 2011, Friedrichs et al., 2013, Beaussart et al., 2014, Smolyakov et al., 2016, Grzeszczuk et al., 2020, Viji Babu et al., 2021).

One of the most common methods for attaching cells to tipless cantilevers is to use poly-L-lysine, concanavalin A, or fibronectin as adhesives (Moreno-Cencerrado et al., 2017). However, concanavalin A may lead to the activation of white blood cells (Chu et al., 2013). The cantilever is coated with the molecular of interest. Then some suspended cells or poorly substrate-adherent cells are cached and allowed to adhere to the cantilever (Viji Babu et al., 2021). Another procedure consists of an amino-functionalization of the AFM cantilever using PEG-NHS linkers, prior to incubation with virus particles (Rankl et al., 2008).

To prevent cell sliding and nonparallel force application to cells due to the mounting angle, wedges made of UV-curable polymer have been mounted on tipless cantilevers (Stewart et al., 2013). Notably, the attachment of cells to cantilevers is not only useful in single-cell force spectroscopy experiments; cellular regulatory processes, accompanied by tiny mass fluctuations, can be monitored by attaching cells to cantilevers and studying their frequency and phase shifts over time (Martínez-Martín et al., 2017).

3.1.2.3 Nonstandard AFM Probes Produced by Lithography and Other Manufacturing Techniques

Lithographic techniques allow the shaping of sharp tips in a reproducible manner. Focused ion beams, focused electron beam-induced deposition, physical sputtering or chemical enhanced etching, as well as the growing of nanostructures on the cantilever using carbon nanotubes, can be used for the tuning of specific AFM probe features (such as the tip geometry or radius of curvature) (Folch et al., 1997, Grow et al., 2002, Ong and Sokolov, 2007, Menozzi et al., 2008, Temiryazev et al., 2016, Onoda et al., 2021).

Dry film photoresist lithography can be further employed to add chemical and physical functionalities to silicon nitride cantilevers; moreover, polymeric tips with specific geometries can be grown on tipless cantilevers (Nilsen et al., 2019). It is noteworthy that cylindrical pillars with flat ends provide a constant and controlled contact area in mechanical measurements, and is often a desirable feature since it simplifies the modelling of the contact (see Chapter 2.2) (Rico et al., 2007, Waters et al., 2012).

3D printing (Alsharif et al., 2018) is an emerging technology to produce custom AFM probes, for example, by 3D direct laser writing, exploiting the two-photon polymerization process (Göring et al., 2016). In this case, polymeric tips with tailored geometry are obtained Figure 3.1.2.3), with a resolution (that can be taken equal to the smallest radius of curvature of the printed tip) as good as 200 nm (Alsharif et al., 2018).

All-diamond and all-metal probes typically maintain their functionalities (conductivity, overall dimensions, and geometry) over time, also in the case of strong tip–sample interaction. For example, an all-metal or doped all-diamond probe will always conduct electricity, despite wear or slight contamination, in contrast to the case of metal-coated tips, where the tip lifetime is typically short, strongly loaded, and are history-dependent, primarily because of the issue of detachment of the thin metallic film. These probes typically are suitable for imaging with good spatial resolution, allowing to couple topographic, electrical and/or mechanical investigation

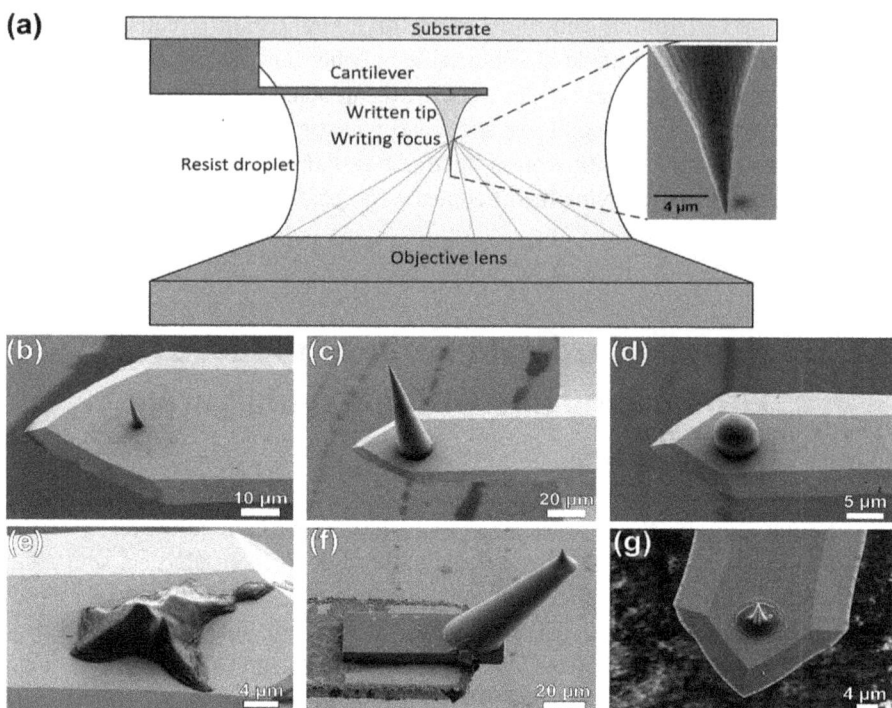

Figure 3.1.2.3: (a) Schematics of the two-photon polymerization process. (b)–(g) Examples of a variety of tips that can be produced (image from Göring et al. (2016)).

(Wood et al., 2015). All-diamond cantilever probes (also conductive) are nowadays commercially available. These probes can be realized by a proximity lithography process (Malavé and Oesterschulze, 2006) and other patented approaches.

All-metal AFM probes can also be fabricated from microstructurally tailored Cu–Hf thin films (Luber et al., 2009). All-metal AFM probes made from platinum and platinum–iridium microwires are commercially available.

Microfluidic cantilevers allow coupling of micro- and nano-pipette analysis to the standard AFM investigation. These cantilevers possess an internal channel that allows measuring local ionic currents, applying and controlling pressure, etc. (Amarouch et al., 2018).

All-plastic cantilevers (made of SU-8 resin), due to the higher internal damping of the polymeric material and the resulting lower quality factor, provide very short response time for high-speed AFM and sensing applications. These cantilevers can be equipped with integrated sharp silicon nitride tips, and also represent suitable platforms for the development of cantilever-based sensors (Adams et al., 2016, Hosseini et al., 2019).

3.1.2.4 Functionalization of Probes

The functionalization of AFM probes, either chemically or morphologically, allows their use as force, mass, or heat sensors (Thundat et al., 1994, Raiteri et al., 2001, Barattin and Voyer, 2008).

Chemical functionalization (Noy et al., 1995, Ozkan et al., 2018) aims at attaching small molecules or molecular constructs to sense specific interactions between the tip chemical end-group and suitable counter-molecules attached to a flat, clean surface. Using functionalized tips, receptor–ligand interactions, specific adhesion phenomena, and protein unfolding events can be studied using the AFM in the force spectroscopy mode (Zlatanova et al., 2000, Dufrêne et al., 2011, Yang et al., 2020). An example of functionalization is the attachment of biomolecules to the AFM probes via amino functionalization of the silicon cantilevers through silane (-SiOx) or gold-coated canti-levers, through thiol groups (-SH) and linkers. Those linkers can be polyethylene gly-col polymers (PEG) (Zimmermann et al., 2010), but also elastin-like polypeptide (ELP) (Ott et al., 2017, Liu et al., 2020). From this base, multiple biomolecules could be at-tached and studied: antibodies/antigens (Kienberger et al., 2005, Ebner et al., 2007, Caneva Soumetz et al., 2010, Liu et al., 2019); peptides (Lehenkari and Horton, 1999); collagen or ECM proteins (Lee et al., 2013, Hong et al., 2015, Li et al., 2020).

Morphological functionalization of probes aims to study how the surface morphol-ogy and the presence of surface peculiar nanotopographic features, a common aspect of biosurfaces and interfaces, affect the interaction of biological entities (cells, pro-teins) with their microenvironment. Typically, for this purpose, a thin film of a mate-rial that confers a peculiar surface nanoscale morphology to the tip is deposited on it (Figure 3.1.2.2I,L) (Chighizola et al., 2020). Single nanoparticles can also be attached to the tip (Marcuello et al., 2018) as well as nanotubes (Woolley et al., 2000, Hafner et al., 2001), or both (Yashchenok et al., 2013). The coating process typically also changes the chemical properties of the tip, depending on the material deposited, and can also make available other interaction channels, for instance, the presence of high aspect ratio gold features on the tip can be used to enhance the Raman scattering at the tip–sample interface (Malavé and Oesterschulze, 2006). Recently, the deposition of thin nanostructured zirconia films on large spherical probes (Figure 3.1.2.2I,L), mim-icking the extracellular matrix nanotopography allowed us to study integrin-related mechanotransductive processes in PC12 cells (Chighizola et al., 2020, 2021a).

References

Adams, J. D., B. W. Erickson, J. Grossenbacher, J., . J. Brugger, A. Nievergely and G. E. Fantner (2016). "Harnessing the damping properties of materials for high-speed atomic force microscopy." Nature Nanotechnology **11**: 147–151.

Albrecht, T. R., S. Akamine, T. E. Carver and C. F. Quate (1990). "Microfabrication of cantilever styli for the atomic force microscope. "." Journal of Vacuum Science and Technology **8**: 3386–3396.

Alcaraz, J., J. Otero, I. Jorba and D. Navajas (2018). "Bidirectional mechanobiology between cells and their local extracellular matrix probed by atomic force microscopy." Seminars in Cell & Developmental Biology **73**: 71–81.

Alessandrini, A. and P. Facci (2005). "AFM: A versatile tool in biophysics." Measurement Science and Technology **16**: 65–92.

Alsharif, N., A. Burkatovsky, C. Lissandrello, K. M. Jones, A. E. White and K. A. Brown (2018). "Design and realization of 3D printed AFM probes." Small **14**: e1800162.

Amarouch, M. Y., J. E. Hilaly and D. Mazouzi (2018). "AFM and FluidFM technologies: Recent applications in molecular and cellular biology." Scanning **2018**: 7801274.

Barattin, R. and N. Voyer (2008). "Chemical modifications of AFM tips for the study of molecular recognition events." Chemical Communications **13**: 1513–1532.

Beaussart, A., S. El-Kirat-Chatel, R. M. A. Sullan, S. Alsteens, P. Herman, S. Derclaye and Y. F. Dûfrene (2014). "Quantifying the forces guiding microbial cell adhesion using single-cell force spectroscopy." Nature Protocols **9**: 1049–1055.

Becerra, N., H. Andrade, B. Lopez, L. M. Restrepo and R. Raiteri (2015). "Probing poly(N-isopropylacrylamide-co-butylacrylate)/cell interactions by atomic force microscopy." Journal of Biomedical Materials Research **103**: 145–153.

Benoit, M., D. Gabriel, G. Gerisch and H. E. Gaub (2000). "Discrete interactions in cell adhesion measured by single-molecule force spectroscopy." Nature Cell Biology **2**: 313–317.

Braunsmann, C., J. Seifert, J. Rheinlaender and T. E. Schäffer (2014). "High-speed force mapping on living cells with a small cantilever atomic force microscope." The Review of Scientific Instruments **85**: 073703.

Caneva Soumetz, F., J. F. Saenz, L. Pastorino, C. Ruggiero, D. Nosi and R. Raiteri (2010). "Investigation of integrin expression on the surface of osteoblast-like cells by atomic force microscopy." Ultramicroscopy **110**: 330–338.

Carl, P. and H. Schillers (2008). "Elasticity measurement of living cells with an atomic force microscope: Data acquisition and processing." Pflügers Archiv: European Journal of Physiology **457**: 551–559.

Chand, A., M. B. Viani, T. E. Schäffer and P. K. Hansma (2000). "Microfabricated small metal cantilevers with silicon tip for atomic force microscopy." Journal of Microelectromechanical Systems **9**: 112–116.

Chaudhuri, O., S. T. Koshy, C. Branco Da Cunha, J. W. Shin, C. S. Verbeke, K. H. Allison and D. J. Mooney (2014). "Extracellular matrix stiffness and composition jointly regulate the induction of malignant phenotypes in mammary epithelium." Nature Materials **13**: 970–978.

Chighizola, M., T. Dini, S. Marcotti, M. D'Urso, C. Piazzoni, F. Borghi, A. Previdi, L. Ceriani, C. Folliero, B. Stramer, C. Lenardi, P. Milani, A. Podesta and C. Shulte. "The glycocalyx affects the mechanotransductive perception of the topographical microenvironment." Journal of Nanobiotechnology **20**: 418 (2022). https://doi.org/10.1186/s12951-022-01585-5

Chighizola, M., A. Previdi, T. Dini, C. Piazzoni, C. Lenardi, P. Milani, C. Shulte and A. Podesta (2020). "Adhesion force spectroscopy with nanostructured colloidal probes reveals nanotopography-dependent early mechanotransductive interactions at the cell membrane level." Nanoscale 12: 14708–14723.

Chighizola, M., L. Puricelli, L. Bellon and A. Podestà (2021b). "Large colloidal probes for atomic force microscopy: Fabrication and calibration issues." Journal of Molecular Recognition : JMR 34: e2879.

Chu, C., E. Celik, F. Rico and V. T. Moy (2013). "Elongated membrane tethers, individually anchored by high affinity α4β1/VCAM-1 complexes, are the quantal units of monocyte arrests." PLoS One 8: e64187.

D'Sa, D. J., H. K. Chan and W. Chrzanowski (2014). "Attachment of micro- and nano-particles on tipless cantilevers for colloidal probe microscopy." Journal of Colloid and Interface Science 426: 190–198.

Dimitriadis, E. K., F. Horkay, J. Maresca, B. Kachar and R. S. Chadwick (2002). "Determination of elastic moduli of thin layers of soft material using the atomic force microscope." Biophysical Journal 82: 2798–2810.

Ditscherlein, L. and U. A. Peuker (2017). "Note: Production of stable colloidal probes for high-temperature atomic force microscopy applications." Review of Scientific Instruments 88: 046107.

Ducker, W. A., T. J. Senden and R. M. Pashley (1991). "Direct measurement of colloidal forces using an atomic force microscope." Nature 353: 239–241.

Ducker, W. A., T. J. Senden and R. M. Pashley (1992). "Measurement of forces in liquids using a force microscope." Langmuir 8: 1831–1836.

Dufrêne, Y. F., E. Evans, A. Engel, J. Helenius, H. E. Gaub and D. J. Mueller (2011). "Five challenges to bringing single-molecule force spectroscopy into living cells." Nature Methods 8: 123–127.

Eaton, P. J. and P. West (2010). Atomic force microscopy. Oxford University Press.

Ebner, A., P. Hinterdorfer and H. J. Gruber (2007). "Comparison of different aminofunctionalization strategies for attachment of single antibodies to AFM cantilevers." Ultramicroscopy 107: 922–927.

Ebner, A., L. Wildling and H. J. Gruber (2019). "Functionalization of AFM tips and supports for molecular recognition force spectroscopy and recognition imaging." Methods in Molecular Biology (Clifton, N.J.) 1886: 117–151.

Edwards, D. T., J. K. Faulk, M. A. LeBlanc and T. T. Perkins (2017). "Force spectroscopy with 9-µs resolution and Sub-pN stability by tailoring AFM cantilever geometry." Biophysical Journal 113: 2595–2600.

Folch, A., M. S. Wrighton and M. A. Schmidt (1997). "Microfabrication of oxidation-sharpened silicon tips on silicon nitride cantilevers for atomic force microscopy." Journal of Microelectromechanical Systems 6: 303–306.

Friedrichs, J., J. Helenius and D. J. Muller (2010). "Quantifying cellular adhesion to extracellular matrix components by single-cell force spectroscopy." Nature Protocols 5: 1353–1361.

Friedrichs, J., K. R. Legate, R. Schubert, M. Bharadwaj, C. Werner, D. J. Mueller and M. Benoit (2013). "A practical guide to quantify cell adhesion using single-cell force spectroscopy." Methods 60: 169–178.

Gahan, P. B. (2004). Atomic force microscopy in cell biology. Jena, B. P. and Horber, J. K. H., Eds. Academic Press.

Garcia, P. D. and R. Garcia (2018). "Determination of the elastic moduli of a single cell cultured on a rigid support by force microscopy." Biophysical Journal 114: 2923–2932.

Gavara, N. and R. S. Chadwick (2012). "Determination of the elastic moduli of thin samples and adherent cells using conical atomic force microscope tips." Nature Nanotechnology 7: 733–736.

Gavara, N. (2017). "A beginner's guide to atomic force microscopy probing for cell mechanics." Microscopy Research and Technique **80**: 75–84.

Göring, G., P. I. Dietrich, M. Blaicher, S. Sharma, J. G. Korvink, T. Schimmel, C. Koos and H. Hoelscher (2016). "Tailored probes for atomic force microscopy fabricated by two-photon polymerization." Applied Physics Letters **109**: 63101.

Grow, R. J., S. C. Minne, S. R. Manalis and C. F. Quate (2002). "Silicon nitride cantilevers with oxidation-sharpened silicon tips for atomic force microscopy." Journal of Microelectromechanical Systems **11**: 317–321.

Grzeszczuk, Z., A. Rosillo, O. Owens and S. Bhattacharjee (2020). "Atomic Force Microscopy (AFM) as a surface mapping tool in microorganisms resistant toward antimicrobials: A mini-review." Frontiers in Pharmacology **11**: 517165.

Gupta, A., J. P. Denton, H. McNally and R. Bashir (2003). "Novel fabrication method for surface micromachined thin single-crystal silicon cantilever beams." Journal of Microelectromechanical Systems **12**: 185–192.

Hafner, J., C. L. Cheung, A. Woolley and C. Lieber (2001). "Structural and functional imaging with carbon nanotube AFM probes." Progress in Biophysics and Molecular Biology **77**: 73–110.

Harjumäki, R., X. Zhang, R. W. N. Nugroho RWN, M. Farooq, Y. R. Lou, M. Yliperttula, J. J. Valle-Delgado and M. Oesterberg (2020). "AFM force spectroscopy reveals the role of integrins and their activation in cell-biomaterial interactions." ACS Applied Bio Materials **3**: 1406–1417.

Helenius, J., C.-P.-P. Heisenberg, H. E. Gaub and D. J. Muller (2008). "Single-cell force spectroscopy." Journal of Cell Science **121**: 1785–1791.

Hong, Z., K. J. Reeves, Z. Sun, Z. Li, N. J. Brown and G. A. Meininger (2015). "Vascular smooth muscle cell stiffness and adhesion to collagen I modified by vasoactive agonists." PLoS One **10**: e0119533.

Hook, M. S., P. G. Hartley and P. J. Thistlethwaite (1999). "Fabrication and characterization of spherical zirconia particles for direct force measurement using the atomic force microscope." Langmuir **15**: 6220–6225.

Hosseini, N., M. Neuenschwander, O. Peric, S. H. Andany, J. D. Adams and G. E. Fantner (2019). "Integration of sharp silicon nitride tips into high-speed SU8 cantilevers in a batch fabrication process." Beilstein Journal of Nanotechnology **10**: 2357–2363.

Indrieri, M., A. Podestà, G. Bongiorno, D. Marchesi and P. Milani (2011). "Adhesive-free colloidal probes for nanoscale force measurements: Production and characterization." Review of Scientific Instruments **82**: 023708.

Jorba, I., G. Beltrán, B. Falcones, B. Suki, R. Farre, J. M. Garcia-Aznar and D. Navajas (2019). "Nonlinear elasticity of the lung extracellular microenvironment is regulated by macroscale tissue strain." Acta Biomaterialia **92**: 265–276.

Jorba, I., J. J. Uriarte, N. Campillo, R. Farre and D. Navajas (2017). "Probing micromechanical properties of the extracellular matrix of soft tissues by atomic force microscopy." Journal of Cellular Physiology **232**: 19–26.

Kienberger, F., G. Kada, H. Mueller and P. Hinterdorfer (2005). "Single molecule studies of antibody–antigen interaction strength versus intra-molecular antigen stability." Journal of Molecular Biology **347**: 597–606.

Krause, O. and P. Russell (2008). "AFM probe manufacturing." NanoWorld Services GmbH.

Kubiak, A., M. Chighizola, C. Schulte, N. Bryniarska, J. Wesołowska, M. Pudełek, M. Lasota, D. Ryszawy, A. Basta-Kaim, P. Laidler, A. Podesta and M. Lekka (2021). "Stiffening of DU145 prostate cancer cells driven by actin filaments-microtubule crosstalk conferring resistance to microtubule-targeting drugs." Nanoscale **13**: 6212–6226.

Kuznetsov, V. and G. Papastavrou (2012). "Note: Mechanically and chemically stable colloidal probes from silica particles for atomic force microscopy." Review of Scientific Instruments **83**: 116103.

Laat, M. L. C., H. H. P. Garza and M. K. Ghatkesar (2016). "In situ stiffness adjustment of AFM probes by two orders of magnitude." Sensors **16**: 523.

Lai, Y., H. Zhang, Y. Sugano, H. Xie and P. Kallio (2019). "Correlation of surface morphology and interfacial adhesive behavior between cellulose surfaces: Quantitative measurements in peak-force mode with the colloidal probe technique." Langmuir **35**: 7312–7321.

Lee, S., Y. Yang, D. Fishman, M. M. Banaszak and S. Hong (2013). "Epithelial–mesenchymal transition enhances nanoscale actin filament dynamics of ovarian cancer cells." The Journal of Physical Chemistry. B **117**: 9233–9240.

Lehenkari, P. P. and M. A. Horton (1999). "Single integrin molecule adhesion forces in intact cells measured by atomic force microscopy." Biochemical and Biophysical Research Communications **259**: 645–650.

Li, Z., T. Liu, J. Yang, J. Lin and S. X. Xin (2020). "Characterization of adhesion properties of the cardiomyocyte integrins and extracellular matrix proteins using atomic force microscopy." Journal of Molecular Recognition : JMR **33**: e2823.

Liu, H., V. Schittny and M. A. Nash (2019). "Removal of a conserved disulfide bond does not compromise mechanical stability of a VHH antibody complex." Nano Letters **19**: 5524–5529.

Liu, Z., H. Liu, A. M. Vera, R. C. Bernardi, P. Tinnefeld and M. A. Nash (2020). "High force catch bond mechanism of bacterial adhesion in the human gut." Nature Communications **11**: 4321.

Luber, E. J., B. C. Olsen, C. Ophus, V. Radmilovic and D. Mitlin (2009). "All-metal AFM probes fabricated from microstructurally tailored Cu-Hf thin films." Nanotechnology **20**: 345703.

Mak, L. H., M. Knoll, D. Weiner, A. Gorschlüter, A. Schirmeisen and H. Fuchs (2006). "Reproducible attachment of micrometer sized particles to atomic force microscopy cantilevers." Review of Scientific Instruments **77**: 46104.

Malavé, A. and E. Oesterschulze (2006). "All-diamond cantilever probes for scanning probe microscopy applications realized by a proximity lithography process." Review of Scientific Instruments **77**: 043708.

Marcuello, C., L. Foulon, B. Chabbert, M. Molinari and V. Aguie-Beghin (2018). "Langmuir–blodgett procedure to precisely control the coverage of functionalized AFM cantilevers for SMFS measurements: Application with cellulose nanocrystals." Langmuir **34**: 9376–9386.

Marsal, M., I. Jorba, E. Rebollo, T. Luque, D. Navajas and E. Martin-Blanco (2017). "AFM and microrheology in the zebrafish embryo yolk cell." Journal of Visualized Experiments **129**: e56224.

Martínez-Martín, D., G. Fläschner, B. Gaub, S. Martin, R. Newton, C. Beerli, J. Mercer, C. Gerber and D. J. Mueller (2017). "Inertial picobalance reveals fast mass fluctuations in mammalian cells. "." Nature **550**: 500–505.

Menozzi, C., L. Calabri, P. Facci, P. Pingue, F. Dinelli and P. Baschieri (2008). "Focused ion beam as tool for atomic force microscope (AFM) probes sculpturing." Journal of Physics: Conference Series **126**: 012070.

Moldovan, N., Z. Dai, H. Zeng, J. A. Carlisle, T. D. B. Jacobs, V. Vahdat, D. S. Grierson, J. Liu, K. T. Turner and R. W. Carpick (2012). "Advances in manufacturing of molded tips for scanning probe microscopy." Journal of Microelectromechanical Systems **21**: 431–442.

Moreno-Cencerrado, A., J. Iturri, I. Pecorari, M. D. M. Vivanco, O. Sbaizero and J. L. Toca-Herrera (2017). "Investigating cell-substrate and cell–cell interactions by means of single-cell-probe force spectroscopy." Microscopy Research and Technique **80**: 124–130.

Müller, D. J. and Y. F. Dufrêne (2008). "Atomic force microscopy as a multifunctional molecular toolbox in nanobiotechnology." Nature Nanotechnology **3**: 261–269. https://doi.org/10.1142/9789814287005_0028.

Nebuloni, M., L. Albarello, A. Andolfo, C. Magagnotti, L. Genovese, I. Locatelli, G. Tonon, E. Longhi, P. Zerbi, R. Allevi, A. Podestà, L. Puricelli, P. Milani, A. Soldarini, A. Salonia and M. Alfano (2016). "Insight on colorectal carcinoma infiltration by studying perilesional extracellular matrix." Scientific Reports **6**: 22522.

Neto, C. and V. S. J. Craig (2001). "Colloid probe characterization: Radius and roughness determination." Langmuir **17**: 2097–2099.

Nilsen, M., F. Port, M. Roos, K. E. Gottschalk and S. Strehle (2019). "Facile modification of freestanding silicon nitride microcantilever beams by dry film photoresist lithography." Journal of Micromechanics and Microengineering **29**: 025014.

Noy, A., C. D. Frisbie, L. F. Rozsnyai, M. S. Wrighton and C. M. Lieber (1995). "Chemical force microscopy: Exploiting chemically-modified tips to quantify adhesion, friction, and functional group distributions in molecular assemblies." Journal of the American Chemical Society **117**: 7943–7951.

Noy, A., D. V. Vezenov and C. M. Lieber (1997). "Chemical force microscopy"." Annual Review of Materials Research **27**: 381–421.

Ong, Q. K. and I. Sokolov (2007). "Attachment of nanoparticles to the AFM tips for direct measurements of interaction between a single nanoparticle and surfaces." Journal of Colloid and Interface Science **310**: 385–390.

Onoda, J., T. Hasegawa and Y. Sugimoto (2021). "In situ reproducible sharp tips for atomic force microscopy." Physical Review Applied **15**: 034079.

Ott, W., M. A. Jobst, M. S. Bauer, E. Durner, L. F. Milles, M. A. Nash and H. E. Gaub (2017). "Elastin-like polypeptide linkers for single-molecule force spectroscopy." ACS Nano **11**: 6346–6354.

Ozkan, A. D., A. E. Topal, F. B. Dikecoglu, M. O. Guler, A. Dana and A. B. Tekinay (2018). "Probe microscopy methods and applications in imaging of biological materials." Seminars in Cell and Developmental Biology **73**: 153–164.

Pedersen, H. G. (1999). "Aluminum oxide probes for AFM force measurements: Preparation, characterization, and measurements." Langmuir **15**: 3015–3017.

Plodinec, M., M. Loparic M. and U. Aebi (2010). "Preparation of microsphere tips for atomic force microscopy (AFM)." Cold Spring Harbor Protocols **5**.

Puricelli, L., M. Galluzzi, C. Schulte, A. Podesta and P. Milani (2015). "Nanomechanical and topographical imaging of living cells by atomic force microscopy with colloidal probes." Review of Scientific Instruments **86**: 033705.

Qu, W., J. M. M. Hooymans, J. de Vries, H. C. van der Mei and H. J. Busscher (2011). "Force analysis of bacterial transmission from contact lens cases to corneas, with the contact lens as the intermediary." Investigative Ophthalmology & Visual Science **52**: 2565–2570.

Raiteri, R., M. Grattarola, H.-J. Butt and P. Skládal (2001). "Micromechanical cantilever-based biosensors." Sensors & Actuators, B: Chemical **79**: 115–126.

Rankl, C., F. Kienberger, L. Wildling, J. Wruss, H. J. Gruber, D. Blaas and P. Hinterdorfer (2008). "Multiple receptors involved in human rhinovirus attachment to living cells." Proceedings of the National Academy of Sciences of the United States of America **105**: 17778–17783.

Rianna, C. and M. Radmacher (2016). "Cell mechanics as a marker for diseases: Biomedical applications of AFM." AIP Conference Proceedings **1790**: 020057.

Rico, F., P. Roca-Cusachs, R. Sunyer, R. Farre and D. Navajas (2007). "Cell dynamic adhesion and elastic properties probed with cylindrical atomic force microscopy cantilever tips." Journal of Molecular Recognition : JMR **20**: 459–466.

Rigato, A., A. Miyagi, S. Scheuring and F. Rico (2017). "High-frequency microrheology reveals cytoskeleton dynamics in living cells." Nature Physics **13**: 771–775.

Rotsch, C. and M. Radmacher (2000). "Drug-induced changes of cytoskeletal structure and mechanics in fibroblasts: An atomic force microscopy study." Biophysical Journal **78**: 520–535.

Sader, J. E. (2003). "Susceptibility of atomic force microscope cantilevers to lateral forces." Review of Scientific Instruments **74**: 2438–2443.

Sader, J. E. and R. C. Sader (2003). "Susceptibility of atomic force microscope cantilevers to lateral forces: Experimental verification." Applied Physics Letters **83**: 3195–3197.

Schillers, H., C. Rianna, J. Schäpe, et al. (2017). "Standardized nanomechanical atomic force microscopy procedure (SNAP) for measuring soft and biological samples." Scientific Reports **7**: 5117.

Slattery, A. D., A. J. Blanch, V. Ejov, J. S. Quinton and C. T. Gibson (2014). "Spring constant calibration techniques for next-generation fast-scanning atomic force microscope cantilevers." Nanotechnology **25**: 335705.

Slattery, A. D., A. J. Blanch, C. J. Shearer, A. J. Stapleton, R. V. Goreham, S. L. Harmer, J. S. Quinton and C. T. Gibson (2019). "Characterization of the material and mechanical properties of atomic force microscope cantilevers with a plan-view trapezoidal geometry." Applied Sciences **9**: 2604.

Smith, D. P. E. (1995). "Limits of force microscopy." Review of Scientific Instruments **66**: 3191–3195.

Smolyakov, G., B. Thiebot, C. Campillo, S. Labdi, C. Severac, J. Pelta and E. Dague (2016). "Elasticity, adhesion, and tether extrusion on breast cancer cells provide a signature of their invasive potential." ACS Applied Materials & Interfaces **8**: 27426–27431.

Stewart, M. P., A. W. Hodel, A. Spielhofer, C. J. Cattin, D. J. Mueller and J. Helenius (2013). "Wedged AFM-cantilevers for parallel plate cell mechanics." Methods **60**: 186–194.

Tang, Y., J. Fang, X. Yan and H. F. Ji (2004). "Fabrication and characterization of SiO2 microcantilever for microsensor application." Sensors & Actuators, B: Chemical **97**: 109–113.

Temiryazev, A., S. I. Bozhko, A. E. Robinson and M. Temiryazeva (2016). "Fabrication of sharp atomic force microscope probes using in situ local electric field induced deposition under ambient conditions." Review of Scientific Instruments **87**: 113703.

Thundat, T., R. J. Warmack, G. Y. Chen and D. P. Allison (1994). "Thermal and ambient-induced deflections of scanning force microscope cantilevers." Applied Physics Letters **64**: 2894.

Trache, A. and G. A. Meininger (2008). "Atomic force microscopy (AFM)." Current Protocols in Microbiology **8**: 1–17.

Viani, M. B., T. E. Schäffer, A. Chand, M. Rief, H. E. Gaub and P. K. Hansma (1999). "Small cantilevers for force spectroscopy of single molecules." Journal of Applied Physics **86**: 2258.

Viji Babu, P. K., U. Mirastschijski, G. Belge and M. Radmacher (2021). "Homophilic and heterophilic cadherin bond rupture forces in homo- or hetero-cellular systems measured by AFM-based single-cell force spectroscopy." European Biophysics Journal : EBJ **50**: 543–559.

Vinogradova, O. I., G. E. Yakubov and H. J. Butt (2001). "Forces between polystyrene surfaces in water-electrolyte solutions: Long-range attraction of two types?"." The Journal of Chemical Physics **114**: 8124–8131.

Voigtländer, B. (2015). Scanning probe microscopy. Heidelberg, Springer Berlin Heidelberg.

Waters, C. M., E. Roan and D. Navajas (2012). "Mechanobiology in lung epithelial cells: Measurements, perturbations, and responses." Comprehensive Physiology **2**: 1–29.

Wojcikiewicz, E. P., M. H. Abdulreda, X. Zhang and V. T. Moy (2006). "Force spectroscopy of LFA-1 and its ligands, ICAM-1 and ICAM-2." Biomacromolecules **7**: 3188–3195.

Wood, D., I. Hancox, T. S. Jones and N. R. Wilson (2015). "Quantitative nanoscale mapping with temperature dependence of the mechanical and electrical properties of poly(3-hexylthiophene) by conductive atomic force microscopy." The Journal of Physical Chemistry C **119**: 11459–11467.

Woolley, A. T., C. L. Cheung, J. H. Hafner and C. M. Lieber (2000). "Structural biology with carbon nanotube AFM probes." Chemistry & Biology **7**: R193–R204.

Yang, B., Z. Liu, H. Liu and M. A. Nash (2020). "Next generation methods for single-molecule force spectroscopy on polyproteins and receptor-ligand complexes." Frontiers in Molecular Biosciences **7**: 85.

Yapici, M. K. and J. Zou (2009). "Microfabrication of colloidal scanning probes with controllable tip radii of curvature." Journal of Micromechanics and Microengineering **19**: 105021.

Yashchenok, A., A. Masic, D. Gorin, B. Sup Shim, N. A. Kotov, P. Fratzl, H. Moehwald and A. Skirtach (2013). "Nanoengineered colloidal probes for Raman-based detection of biomolecules inside living cells." Small **9**: 351–356.

Yu, Q., G. Qin, C. Darne, C. Caib, W. Wosika and -S.-S. Pei (2006). "Fabrication of short and thin silicon cantilevers for AFM with SOI wafers." Sensors and Actuators A: Physical **126**: 369–374.

Yuan, C. C., D. Zhang and Y. Gan (2017). "Invited review article: Tip modification methods for tip-enhanced Raman spectroscopy (TERS) and colloidal probe technique: A 10-year update (2006–2016) review." Review of Scientific Instruments **88**: 031101.

Zimmermann, J. L., T. Nicolaus, G. Neuert and K. Blank (2010). "Thiol-based, site-specific and covalent immobilization of biomolecules for single-molecule experiments." Nature Protocols **5**: 975–985.

Zlatanova, J., S. M. Lindsay and S. H. Leuba (2000). "Single molecule force spectroscopy in biology using the atomic force microscope." Progress in Biophysics and Molecular Biology **74**: 37–61.

Matteo Chighizola, Jorge Rodriguez-Ramos, Felix Rico,
Manfred Radmacher, Alessandro Podestà

3.1.3 AFM Calibration Issues

Atomic force microscopy (AFM) is a powerful tool to investigate molecular interactions and mechanical properties at biointerfaces, with nanometric spatial resolution and pN force sensitivity (Binnig et al., 1986, Martin et al., 1987). However, to fully exploit the quantitative spectroscopic and imaging capabilities of an atomic force microscope, the instrument must be accurately calibrated. The calibration regards several components of an AFM.

First of all, the scaling factors of the force and deflection detection apparatus, that is, the cantilever spring constant (intrinsic and effective) and the deflection sensitivity of the optical beam deflection system must be determined.

The displacement of the sample with respect to the AFM tip must be controlled with sub-nanometer accuracy, which requires the proper calibration of the sensitivities of the piezoelectric scanners of the instrument along the three directions x,y,z.

The tip radius of curvature and the tip geometry are important parameters in the modelling of tip–sample interactions, as in the case of indentation experiments; these parameters must be accurately characterized.

This chapter presents a critical overview of the most important calibration procedures for AFM and discusses the related issues and how to mitigate them.

Acknowledgment: M.C. acknowledges the support of the European Union's Horizon 2020 research and innovation program under the FET Open grant agreement no. 801126, project EDIT.

Matteo Chighizola, Alessandro Podestà, Dipartimento di Fisica "Aldo Pontremoli" and CIMaINa, Università degli Studi di Milano, Milano, Italy
Jorge Rodriguez-Ramos, Felix Rico, Aix-Marseille University, CNRS, INSERM, LAI, Turing Centre for Living Systems
Manfred Radmacher, Institute of Biophysics, University of Bremen, Bremen, Germany

https://doi.org/10.1515/9783110640632-007

Table of Symbols

Symbol	Units	Comments
A	N/(m Hz$^{1/3}$)	AFM probe-specific A-factor for the CGI Sader method
A_0	m^2/Hz	White noise in PSD
b	m	Cantilever width
B	m^2/Hz	Amplitude of the PSD of the cantilever deflection (measured in m) for the SHO model
B_v	V^2/Hz	Amplitude of the PSD of the raw cantilever deflection signal (measured in V) for the SHO model
C_θ	–	Correction factor, which converts the intrinsic spring constant into the effective spring constant
d	m	Deflection of cantilever, measured in the direction perpendicular to the sample surface
$d_\perp$	m	Deflection of cantilever, measured in the direction perpendicular to the cantilever axis
$d_c(x_c)$, $d_{c,\mathrm{dyn}}(x_c)$	m	Static and dynamic deflection profiles of the cantilever, in the cantilever reference frame
F	N	Force applied on the AFM tip, in the direction perpendicular to the sample surface
f	Hz	Frequency (in PSD)
f_R	Hz	Resonance frequency of the cantilever (first mode)
H	m	Height of the (sharp) AFM tip
invOLS	V/m	Inverse optical lever sensitivity, another name for the deflection sensitivity S
k	N/m	Intrinsic spring constant of the cantilever at full length
k_{LP}	N/m	Intrinsic spring constant of the cantilever at the loading point (tip location)
k_{eff}	N/m	Effective spring constant of the cantilever in the measurement configuration (cantilever tilt, tip backshift, tip height, etc.)
k_B	J/K	Boltzmann constant
ΔL	m	Backshift of the loading point (tip location) with respect to the end of the cantilever
L	m	Cantilever length
PSD	m^2/Hz	Power spectral density, in AFM used for the spectral density of the cantilever oscillation

(continued)

Symbol	Units	Comments
PSD_V	V^2/Hz	Power spectral density of the raw deflection signal of the cantilever (photodetector output)
PSD^{SHO}	m^2/Hz	Power spectral density of the SHO model
PSD_V^{SHO}	V^2/Hz	Power spectral density for the raw deflection signal of the cantilever (photodetector output) of the SHO model
P_V	V^2	Power in the raw PSD_V obtained by integration of the first resonance peak
Q	–	Quality factor of (first) resonance peak
R	m	Radius of the AFM tip
Re	–	Reynolds number
S	m/V	Deflection sensitivity, the sensitivity of the optical beam deflection system (also known as invOLS)
T	K	Absolute temperature
ΔV	V	Raw deflection signal from the split photodiode
x_{laser}	m	Position of the laser along the cantilever axis
Z_p	m	z-Piezo displacement
β	–	Fraction of squared oscillation amplitude of the cantilever in the first dynamic mode to be used in the Sader and thermal noise methods
λ		Correction factor for the deflection sensitivity (SNAP procedure)
χ_0	–	Correction factor between dynamic and static deflection sensitivity for the rectangular cantilever, with the laser aligned at the end of the cantilever, for a tilt angle of $0°$
χ_{eff}	–	Correction factor between dynamic and static deflection sensitivity for a given loading configuration
$\Lambda(Re)$	–	Hydrodynamic function of the cantilever
ω_R	Hz	Angular resonance frequency
μ	Pa s	Dynamic viscosity of surrounding medium
ρ	kg/m^3	Density of surrounding medium
θ	rad	Mounting or tilt angle of cantilever

Acronyms

OBD	Optical beam deflection
OLD	Optical lever detection
SHO	Simple harmonic oscillator

3.1.3.1 Measuring Force and Deflection with the AFM with the Optical Beam Deflection Method

The force sensing capability of AFM is based on the use of probes consisting of an elastic microlever (cantilever) with an interacting tip at its apex (Figure 3.1.3.1).

In AFM force measurements, the tip–sample force (F) perpendicular to the sample surface must be measured as a function of either the tip–sample separation or the sample deformation.

The force F perpendicular to the sample surface causes a vertical deflection d of the cantilever as small as 1/10 of a nanometer that can be detected through an optical beam deflection (OBD) apparatus, also known in the literature as the optical lever detection apparatus (OLD) (Figure 3.1.3.1a) (Meyer and Amer, 1988, 1990a, 1990b, Putman et al., 1992, Alexander et al. 1989).

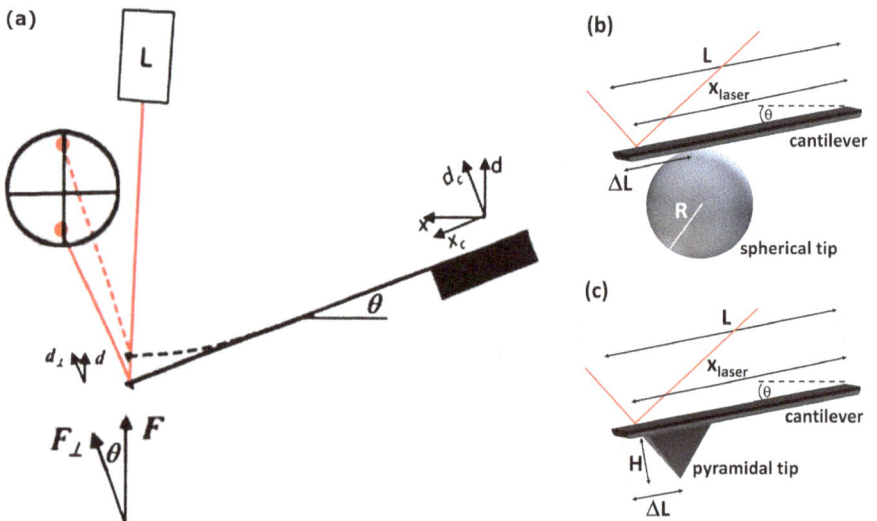

Figure 3.1.3.1: (a) The AFM cantilever and the OBD apparatus. (b, c) Spherical and sharp (pyramidal) tips, with the relevant lengths highlighted.

Force F and deflection d are proportional, according to Hooke's law:

$$F = k_{eff}d \tag{3.1.3.1}$$

In eq. (3.1.3.1), the spring constant k_{eff} (with units of N/m) is termed effective rather than intrinsic because, due to the cantilever mounting angle or tilt θ (usually $\theta = 10°–15°$), force and deflection are perpendicular to the sample surface rather than to the cantilever axis.

The intrinsic spring constant k of the cantilever, in turn, relates the deflection $d_\perp$ and the force $F_\perp$, respectively, perpendicular to the cantilever axis:

$$F_\perp = kd_\perp \tag{3.1.3.2}$$

The effective spring constant k_{eff} depends on both the elastic properties of the cantilever and the instrument configuration (i.e., the tilt of the cantilever, but also the tip height and the loading point position). In the simplest case of a negligibly small tip at the very end of the cantilever, given that $d_\perp = d/\cos(\theta)$ and $F_\perp = F\cos(\theta)$, one obtains:

$$k_{eff} = k/\cos^2(\theta) \tag{3.1.3.3}$$

This formula can be far more complex for other configurations (see eq. (3.1.3.26)), and in general, a correction factor C_θ must be determined, so that:

$$k_{eff} = C_\theta k \tag{3.1.3.4}$$

The cantilever deflection is measured in commercial AFMs using the OBD method, schematically represented in Figure 3.1.3.1. A laser beam is aligned on the cantilever in correspondence with the tip and reflected onto a segmented photodiode for detection. A small cantilever deflection causes a much larger vertical displacement of the laser spot on the detector (the geometrical amplification factor, proportional to the cantilever–detector distance, can be as large as 1,000 (Butt et al., 1995)). Upon suitable conversion and amplification, a voltage output ΔV proportional to the laser spot displacement, and therefore to the vertical cantilever deflection d is generated.

The cantilever deflection can therefore be measured using the OBD system as

$$d = S\Delta V \tag{3.1.3.5}$$

where the calibration factor S (with units of nm/V) is called deflection sensitivity (also known as inverse optical lever sensitivity, or invOLS).

The force F can therefore be calculated as

$$F = k_{eff}S\Delta V \tag{3.1.3.6}$$

The deflection sensitivity is also important for calculating the actual tip–sample distance (or indentation, in the contact region of the force–distance curve) since the latter distance is obtained by adding the cantilever deflection d to the z-piezo displacement Z_p (Butt et al., 2005).

To carry out quantitative force measurements by AFM, using eqs. (3.1.3.5) and (3.1.3.6), it is essential to calibrate the value of the spring constants (intrinsic or effective) of the cantilever as well as of the deflection sensitivity of the OBD system.

3.1.3.2 Calibration of the Cantilever Spring Constant

Several approaches have been developed to characterize the spring constant of a cantilever (Heim et al., 2014, Hutter and Bechhoefer, 1993, Cleveland et al., 1993, Kim et al., 2007; Sader et al., 2016, Craig and Neto, 2001; Butt and Jaschke, 1995, Gates and Pratt, 2012, Langlois et al., 2007; Chung et al., 2008, Clifford and Seah, 2009; Sader et al., 2012, Sikora, 2016). Here, we describe the Sader method and the thermal noise method, which are widely used due to their accuracy, simplicity and minimal, or null, invasiveness for the tip. As we will show, despite the simplicity of the thermal noise method, several potential issues can affect its accuracy and need to be considered. Nevertheless, when properly implemented, the Sader and thermal noise method agree well with each other (Burnham et al., 2003, Ohler, 2007) and with the results of finite-element analyses (FEA) or interferometric measurements (Ohler, 2007, Gates and Pratt, 2012).

Despite the considerable effort dedicated on the development and optimization of reliable calibration procedures for AFM, this is still an open field of research, considering that probes with unconventional shapes and mass distributions are continuously designed to support novel imaging and force spectroscopy modes (see Chapter 3.1.2).

3.1.3.2.1 The Sader Method

The Sader method is based on the modelling of the dynamical behavior of an AFM probe oscillating upon thermal excitation in a viscous fluid (Sader et al., 2016, 1999; 2005, 2012), while the method is usually applied to cantilevers oscillating in air, it was shown to work also reliably in fluids (Sumbul et al., 2020, Burnham et al., 2003).

The method requires, in its original formulation, the planar dimensions of the cantilever, the resonance frequency $f_R = \omega_R/2\pi$ (ω_R is the resonant angular frequency) and the quality factor Q of the first flexural mode. The intrinsic spring constant at the end of the cantilever ($x_C = L$) is then determined as

$$k = \beta \rho b^2 L \Lambda(\mathrm{Re}) \omega_R^2 Q \qquad (3.1.3.7)$$

being ρ the density of the surrounding fluid, b and L the width and length of the cantilever, respectively, and $\Lambda(\mathrm{Re})$ the hydrodynamic function (John E. Sader et al., 2005, 2012), which depends on the Reynolds number $\mathrm{Re} = \rho\, b^2 \omega_R/(4\mu)$, where μ is the dynamic viscosity of the fluid. One of the advantages of the Sader method is that it does not require the determination of the cantilever thickness since only the planar dimensions of the cantilever are required, which can be easily measured using optical microscopy for most commercial cantilevers.

The factor β in eq. (3.1.3.7) is the fraction of the total quadratic oscillation amplitude in the first normal mode (Butt and Jaschke, 1995, Sader et al., 2014) or, equivalently, the ratio of static and dynamic (first mode) spring constants of the cantilever, both constants measured at $x_C = L$. For a rectangular cantilever with a negligibly small tip at the end, $\beta = 0.971$, while for more complex geometries, it must be calculated by FEA (Rodriguez-Ramos and Rico, 2021; Sader et al., 2012, 2014) or measured by interferometry (Gates et al., 2013).

The values of Q and f_R can be easily obtained from the power spectral density (PSD) of the cantilever oscillation, defined as the squared modulus of the Fourier transform of the deflection $d(t)$ (or $d_\perp(t)$), with units of m^2/Hz or V^2/Hz, depending on whether it is measured by interferometry or the OBD method (see Pottier and Bellon (2017), Labuda (2016), and Sader et al. (2011) for details about both acquisition, calculation, and distortions of the PSD). Indeed, the oscillation versus time signal contains several disturbances that cannot be easily decoupled from the true signal since they are distributed across a wide frequency range. In turn, the PSD consists of a numerable series of resonance peaks corresponding, in the absence of cross talk between vertical and lateral segments (Piner and Ruoff, 2002, Hoffmann et al., 2007, Onal et al., 2008), to the normal flexural modes of the cantilever (Figure 3.1.3.2), and the contribution of the true oscillation of the cantilever can be easily distinguished from noise.

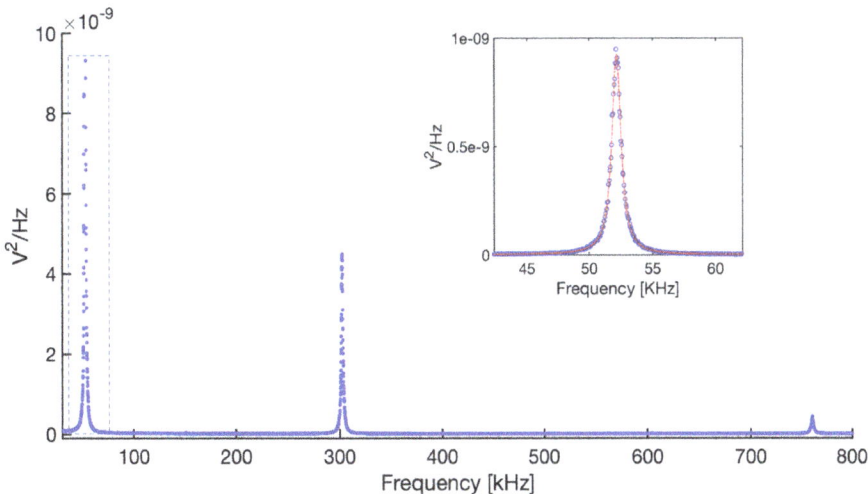

Figure 3.1.3.2: Power spectral density of a contact mode cantilever showing the resonance peaks of the first three flexural modes. The inset shows the first resonance peak with the superimposed fit of the SHO model (eq. (3.1.3.8)).

For example, if the simple harmonic oscillator (SHO) model is used, the PSD is given by eq. (3.1.3.8), where A_0 is a white noise offset, Q and f_R are the quality factor and the resonance frequency of the first flexural mode of the cantilever, respectively, and B is the oscillation amplitude at resonance; both A_0 and B have units of m/√Hz (or V/√Hz if the OBD system is used) (Pirzer and Hugel, 2009):

$$\text{PSD}^{\text{SHO}}(f) = A_0{}^2 + \frac{B^2 f_R{}^4}{Q^2}\left[\left(f^2 - f_R{}^2\right)^2 + f^2 f_R{}^2/Q^2\right]^{-1} \tag{3.1.3.8}$$

Other models can be used; for example, some authors found that the Lorentzian PSD better reproduces the experimental PSD of small-Q cantilevers in liquids (Pirzer and Hugel, 2009), However, recent comparison of the different fitting models suggests that the SHO is more consistent also in liquid and for low Q-factor cantilevers (Sumbul et al., 2020). Fitting eq. (3.1.3.8) or another suitable resonance model to the first peak of the measured thermal spectrum, it is possible to obtain the values of Q and f_R to be used in eq. (3.1.3.7).

Sader later showed that it is possible to determine the spring constant of an uncalibrated cantilever using the parameters of a calibrated reference cantilever, with the same (nominal) geometry and dimensions. This is a consequence of the fact that, for any given cantilever type with the same plan view dimensions (Sader and Friend, 2015), the intrinsic spring constant can be approximated as

$$k = AQf_R{}^{1.3} \tag{3.1.3.9}$$

where A is a constant factor recapitulating the quantities appearing in eq. (3.1.3.7). Thus, a global calibration initiative was proposed by Sader et al. (2016) to standardize the calibration of AFM cantilevers using an online, user-contributed database of A factors calculated from reference cantilevers of specific types and then averaged. A web-based applet is available at the following URL: https://sadermethod.org/. The accuracy of this universal A-coefficient for a particular cantilever, that is, of the calibration method, improves as the number of reference cantilevers uploaded by the community increases. Importantly, reporting the A-factor in publications would allow for a potential future recalibration of data.

3.1.3.2.2 The Thermal Noise Method

The thermal noise method (Butt and Jaschke, 1995, Gates and Pratt, 2012, Heim et al., 2014, Hutter and Bechhoefer, 1993) is based on Boltzmann's equipartition theorem:

$$\frac{1}{2} k < d_\perp^2 > = \frac{1}{2} k_B T \tag{3.1.3.10}$$

which relates the mean squared oscillation amplitude $<d_\perp^2>$ of a simple spring with the thermal energy in thermodynamic equilibrium. For a cantilever, which has several bending modes, eq. (3.1.3.10) will hold for each individual mode as can be seen in the PSD above.

The intrinsic cantilever spring constant k is then calculated from eq. (3.1.3.10) as

$$k = \frac{k_B T}{<d_\perp^2>} \qquad (3.1.3.11)$$

The elegance and the simplicity of eqs. (3.1.3.10) and (3.1.3.11) hide several details, which become evident when one tries to implement the thermal noise method. Here, we discuss two main approaches for implementing eq. (3.1.3.11) through the measurement of the cantilever deflection: the first is based on interferometry, while the second (the standard approach) is based on the OBD method, typically used in commercial AFMs.

Since it is convenient to focus only on the first flexural normal mode of the cantilever, which encompasses 98.5% of the total oscillation amplitude, the correction factor β introduced in eq. (3.1.3.7) must be used to account for the fraction of the total oscillation stored in this mode: $<d_{\perp,1}^2> = \beta <d_\perp^2>$, where $<d_{\perp,1}^2>$ is the measured oscillation amplitude of the first mode.

Equation (3.1.3.11) is then replaced by eq. (3.1.3.12):

$$k = \beta \frac{k_B T}{<d_{\perp,1}^2>} \qquad (3.1.3.12)$$

It was noted already that the cantilever oscillation is distributed among its normal modes, which are represented in the PSD by resonance peaks (Figure 3.1.3.2).

The interferometric and OBD-based thermal noise methods differ in how the oscillation amplitude $<z_{\perp,1}^2>$ is measured. This is described in the following sections.

3.1.3.2.2.1 Interferometric Thermal Noise Method

Interferometric techniques (Gates et al., 2015, Ohler, 2007, Paolino et al., 2013, Gates and Pratt, 2012) permit to directly measure with great accuracy the cantilever deflection in metric units. Compared to the OBD apparatus, interferometry allows to perform measurements with significantly lower noise, at sufficiently large bandwidth (above 1 MHz) and with a high dynamic range of deflection (from below 0.1 to above 500 nm); moreover, interferometric techniques offer the advantage of not requiring the calibration of the deflection sensitivity, the major drawback of the OBD setup (Laurent et al., 2013). At present, interferometric techniques are considered the gold standard for the calibration (through the thermal noise method) of AFM cantilevers. Precision and accuracy, tested against SI traceable technique (like the NIST electrostatic force balance), are well below 2%, typically around 1% (Gates and Pratt, 2012).

In interferometric measurements, the cantilever is not tilted; therefore, the displacement $d_\perp$ in the direction perpendicular to the cantilever is directly measured.

The fraction of the total cantilever oscillation of a specific normal mode corresponds to the area below the corresponding peak in the PSD (Figure 3.1.3.2). Fitting the PSD model of the SHO (eq. (3.1.3.8)) to the first peak of the measured thermal spectrum provides the values of the parameters B, f_R, and Q, and these, in turn, allow to calculate the oscillation amplitude $<d^2_{\perp,1}>$ of the first mode of the cantilever as (Pirzer and Hugel, 2009):

$$< d^2_{\perp,1} > = \int_0^\infty PSD^{SHO}(f)df = \frac{\pi B^2 f_R}{2Q} \tag{3.1.3.13}$$

According to eq. (3.1.3.12), the intrinsic spring constant can thus be calculated as

$$k = \beta \frac{2Q}{\pi B^2 f_R} k_B T \tag{3.1.3.14}$$

Besides requiring the knowledge of the value of the correction factor β, which can be calculated analytically only in the case of the ideal rectangular cantilever, it should be considered that interferometry provides the value of the intrinsic spring constant at the laser location. Therefore, on the one hand, the precise positioning of the laser along the cantilever axis is required (at odd with the case of the thermal noise method based on OBD), and on the other hand, it is important to precisely determine the distance between the laser position and the tip location (the loading point); this latter distance, together with other geometrical parameters, like the tip height, the cantilever tilt angle, and the cantilever length, is necessary to evaluate the correction factor C_θ (eq. (3.1.3.4) and Section 3.1.3.2.5) for the determination of the effective spring constant.

Besides the greater accuracy and precision provided by interferometry over the OBD system and the strong advantage of not requiring the characterization of the invOLS, also this calibration technique requires the calculation (typically by FEA) of both β and C_θ correction factors.

Last but not the least, at present, interferometric devices for the calibration of AFM cantilevers are either very expensive (when commercially available) or a complex, research-oriented, custom-built apparatus. Thus, there is a clear need to develop cost-effective, simplified interferometric devices for the calibration of AFM cantilevers to be installed and used routinely in research laboratories.

3.1.3.2.2.2 OBD-Based Thermal Noise Method

Commercial AFMs use the OBD method to measure the cantilever deflection, with the laser typically aligned close to the end of the cantilever; this implies that in addition to the modal correction factor β, other factors must be known.

First, the raw output of the photodetector in volts must be converted into a deflection in the physical units of meters via the deflection sensitivity S (eq. (3.1.3.5)), which implies that, with respect to the Sader and the interferometric approaches, an additional calibration step is required. Moreover, the deflection sensitivity S accounts for the static deflection of a cantilever in the experimental loading configuration (i.e., with the cantilever tilted, the tip of finite height located at a finite distance ΔL from the cantilever end, see Figure 3.1.3.1b,c), while the thermal noise is a dynamical oscillation measured at the laser position, in the direction perpendicular to the cantilever axis. As a consequence, an additional correction factor χ_{eff} is required (Sader et al., 2014, Cook et al., 2006; Butt and Jaschke, 1995). The squared oscillation amplitude of the first flexural mode measured by the OBD system is, therefore:

$$< d_{\perp,1}^2 > = \chi_{\text{eff}}^2 S^2 P_V \tag{3.1.3.15}$$

where P_V is the area below the first resonant peak of the PSD of the raw output of the photodetector (PSD$_V(f)$, in units of V^2/Hz).

Equation (3.1.3.12) is thus replaced by eq. (3.1.3.16) (Cook et al., 2006, Edwards et al., 2008, Hutter, 2005):

$$k = \frac{\beta}{\chi_{\text{eff}}^2} \frac{k_B T}{S^2 P_V} \tag{3.1.3.16}$$

If the SHO model is used, PSD$_V^{\text{SHO}}(f)$ is given by eq. (3.1.3.8), but with an amplitude B_V with units V/√Hz. Adapting eq. (3.1.3.13), one obtains: $P_V = \left(\pi B_V^2 f_R\right)/(2Q)$. Equation (3.1.3.16) thus becomes

$$k = \frac{\beta}{\chi_{\text{eff}}^2} \frac{2Q}{S^2 \pi B_V^2 f_R} k_B T \tag{3.1.3.17}$$

As pointed out by Butt and Jaschke (1995) in their seminal work, the OBD method measures directly the cantilever angular deflection, or inclination, rather than its vertical displacement; therefore, the correction factor χ_{eff}, originally defined as the ratio of the dynamic to the static deflection sensitivity (Proksch et al., 2004), is also equivalent to the ratio of the slopes of the statically and thermally dynamically loaded cantilevers, calculated in the frame of reference (x_c, d_c) of the cantilever (Figure 3.1.3.1a) (Schäffer and Fuchs, 2005; Butt and Jaschke, 1995, Cook et al., 2006):

$$\chi_{\text{eff}} = \frac{\frac{d}{dx_c}(d_c(x_c))|_{x_c = L - \Delta L}}{\frac{d}{dx_c}(d_{c,\text{dyn}}(x_c))|_{x_c = x_{\text{laser}}}} \tag{3.1.3.18}$$

In eq. (3.1.3.18), $d_c(x_c)$ and $d_{c,\text{dyn}}(x_c)$ represent the static and dynamic (first normal mode) profiles, respectively, of the cantilever, and x_{laser} is the laser position. For the first normal mode of an ideal, not tilted, rectangular cantilever, with a negligibly small tip and laser spot located at the end of the cantilever, $\chi_{eff} = \chi_0/\cos(\theta)$, with $\chi_0 = 1.09$ (the case of the finite spot size of the laser is treated in Schäffer and Fuchs (2005) and Proksch et al. (2004)).

In the general case of non-standard loading configurations and for exotically shaped cantilevers, with uneven mass distribution, including those with bulky and tall integrated tips (see Chapter 3.1.2 for an overview), not only the modal factor β, but also the factor χ_{eff} must be calculated by FEA (Rodriguez-Ramos and Rico, 2021), or by directly measuring the modal cantilever shapes by interferometry(Gates et al., 2013, Laurent et al., 2013).

For AFM probes based on the rectangular geometry (Figure 3.1.3.1b,c), assuming that the laser is positioned at the cantilever end, the following equation can be used, where the tip radius R must be replaced by the tip height H in the case of nonspherical probes (Edwards et al., 2008, Hutter, 2005, Chighizola et al., 2021):

$$\chi_{eff} = \chi_0 \left\{ \left(1 - \frac{\Delta L}{L}\right) \frac{\left[1 - \frac{3}{2}\frac{R/L}{(1-\frac{\Delta L}{L})}\tan\theta\right]}{\left[1 - 2\frac{R/L}{(1-\frac{\Delta L}{L})}\tan\theta\right]} \cos\theta \right\}^{-1} \tag{3.1.3.19}$$

and eq. (3.1.3.16) becomes

$$k = \frac{\beta}{\chi_0^2} \left\{ \left(1 - \frac{\Delta L}{L}\right)^2 \frac{\left[1 - \frac{3}{2}\frac{R/L}{(1-\frac{\Delta L}{L})}\tan\theta\right]^2}{\left[1 - 2\frac{R/L}{(1-\frac{\Delta L}{L})}\tan\theta\right]} \cos^2(\theta) \right\} \frac{k_B T}{S^2 P_V} \tag{3.1.3.20}$$

If the SHO model is used to fit the PSD, then eq. (3.1.3.20) becomes

$$k = \frac{\beta}{\chi_0^2} \left\{ \left(1 - \frac{\Delta L}{L}\right)^2 \frac{\left[1 - \frac{3}{2}\frac{R/L}{(1-\frac{\Delta L}{L})}\tan\theta\right]^2}{\left[1 - 2\frac{R/L}{(1-\frac{\Delta L}{L})}\tan\theta\right]} \cos^2(\theta) \right\} \frac{2Q}{S^2 \pi B_V^2 f_R} k_B T \tag{3.1.3.21}$$

The $\cos^2(\theta)$ term in eqs. (3.1.3.20) and (3.1.3.21) accounts for the tilt θ of the cantilever (Gates, 2017); the term in square brackets accounts for the torque applied to the cantilever because of the finite height of the tip; the first term $(1 - \Delta L/L)^2$ accounts for the fact that the static deflection sensitivity is measured at the loading point rather than at the end of the cantilever (Edwards et al., 2008, Hutter, 2005). A special case of eq. (3.1.3.20) was first proposed by Hutter (2005) under the hypothesis $\Delta L = 0$; in this case, eq. (3.1.3.20) simplifies to

$$k = \frac{\beta}{\chi_0^2} \left\{ \frac{\left[1 - \frac{3}{2}\frac{R/L}{(1-\frac{\Delta L}{L})}\tan\theta\right]^2}{\left[1 - 2\frac{R/L}{(1-\frac{\Delta L}{L})}\tan\theta\right]} \cos^2(\theta) \right\} \frac{k_B T}{S^2 P_V} \tag{3.1.3.22}$$

and an expression similar to eq. (3.1.3.21) but without the $(1 - \Delta L/L)^2$ term can be obtained for the SHO fit.

For a rectangular cantilever, with a negligibly small tip and laser spot located at the end of the cantilever, $\beta/\chi_0^2 = 0.817$, and the static intrinsic spring constant can be calculated as

$$k = 0.817 \cos^2(\theta) \frac{k_B T}{S^2 P_V} \tag{3.1.3.23}$$

Fitting the SHO model to the measured PSD_V, one obtains:

$$k = 0.817 \frac{2Q}{S^2 \pi B_V^2 f_R} \cos^2(\theta) k_B T \tag{3.1.3.24}$$

3.1.3.2.3 The Added Mass Effect

The implementation of the thermal noise method based on the standard rectangular geometry of the cantilever assumes that the mass of the cantilever is uniformly distributed. For this statement to be valid, the mass of the tip must be negligible compared to the mass of the cantilever. However, when large tips, for example, spheres, are attached at the end of a cantilever, this hypothesis is not always satisfied. Since the mass of the microsphere scales up with the cube of radius R, it can easily reach values comparable to or larger than the mass of the cantilever. For example, this is the case for a glass sphere with a radius $R > 10$ µm, attached to a silicon nitride cantilever with $L = 200$ µm.

Using interferometric approaches, it was recently demonstrated that the cantilever oscillation dynamics changes if a mass is attached to the free end (Gates et al., 2013, Laurent et al., 2013). Among other effects, the large attached mass of the tip and/or the non-uniform distribution of the mass across the probe length induces a change in the modal shapes of the flexural normal modes. As a consequence, both factors β and χ_{eff} change; if those two factors are not corrected, the resulting spring constant can be underestimated by a significant factor (Chighizola et al., 2021).

For the rectangular geometry, the added mass effect can be accounted for according to the model developed by Laurent et al. (2013) and Chighizola et al. (2021) (an online tool is available for the $\Delta L = 0$ case: https://perso.ens-lyon.fr/ludovic.bel lon/wp/tools/colloidal-probe-calibrator/, with the change of notation: $\hat{\alpha} \equiv \beta$ and $\beta \equiv \hat{\alpha}/\chi_0^2$).

If the factors β and χ_{eff}^2 are numerically calculated by FEA, then the added mass effect is automatically taken into account, together with all other effects related to the nonideality of the probe geometry (such as the backshift of the loading point, see Section 3.1.3.2.4, or the deviation from the rectangular geometry).

3.1.3.2.4 Effect of the Backshift of the Loading Point of the Probe

Usually, it is assumed that the loading point, that is, the point where the surface force is applied to the probe, is located at the free end ($x_C = L$) of the lever. In general, however, the loading point is horizontally displaced from the cantilever end by a distance ΔL. Consequently, the intrinsic spring constant of the rectangular cantilever increases cubically as the loading point is back shifted from the cantilever end by a distance ΔL (Sader and White, 1993; Sader et al., 1995):

$$k_{LP} = \left(\frac{L}{L - \Delta L}\right)^3 k \tag{3.1.3.25}$$

where k_{LP} is the intrinsic spring constant at the loading point.

In the case of rectangular, axisymmetric, and relatively long cantilevers, eq. (3.1.3.25) is valid for arbitrary offset ΔL, not only in the limit $\Delta L/L \ll 1$, while eq. (3.1.3.25) is reasonably accurate for triangular cantilevers, as long as $\Delta L/L$ and b/L are small, where b represents the width of the single arms of the cantilever (Sader, 1995).

For AFM probes of arbitrary shape, the correction factor to obtain the static spring constant at the loading point must be calculated numerically via FEA (Rodriguez-Ramos and Rico, 2021). The correction factor C_θ for the effective spring constant (eq. (3.1.3.4), see also the next section) typically considers the actual position of the loading point.

3.1.3.2.5 The Effective Spring Constant

In the presence of a cantilever tilt θ and of a nonnegligible R/L or H/L ratio, a torque, proportional to the component of the force F perpendicular to the tip axis, causes an additional deflection of the cantilever (an additional torque can be caused by friction forces, here neglected (Attard et al., 1999, Stiernstedt et al., 2005, 2006)). In the general case when the ratios R/L and $\Delta L/L$ are not negligible, the equation for the effective spring constant k_{eff} (eq. (3.1.3.3)) must be replaced by the following equation (Edwards et al., 2008, Hutter, 2005):

$$k_{eff} = \left\{\left[1 - \frac{3R/L}{2(1 - \Delta L/L)}\tan\theta\right]\cos^2(\theta)\right\}^{-1} \left(\frac{L}{L - \Delta L}\right)^3 k \tag{3.1.3.26}$$

which accounts for both the torque effect and the backshift of the loading point. The laser is supposed to be aligned between the loading point and the end of the cantilever (in this region, the inclination of the loaded cantilever is constant). In eq. (3.1.3.26), the radius R of the spherical probe must be replaced by the tip height H in case a standard sharp tip is used (Edwards et al., 2008).

For a rectangular cantilever, the correction factor C_θ for the effective spring constant (eq. 3.1.3.4) is

$$C_\theta = \left\{ \left[1 - \frac{3R/L}{2(1 - \Delta L/L)} \tan\theta \right] \cos^2(\theta) \right\}^{-1} \left(\frac{L}{L - \Delta L} \right)^3 \qquad (3.1.3.27)$$

For negligibly small R/L or H/L ratios (i.e., negligible torque), eq. (3.1.3.26) simplifies to

$$k_{\text{eff}} = \left(\frac{L}{L - \Delta L} \right)^3 k / \cos^2(\theta) \qquad (3.1.3.28)$$

In the case of also negligible $\Delta L/L$, eq. (3.1.3.28) further simplifies to eq. (3.1.3.3), with the simple correction factor $C_\theta = 1/\cos^2(\theta)$.

As soon as the geometry of the cantilever deviates from the ideal rectangular one, the factor C_θ must be calculated by FEA (Rodriguez-Ramos and Rico, 2021).

3.1.3.2.6 Calibration of the Deflection Sensitivity of the OBD System

The thermal method based on the OBD system requires the accurate calibration of the deflection sensitivity; since the latter parameter enters as a square in the formula for the spring constant, a small relative error in the deflection sensitivity leads to a twice as large relative error in the value of k.

3.1.3.2.6.1 Standard Calibration Method of the Deflection Sensitivity

The calibration of the deflection sensitivity S is typically performed by making contact with the AFM tip on a very rigid surface and collecting a series of raw photodiode signal versus z-piezo displacement curves (Figure 3.1.3.3). Assuming that neither the tip nor the surface are deformed, the deflection sensitivity S is calculated as the inverse of the mean slope of the contact region of the ΔV versus Z_p curves (which justifies the name invOLS), or equivalently as the mean of the inverse slopes, which leads to the same relative error.

Despite the simplicity of this approach, there are some drawbacks in the standard calibration method of the deflection sensitivity:

- A clean and rigid substrate is required, which is not always available, for example when a soft sample at full coverage (a confluent cell layer, a tissue slice) is studied.
- The tip–surface contact can reduce the tip sharpness or damage the tip functionalization, if any.

- When tips with large radii and/or stiff cantilevers are used, especially on adhesive surfaces, friction forces can produce a torque, which can influence the measured deflection and can result in an apparent deflection sensitivity (Attard et al., 1999, Stiernstedt et al., 2005, 2006, Warmack et al., 1994).
- The z-piezo must be properly calibrated (see Chapter 3.1.1); otherwise, the measured deflection sensitivity will be systematically rescaled.
- The vertical deflection signal must be free of artifacts, for instance, due to crosstalk between vertical and lateral segments of the photodetector (Piner and Ruoff, 2002, Onal et al., 2008, Hoffmann et al., 2007); these effects can be important when a large deflection interval is probed, as during the acquisition of a force curve.
- The deflection signal should be measured well within the linearity range of the photodetector, which, depending on the system, can be as small as one-third of the total range.

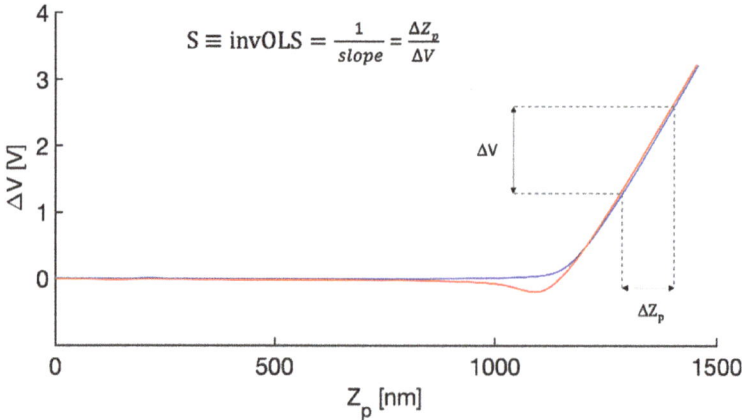

Figure 3.1.3.3: A typical force curve acquired on a rigid substrate. In blue and red, the approaching and retracting portions of the force curve, respectively.

The first point is critical because it can force to perform separate experiments for calibrating the deflection sensitivity and for measuring the sample; ideally, one should try to perform both tasks without changing the experimental setup (including the laser alignment on the cantilever, the thermalization of the liquid medium, among others). Therefore, in practice, it is wise, whenever possible, to leave a portion of the substrate uncovered and clean to calibrate the deflection sensitivity.

Because of friction, the tip can get pinned on the surface instead of sliding (Weafer et al., 2012). The friction, and therefore the torque, are stronger when the effective tip height is larger, as in the case of large spherical probes, and when the normal force is higher, as in the case stiffer cantilevers are used. These two conditions are often met since stiffer cantilevers must be used, in combination with large spheres, in order to achieve reasonably high indentations.

It turns out that, as long as the tip sharpness or functionalization is not an issue, the standard calibration method for the deflection sensitivity is more reliable when sharp tips are used, provided the maximum load is kept reasonably low (also to avoid detector non-linearity), and the first part of the force curve after tip–surface contact is discarded. When torque and friction-related effects are important and the use of large tips cannot be avoided, Chung et al.(Chung et al., 2009) suggested calculating the mean of the slopes of the loading and unloading portions of the force curve to subtract the deflection artifact effectively and then inverting the mean slope to obtain the deflection sensitivity.

Other points in the above list refer to the calibration and linearity of both the z-piezo and the deflection signals, which directly affect the slope of the force curve, and therefore the deflection sensitivity. Whenever possible, the output of a well-calibrated displacement z-sensor should be recorded and used as z-piezo displacement axis or used to operate the z-scanner in close-loop mode (see Chapter 3.1.1).

3.1.3.2.6.2 Contactless Calibration of the Deflection Sensitivity

To avoid many of the possible artifacts and problems noted above on the conventional contact method to determine the deflection sensitivity, a contactless approach was suggested (Higgins et al., 2006). The method is based on the inversion of eq. (3.1.3.17) and on the general idea that the intrinsic spring constant of the cantilever is not supposed to change; it requires independent knowledge of the intrinsic spring constant k of the cantilever, for example calibrated using either Sader or interferometric methods, and of the parameters obtained by fitting the SHO model to the thermal spectrum:

$$S = \sqrt{\frac{\beta}{\chi_{\text{eff}}^2} \frac{2Q}{k\pi B_V^2 f_{\text{R}}} k_{\text{B}} T} \qquad (3.1.3.29)$$

The method was first applied assuming that the spring constant calibrated in air did not change in liquid, a reasonable assumption that has been later confirmed (Higgins et al., 2006, Sumbul et al., 2020). Apart from the advantage of avoiding damage on the sample and possible artifacts due to contaminated tip or sample or to photodiode nonlinearities, the uncertainty in the spring constant propagates as half to determine the deflection sensitivity.

A similar approach, based on the same assumptions that led to eq. (3.1.3.29), can be used when the thermal noise procedure directly provides the spring constant rather than the parameters of the SHO fit. If an incorrect value S_{temp} of the deflection sensitivity is used, an incorrect estimation of the spring constant k_{temp} is obtained. The correction factor λ for the deflection sensitivity, such that:

$$S = \lambda\, S_{\text{temp}} \qquad\qquad (3.1.3.30)$$

can be calculated as

$$\lambda = \sqrt{\frac{k_{\text{temp}}}{k}} \qquad\qquad (3.1.3.31)$$

where k is the known true value of the intrinsic spring constant of the cantilever, that is, a value obtained by means of an accurate calibration method. This method is known as the SNAP procedure (Schillers et al., 2017). The acquisition of thermal spectra during the experiments and the calculation of the deflection sensitivity through eqs. (3.1.3.29) or (3.1.3.30, 3.1.3.31) represent good practices to monitor possible variations of the deflection sensitivity due, for example, to the displacement of the laser on the cantilever.

Due to its robustness, the contactless methods have shown to be more reliable than the standard method in a series of measurements across laboratories to determine the mechanics of soft biological samples, reducing the variability to 1%, much lower than using the conventional contact approach (Schillers et al., 2017). This method is now available on some commercial AFM systems. Thus, the advantages of this method are clear, and we recommend it for robust calibration of the deflection sensitivity, provided an accurate spring constant is known.

3.1.3.3 Calibration of the Tip Radius and Geometry

It is important to know both the tip geometry and its characteristic dimensions, like the opening angle, for conical or pyramidal tips, and the radius of curvature because the contact mechanics models depend on these parameters (see Chapter 2.2).

Unfortunately, the quantitative characterization of the tip geometry is not straightforward, requiring typically access to an electron microscopy facility. Optical microscopy can hardly provide accurate results except for large spherical probes with a radius $R > 10\,\mu m$.

In the case of the spherical geometry, however, or as long as one is interested mainly in the characterization of the radius of curvature of not extremely sharp probes, it is possible to the well-known phenomenon of tip–sample convolution (Montelius and Tegenfeldt, 1993, Keller, 1991, Villarrubia, 1997, Marques-Moros et al., 2020, Westra et al., 1993), which is usually considered as a resolution-limiting artifact, but which in this case turns into an advantage. In the case of extreme tip–sample convolution regime, when a probe with a large radius scans very high-aspect ratio surface features, the accurate reverse imaging of the probe apical part is obtained, which allows to directly and quantitatively estimate both the geometry and dimensions of the apical part of the tip; in the general case, however, the convoluted image will contain the information of both the surface feature and the tip radii.

Polymeric particles can be used to produce suitable convoluted images upon scanning of relatively sharp tips (Van Cleef et al., 1996, Nagy et al., 1996, Colombi et al., 2009). Sharp rectangular steps can also be used (Yan et al., 2016, Markiewicz and Cynthia Goh, 1995, Hübner et al., 2003). These approaches typically, but not always, provide 1D topographic profiles that can be analyzed to extract the value of the curvature radius of the tip. When the convolution regime is stronger, the reverse AFM image represents a truly 3D reconstruction of the apical part of the tip (Neto and Craig, 2001), which can be fitted using 3D models, like the spherical cap model (Indrieri et al., 2011), or can feed blind tip reconstruction algorithms (Flater et al., 2014, Villarrubia, 1997), for the estimation of the radius. If several high aspect ratio surface features are present, each will provide an independent replica of the tip apical portion to the benefit of statistics. The typical accuracy of the determination of the radius of spherical tips by reverse imaging is 1% (Indrieri et al., 2011).

Reverse imaging for the calibration of the spherical probe radius has some limitations. Indeed, for radii comparable or larger than the typical separation of spikes in the calibration grating (typically a few micrometers), the spherical tip cannot penetrate deeply in between the spikes, and only a small portion of it will be imaged. While this is not an issue for force spectroscopy measurements, where the tip–surface interaction takes place mostly in the apical portion of the tip, it could be a limitation in indentation measurements; in this case, the tip radius should be characterized up to distances from the apex comparable to the maximum indentation, which can be as large as a fewmicrometers. At present, commercial spiked gratings cannot provide sufficiently large planar spacings to accomplish this task.

Scanning electron microscopy is the technique of choice to investigate the geometrical details of new probes, with uncommon shape, especially when the manufacturer does not provide exhaustive information.

References

Alexander, S., L. Hellemans, O. Marti, J. Schneir, V. Elings, P. K. Hansma, M. Longmire and J. Gurley (1989). "An atomic-resolution atomic-force microscope implemented using an optical lever." Journal of Applied Physics **65**(1): 164–167. https://doi.org/10.1063/1.342563.

Attard, P., A. Carambassis and M. W. Rutland (1999). "Dynamic surface force measurement. 2. Friction and the atomic force microscope." Langmuir **15**(2): 553–563. https://doi.org/10.1021/la980848p.

Binnig, G., C. F. Quate and C. Gerber (1986). "Atomic force microscope." Physical Review Letters **56**(9): 930–933. https://doi.org/10.1103/PhysRevLett.56.930.

Burnham, N. A., X. Chen, C. S. Hodges, G. A. Matei, E. J. Thoreson, C. J. Roberts, M. C. Davies and S. J. B. Tendler (2003). "Comparison of calibration methods for atomic-force microscopy cantilevers." Nanotechnology **14**. https://doi.org/10.1088/0957-4484/14/1/301.

Butt, H. J. and M. Jaschke (1995). "Calculation of thermal noise in atomic force microscopy." Nanotechnology **6**(1): 1–7. https://doi.org/10.1088/0957-4484/6/1/001.

Butt, H. J., B. Cappella and M. Kappl (2005). "Force measurements with the atomic force microscope: Technique, interpretation and applications." Surface Science Reports **59**(1–6): 1–152. https://doi.org/10.1016/j.surfrep.2005.08.003.

Butt, H. J., M. Jaschke and W. Ducker (1995). "Measuring surface forces in aqueous electrolyte solution with the atomic force microscope." Bioelectrochemistry and Bioenergetics **38**(1): 191–201. https://doi.org/10.1016/0302-4598(95)01800-T.

Chighizola, M., L. Puricelli, L. Bellon and P. Alessandro (2021). "Large colloidal probes for atomic force microscopy: Fabrication and calibration issues." Journal of Molecular Recognition **34**(1): e2879. https://doi.org/10.1002/jmr.2879.

Chung, K. H., S. Scholz, G. A. Shaw, J. A. Kramar and J. R. Pratt (2008). "SI traceable calibration of an instrumented indentation sensor spring constant using electrostatic force." Review of Scientific Instruments **79**(9). https://doi.org/10.1063/1.2987695.

Chung, K.-H.-H., G. A. Shaw and J. R. Pratt (2009). "Accurate noncontact calibration of colloidal probe sensitivities in atomic force microscopy." Review of Scientific Instruments **80**(6): 65107. https://doi.org/10.1063/1.3152335.

Cleef, M. V. A. N., S. A. Holt, G. S. Watson and S. Myhra (1996). "Polystyrene spheres on mica substrates: AFM Calibration, tip parameters and scan artefacts." Journal of Microscopy **181**(1): 2–9. https://doi.org/10.1046/j.1365-2818.1996.74351.x.

Cleveland, J. P., S. Manne, D. Bocek and P. K. Hansma (1993). "A nondestructive method for determining the spring constant of cantilevers for scanning force microscopy." Review of Scientific Instruments **64**(2): 403–405. https://doi.org/10.1063/1.1144209.

Clifford, C. A. and M. P. Seah (2009). "Improved methods and uncertainty analysis in the calibration of the spring constant of an atomic force microscope cantilever using static experimental methods." Measurement Science and Technology **20**(12). https://doi.org/10.1088/0957-0233/20/12/125501.

Colombi, P., I. Alessandri, P. Bergese, S. Federici and L. E. Depero (2009). "Self-assembled polystyrene nanospheres for the evaluation of atomic force microscopy tip curvature radius." Measurement Science and Technology **20**(8): 084015. https://doi.org/10.1088/0957-0233/20/8/084015.

Cook, S. M., T. E. Schäffer, K. M. Chynoweth, M. Wigton, R. W. Simmonds, K. M. Lang, K. M. Chynoweth, M. Wigton, R. W. Simmonds and T. E. Schäffer (2006). "Practical implementation of dynamic methods for measuring atomic force microscope cantilever spring constants." Nanotechnology **17**(9): 2135–2145. https://doi.org/10.1088/0957-4484/17/9/010.

Craig, V. S. J. and C. Neto (2001). "In situ calibration of colloid probe cantilevers in force microscopy: Hydrodynamic drag on a sphere approaching a wall." Langmuir **17**(19): 6018–6022. https://doi.org/10.1021/la010424m.

Edwards, S. A., W. A. Ducker and J. E. Sader (2008). "Influence of atomic force microscope cantilever tilt and induced torque on force measurements." Journal of Applied Physics **103**(6): 64513. https://doi.org/10.1063/1.2885734.

Flater, E. E., G. E. Zacharakis-Jutz, B. G. Dumba, I. A. White and C. A. Clifford (2014). "Towards easy and reliable AFM tip shape determination using blind tip reconstruction." Ultramicroscopy **146** (July): 130–143. https://doi.org/10.1016/j.ultramic.2013.06.022.

Gates, R. S. (2017). "Experimental confirmation of the atomic force microscope cantilever stiffness tilt correction." Review of Scientific Instruments **88**(12): 123710. https://doi.org/10.1063/1.4986201.

Gates, R. S., W. A. Osborn and J. R. Pratt (2013). "Experimental determination of mode correction factors for thermal method spring constant calibration of AFM cantilevers using laser doppler vibrometry." Nanotechnology **24**(25): 255706. https://doi.org/10.1088/0957-4484/24/25/255706.

Gates, R. S., W. A. Osborn and G. A. Shaw (2015). "Accurate flexural spring constant calibration of colloid probe cantilevers using scanning laser doppler vibrometry." Nanotechnology **26**(23): 235704. https://doi.org/10.1088/0957-4484/26/23/235704.

Gates, R. S. and J. R. Pratt (2012). "Accurate and precise calibration of AFM cantilever spring constants using laser doppler vibrometry." Nanotechnology **23**(37): 375702. https://doi.org/10.1088/0957-4484/23/37/375702.

Heim, L. O., T. S. Rodrigues and E. Bonaccurso (2014). "Direct thermal noise calibration of colloidal probe cantilevers." Colloids and Surfaces. A, Physicochemical and Engineering Aspects **443** (February): 377–383. https://doi.org/10.1016/j.colsurfa.2013.11.018.

Higgins, M. J., R. Proksch, J. E. Sader, M. Polcik, S. Mc Endoo, J. P. Cleveland and S. P. Jarvis (2006). "Noninvasive determination of optical lever sensitivity in atomic force microscopy." Review of Scientific Instruments **77**(1): 1–5. https://doi.org/10.1063/1.2162455.

Hoffmann, Á., T. Jungk and E. Soergel (2007). "Cross-talk correction in atomic force microscopy." Review of Scientific Instruments **78**(1): 016101. https://doi.org/10.1063/1.2424448.

Hübner, U., W. Morgenroth, H. G. Meyer, T. Sulzbach, B. Brendel and W. Mirandé (2003). "Downwards to metrology in nanoscale: determination of the AFM tip shape with well-known sharp-edged calibration structures." Applied Physics. A, Materials Science & Processing **76**: 913–917. Springer. https://doi.org/10.1007/s00339-002-1975-6.

Hutter, J. L. (2005). "Comment on tilt of atomic force microscope cantilevers: Effect on spring constant and adhesion measurements." Langmuir **21**. https://doi.org/10.1021/la047670t.

Hutter, J. L. and J. Bechhoefer (1993). "Calibration of atomic-force microscope tips." Review of Scientific Instruments **64**(7): 1868–1873. https://doi.org/10.1063/1.1143970.

Indrieri, M., A. Podestà, G. Bongiorno, D. Marchesi and P. Milani (2011). "Adhesive-free colloidal probes for nanoscale force measurements: Production and characterization." Review of Scientific Instruments **82**(2): 023708. https://doi.org/10.1063/1.3553499.

Keller, D. (1991). "Reconstruction of STM and AFM images distorted by finite-size tips." Surface Science **253**(1–3): 353–364. https://doi.org/10.1016/0039-6028(91)90606-S.

Kim, M. S., J. H. Choi, J. H. Kim and Y. K. Park (2007). "SI-traceable determination of spring constants of various atomic force microscope cantilevers with a small uncertainty of 1%." Measurement Science and Technology **18**(11): 3351–3358. https://doi.org/10.1088/0957-0233/18/11/014.

Labuda, A. (2016). "Daniell method for power spectral density estimation in atomic force microscopy." Review of Scientific Instruments **87**(3). https://doi.org/10.1063/1.4943292.

Langlois, E. D., G. A. Shaw, J. A. Kramar, J. R. Pratt and D. C. Hurley (2007). "Spring constant calibration of atomic force microscopy cantilevers with a piezosensor transfer standard." Review of Scientific Instruments **78**(9). https://doi.org/10.1063/1.2785413.

Laurent, J., A. Steinberger and L. Bellon (2013). "Functionalized AFM probes for force spectroscopy: Eigenmode shapes and stiffness calibration through thermal noise measurements." Nanotechnology **24**(22): 225504. https://doi.org/10.1088/0957-4484/24/22/225504.

Markiewicz, P. and M. Cynthia Goh (1995). "Atomic force microscope tip deconvolution using calibration arrays." Review of Scientific Instruments **66**(5): 3186–3190. https://doi.org/10.1063/1.1145549.

Marques-Moros, F., A. Forment-Aliaga, E. Pinilla-Cienfuegos and J. Canet-Ferrer (2020). "Mirror effect in atomic force microscopy profiles enables tip reconstruction." Scientific Reports **10**(1): 1–8. https://doi.org/10.1038/s41598-020-75785-0.

Martin, Y., C. C. Williams and H. K. Wickramasinghe (1987). "Atomic force microscope-force mapping and profiling on a sub 100-Å Scale." Journal of Applied Physics **61**(10): 4723–4729. https://doi.org/10.1063/1.338807.

Meyer, G. and N. M. Amer (1988). "Erratum: Novel optical approach to atomic force microscopy (Applied Physics Letters (1988) 53, (1045))." Applied Physics Letters **53**(24): 2400–2402. https://doi.org/10.1063/1.100425.

Meyer, G. and N. M. Amer (1990a). "Optical-beam-deflection atomic force microscopy: The NaCl (001) surface." Applied Physics Letters **56**(21): 2100–2101. https://doi.org/10.1063/1.102985.

Meyer, G. and N. M. Amer (1990b). "Simultaneous measurement of lateral and normal forces with an optical-beam-deflection atomic force microscope." Applied Physics Letters **57**(20): 2089–2091. https://doi.org/10.1063/1.103950.

Montelius, L. and J. O. Tegenfeldt (1993). "Direct observation of the tip shape in scanning probe microscopy." Applied Physics Letters **62**(21): 2628–2630. https://doi.org/10.1063/1.109267.

Nagy, P., G. I. Márk and B. Erzsébet (1996). "Determination of SPM TIP shape using polystyrene latex balls". In Microbeam and nanobeam analysis. Mikrochimica Acta Supplement. Vienna, Springer Vienna, 425–433. https://doi.org/10.1007/978-3-7091-6555-3_35.

Neto, C. and V. S. J. Craig (2001). "Colloid probe characterization: radius and roughness determination." Langmuir **17**(7): 2097–2099. https://doi.org/10.1021/la001506y.

Ohler, B. (2007). "Cantilever spring constant calibration using laser doppler vibrometry." Review of Scientific Instruments **78**(6): 63701. https://doi.org/10.1063/1.2743272.

Onal, C. D., B. Sümer and M. Sitti (2008). "Cross-talk compensation in atomic force microscopy." Review of Scientific Instruments **79**(10): 103706. https://doi.org/10.1063/1.3002483.

Paolino, P., F. A. Aguilar Sandoval and L. Bellon (2013). "Quadrature phase interferometer for high resolution force spectroscopy." Review of Scientific Instruments **84**(9): 95001. https://doi.org/10.1063/1.4819743.

Piner, R. and R. S. Ruoff (2002). "Cross talk between friction and height signals in atomic force microscopy." Review of Scientific Instruments **73**(9): 3392–3394. https://doi.org/10.1063/1.1499539.

Pirzer, T. and T. Hugel (2009). "Atomic force microscopy spring constant determination in viscous liquids." Review of Scientific Instruments **80**(3): 035110. https://doi.org/10.1063/1.3100258.

Pottier, B. and L. Bellon (2017). "'Noiseless' thermal noise measurement of atomic force microscopy cantilevers." Applied Physics Letters **110**(9): 94105. https://doi.org/10.1063/1.4977790.

Proksch, R., T. E. Schäffer, J. P. Cleveland, R. C. Callahan and M. B. Viani (2004). "Finite optical spot size and position corrections in thermal spring constant calibration." Nanotechnology **15**(9): 1344–1350. https://doi.org/10.1088/0957-4484/15/9/039.

Putman, C. A. J., B. G. de Grooth, N. F. van Hulst and J. Greve (1992). "A detailed analysis of the optical beam deflection technique for use in atomic force microscopy." Journal of Applied Physics **72**. https://doi.org/10.1063/1.352149.

Rodriguez-Ramos, J. and F. Rico (2021). "Determination of calibration parameters of cantilevers of arbitrary shape by finite element analysis." Review of Scientific Instruments **92**(4): 045001. https://doi.org/10.1063/5.0036263.

Sader, J. E., C. T. Riccardo Borgani, D. B. Gibson, M. J. Haviland, J. I. Higgins, J. L. Kilpatrick, et al. (2016). "A virtual instrument to standardise the calibration of atomic force microscope cantilevers." Review of Scientific Instruments **87**(9): 93711. https://doi.org/10.1063/1.4962866.

Sader, J. E., W. M. James, M. Chon and P. Mulvaney (1999). "Calibration of rectangular atomic force microscope cantilevers." Review of Scientific Instruments **70**(10): 3967–3969. https://doi.org/10.1063/1.1150021.

Sader, J. E. and J. R. Friend (2015). "Note: Improved calibration of atomic force microscope cantilevers using multiple reference cantilevers." Review of Scientific Instruments **86**(5): 56106. https://doi.org/10.1063/1.4921192.

Sader, J. E., I. Larson, P. Mulvaney and L. R. White (1995). "Method for the calibration of atomic force microscope cantilevers." Review of Scientific Instruments **66**(7): 3789–3798. https://doi.org/10.1063/1.1145439.

Sader, J. E., L. Jianing and P. Mulvaney (2014). "Effect of cantilever geometry on the optical lever sensitivities and thermal noise method of the atomic force microscope." Review of Scientific Instruments **85**(11): 113702. https://doi.org/10.1063/1.4900864.

Sader, J. E., J. Pacifico, C. P. Green and P. Mulvaney (2005). "General scaling law for stiffness measurement of small bodies with applications to the atomic force microscope." Journal of Applied Physics **97**: 12. https://doi.org/10.1063/1.1935133.

Sader, J. E., J. A. Sanelli, B. D. Adamson, J. P. Monty, S. A. Xingzhan Wei, J. R. Crawford, I. M. Friend, P. Mulvaney and E. J. Bieske (2012). "Spring constant calibration of atomic force microscope cantilevers of arbitrary shape." Review of Scientific Instruments **83**(10): 103705. https://doi.org/10.1063/1.4757398.

Sader, J. E., J. Sanelli, B. D. Hughes, J. P. Monty and E. J. Bieske (2011). "Distortion in the thermal noise spectrum and quality factor of nanomechanical devices due to finite frequency resolution with applications to the atomic force microscope." Review of Scientific Instruments **82**(9). https://doi.org/10.1063/1.3632122.

Sader, J. E. (1995). "Parallel beam approximation for V-shaped atomic force microscope cantilevers." Review of Scientific Instruments **66**(9): 4583–4587. https://doi.org/10.1063/1.1145292.

Sader, J. E. and L. White (1993). "Theoretical analysis of the static deflection of plates for atomic force microscope applications." Journal of Applied Physics **74**(1): 1–9. https://doi.org/10.1063/1.354137.

Schäffer, T. E. and H. Fuchs (2005). "Optimized detection of normal vibration modes of atomic force microscope cantilevers with the optical beam deflection method." Journal of Applied Physics **97**(8): 83524. https://doi.org/10.1063/1.1872202.

Schillers, H., C. Rianna, J. Schäpe, T. Luque, H. Doschke, M. Wälte, J. J. Uriarte, et al. (2017). "Standardized nanomechanical atomic force microscopy procedure (SNAP) for measuring soft and biological samples." Scientific Reports **7**(1): 5117. https://doi.org/10.1038/s41598-017-05383-0.

Sikora, A. (2016). "Quantitative normal force measurements by means of atomic force microscopy towards the accurate and easy spring constant determination." Nanoscience and Nanometrology **2**(1): 8. https://doi.org/10.11648/j.nsnm.20160201.12.

Stiernstedt, J., M. W. Rutland and P. Attard (2005). "A novel technique for the in situ calibration and measurement of friction with the atomic force microscope." Review of Scientific Instruments **76**(8): 1–9. https://doi.org/10.1063/1.2006407.

Stiernstedt, J., M. W. Rutland and P. Attard (2006). "Erratum: A novel technique for the in situ calibration and measurement of friction with the atomic force microscope (Review of Scientific Instruments (2005) 76 (083710))." Review of Scientific Instruments **77**(1): 1. https://doi.org/10.1063/1.2162429.

Sumbul, F., N. Hassanpour, J. Rodriguez-Ramos and F. Rico (2020). "One-step calibration of AFM in liquid." Frontiers in Physics **8**(September): 301. https://doi.org/10.3389/fphy.2020.00301.

Villarrubia, J. S. (1997). "Algorithms for scanned probe microscope image simulation, surface reconstruction, and tip estimation." Journal of Research of the National Institute of Standards and Technology **102**(4): 425. https://doi.org/10.6028/jres.102.030.

Warmack, R. J., X. Y. Zheng, T. Thundat and D. P. Allison (1994). "Friction effects in the deflection of atomic force microscope cantilevers." Review of Scientific Instruments **65**(2): 394–399. https://doi.org/10.1063/1.1145144.

Weafer, P. P., J. P. McGarry, M. H. Van Es, J. I. Kilpatrick, W. Ronan, D. R. Nolan and S. P. Jarvis (2012). "Stability enhancement of an atomic force microscope for long-term force measurement including cantilever modification for whole cell deformation. Review of scientific instruments." **83**. https://doi.org/10.1063/1.4752023.

Westra, K. L., A. W. Mitchell and D. J. Thomson (1993). "Tip artifacts in atomic force microscope imaging of thin film surfaces." Journal of Applied Physics **74**(5): 3608–3610. https://doi.org/10.1063/1.354498.

Yan, Y., B. Xue, H. Zhenjiang and X. Zhao (2016). "AFM tip characterization by using FFT filtered images of step structures." Ultramicroscopy **160**(January): 155–162. https://doi.org/10.1016/j.ultramic.2015.10.015.

Nelda Antonovaite, Massimiliano Berardi, Kevin Bielawski,
Niek Rijnveld

3.2 Fiber-Optics-Based Nanoindenters

3.2.1 Introduction

Although atomic force microscopy (AFM) has been proven to be a widely used device for mechanical characterization of biological and biomimetic materials, especially cells, it has some limitations. Typical AFM devices are designed to work at a very high resolution, axial and lateral, which limits the size of the scanned area, usually to a few hundreds of μm, and vertical displacement range to a few tens of μm while many biomaterials, such as hydrogels and tissues, often need to be tested over an area on the order of millimeters. Furthermore, operating an AFM system requires a considerable amount of knowledge, skill, and experience. This makes AFM less accessible for many research groups, especially those that are new to the field of mechanical testing. In addition, most AFM devices are built as stand-alone systems with little flexibility in modifications, which limits the possibilities to easily combine it with different instruments, such as microelectrode arrays, different types of microscopes, or stretching devices for novel multimode experiments. Finally, laser triangulation-based readout of AFM cantilevers makes the system incompatible with the standard high-number well plates (e.g., 96- and 384-well plates) for high-throughput testing (Dujardin et al., 2019).

To address the new needs of the growing mechanobiology field, a new type of indenters was developed by Optics11 Life, which has a focus on ease of use, compact design, versatility, and is designed for micro- to macro-scale mechanical testing. The user-friendly operation was achieved by implementing an all-optical fiber interferometric readout of the cantilever deflection. The cantilever and the fiber are aligned and integrated into a probe during manufacturing. Such a feature allows one to improve stability and handling while maintaining high precision. Furthermore, with this design, the range of cantilever bending is expanded to 30 μm, which allows measuring over 4 orders of magnitude in sample stiffness (e.g., 0.1–100 kPa) with the same probe. This is advantageous when testing mechanically heterogeneous samples (see Figure 3.2.1A for the measurement range). Probes are precalibrated, reusable, and equipped with spherical tips from 3 to 250 μm, where 3–25 μm radius tips are also extended by 30–100 μm to prevent one from touching the sample with a cantilever rather than the sphere when testing non-flat samples. Furthermore, *XYZ*-stages have a range of several centimeters, enabling measurements of large samples with a non-flat surface profile without the need to adjust the sample position (see example in Figure 3.2.1B–E).

Nelda Antonovaite, Massimiliano Berardi, Kevin Bielawski, Niek Rijnveld, Optics 11 Life, Amsterdam, Netherlands

https://doi.org/10.1515/9783110640632-008

For the displacement of the probe during indentation, a piezo-transducer with 100 μm travel range is used to achieve large indentation depths, as many of the biomaterials have different surface properties in comparison to deeper layers due to interface effects or sample preparation procedures such as slicing. In addition, the combination of long-range stages and piezo-transducer allows performing indentations on extremely adhesive samples such as soft gels. Therefore, this novel nanoindenter configuration has expanded the range of mechanical tests while decreasing the complexity of the operation of the instrument and sample preparation procedures.

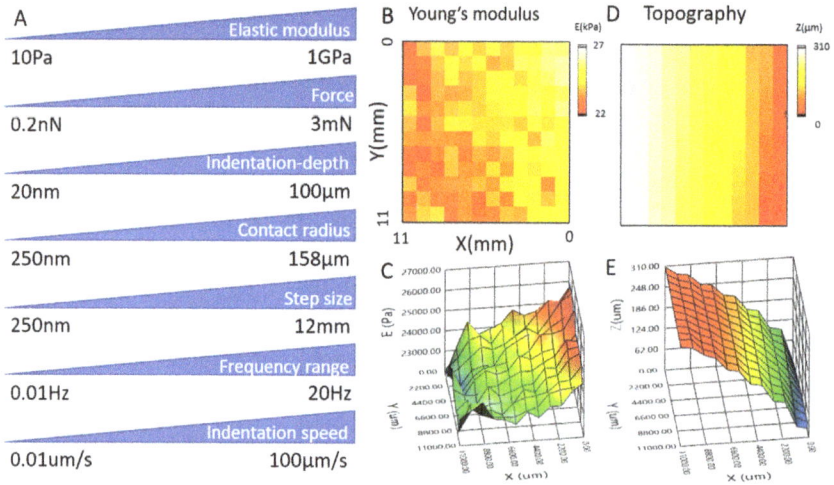

Figure 3.2.1: (A) Measurement range of the system (from top to bottom): the range of elastic modulus that can be measured by cantilevers with spring constants between 0.025 and 250 N/m; applied and sensed force range; the range of reached indentation depth; the range of contact radius $a = \sqrt{hR}$, where h is indentation depth; minimum step size and maximum scanned distance; frequency range of oscillations during dynamic mechanical analysis (DMA); indentation speed during static indentations. (B)–(E) Example of results from the matrix scan on a swelling hydrogel (Petrisoft 25 kPa). The start of the matrix was at X = 0, Y = 0. During the scan, hydrogel swelled by 310 μm and Young's modulus decreased from 27 to 22 kPa. (B) 2D map of Young's modulus E (kPa), (C) surface plot of Young's modulus E (Pa), (D) 2D map of topography Z (μm), and (E) surface plot of topography Z (μm).

The mechanical behavior of biomaterials exhibits nonlinearity and poro-viscoelasticity that results in depth, tip radius, and speed-dependent properties that arise from complex material architecture and solid–fluid interactions. To address this, the Optics11 Life nanoindentation systems were designed to operate in a feedback-control mode where the user can freely define a quasi-static or dynamic indentation profile in indentation depth or load control modes. As a result, in addition to commonly used Young's modulus values from quasi-static measurements, other mechanical parameters can be extracted such as storage, loss moduli, and damping factor as a function of strain and frequency, relaxation time constants, or even permeability and diffusivity from time-dependent measurements (Offeddu et al., 2018). These parameters can be used to investigate the role of mechanics in diseases as they might contain important information about structure and function.

Three fiber-optics-based nanoindenter instruments have been developed to accommodate different types of mechanical experiments: (1) a bench-top stand-alone version suited for testing hydrogels, elastomers, and large tissues (PIUMA Nanoindenter); (2) a microscope-compatible system for small tissue and single-cell characterization or other high-resolution imaging requiring experiments (CHIARO Nanoindenter); (3) a high-throughput nanoindentation platform, which combines automatic mechanical testing of samples inside well plates (up to 384 wells), imaging and incubation (PAVONE Nanoindentation platform) suited for automated testing of hydrogels, biofilms, tissues, single cells, spheroids, organoids, and other biomaterials. These novel nanoindenters were adapted by many research labs: the ones who focus on mechanical testing at various scales by rheometry and AFM, and the ones who just started to investigate the mechanical behavior of their biological samples. We overview the main operation principles, capabilities, and applications of these three nanoindenters to study the role of mechanics in various diseases.

3.2.2 Working Principles

3.2.2.1 Optical Fiber Interferometry

The probe design consists of a glass ribbon glued on top of the ridge carved in a glass ferrule to act as a cantilever (Figure 3.2.2B). Alternatively, microelectromechanical systems produced via lithography are glued directly on top of a glass ferrule (Figure 3.2.2A). The cantilever is coated with gold to increase its reflectivity. An optical fiber is glued perpendicularly to the cantilever into a groove, with the end of the fiber facing the cantilever side opposite to the sphere. The tip of the cantilever is equipped with a sphere. Small spheres of radius 3–25 μm are extended with a glass rod (Figure 3.2.2C and D) and large spheres of radius 50–250 μm are glued directly on the cantilever (Figure 3.2.2B). The dimensions of the probe allow measuring samples at the bottom of standard well plates up to 384 wells (Figure 3.2.2E).

The cantilever displacement is monitored via Fabry–Perot interferometry, where the cavity is defined by the end facet of the optical fiber and the top surface of the cantilever. As shown in Figure 3.2.2F, the measuring fiber is connected to a 2 × 2 90/10 coupler via the 10% arm, while the 90% one is not in use. The two remaining arms are used to connect a monochromatic infrared laser source of wavelength λ=1,550 nm, and a photodiode. The latter detects the changes in light intensity, W, that arise from the interference between the light beams reflecting at refractive index discontinuities, that is, the fiber end and the cantilever top (Chavan et al., 2012). This signal is a function of the cantilever deflection, as that changes the optical path length d traveled by the light in the Fabry–Perot cavity. Assuming that only one reflection occurs, the interference signal can be described with the equation (Chavan et al., 2012):

$$W(d) = W_0 \left[1 + V \cdot \cos\left(\frac{4\pi d}{\lambda} + \varphi_0 \right) \right] \tag{3.2.1}$$

where W_0 is the midpoint (quadrature) interference signal $((W_{max} - W_{min})/2)$, V is the fringe visibility $((W_{max} - W_{min})/(W_{max} + W_{min}))$, and φ_0 is a constant phase shift that depends on the cantilever geometry. From eq. 3.2.1, it is clear that the photodiode's response to a moving cantilever is not linear. Working in quadrature (i.e., by tuning either d or λ) allows to capture an approximately linear response, but this holds only for small displacements ($d \ll \lambda$). On top of this, the readout suffers from cyclic ambiguity.

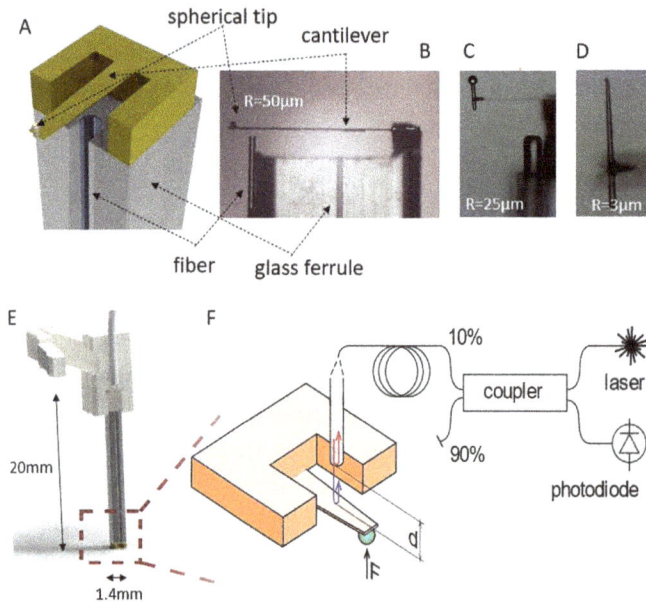

Figure 3.2.2: Optical fiber-based force sensor design. (A) Assembled probe view with components identified. (B)–(D) Side images of probes with different spheres. (E) Probe design. (F) Highlight of the cantilever/fiber positioning and detection scheme. In red and blue, the Fresnel reflections occur at the refractive index discontinuities. d is the Fabry–Perot cavity and F is the point where the load is applied.

Since a typical indentation measurement requires $d > \lambda$, it is important to linearize the photodiode signal. This can be achieved by modulating the laser wavelength around a central value λ_c (Beekmans and Iannuzzi, 2015, Van Hoorn et al., 2016):

$$\lambda(t) = \lambda_c + \delta\lambda \sin(\omega t) \tag{3.2.2}$$

where $\delta\lambda$ is the modulation amplitude and ω is its frequency. Assuming $\varphi_0 = 0$ and given the amplitude $\delta\lambda \ll \lambda$, it is possible to rewrite eq. 3.2.1, given the input wavelength (eq. 3.2.2), as a Taylor expansion around 0:

$$W(t) \approx W_0 \left[1 + V\left[\cos\left(\frac{4\pi d}{\lambda_c}\right) + \left(\frac{4\pi d \cdot \delta\lambda \cdot \sin(\omega t)}{\lambda_c^2}\right)\sin\left(\frac{4\pi d}{\lambda_c}\right)\right]\right] \tag{3.2.3}$$

$$= W_0[1 + V[W_{dc} + W_\omega(t)]]$$

By observing the time dependence of W it is possible to distinguish a DC component (W_{dc}, i.e., the part dependent on d only) and an oscillating one, $W_\omega(t)$, which arises from the wavelength modulation (Figure 3.2.3C and D). These two terms are in quadrature, which allows the linearization of the signal as a function of the cavity, d. The amplitude of W_{dc} can be measured by applying a low-pass filter to the signal, while the W_ω amplitude can be measured with a lock-in amplifier tuned to the wavelength modulation frequency. By rescaling the W_{dc} and W_ω amplitudes and plotting them one against the other, it is possible to obtain a real-time readout describing a circle (Figure 3.2.3E; Van Hoorn et al., 2016).

This gives a unique solution for the displacement over one revolution and allows tracking of large displacements by unwrapping the measured phase, that is, adding one period per each full revolution according to the equation (Beekmans and Iannuzzi, 2015):

$$d = \frac{\lambda_c}{4\pi}\arctan\left(\frac{W_{dc}}{W_\omega}\right) \tag{3.2.4}$$

Because of the real-time readout, this sensing scheme is suitable for the implementation of electromechanical feedback loops, which enable multiple testing protocols. For instance, it is possible to control the piezo-movement to achieve linear force or indentation depth profile in static measurements, or to perform dynamic mechanical analysis in equilibrium or during the loading ramp (Antonovaite et al., 2018, Van Hoorn et al., 2016). Furthermore, it allows automating operations, for example, surface finding based on cantilever deflection.

The calibration of the spring constant of these ferrule-top cantilevers is performed in a quasi-static manner: a piezoelectric transducer presses the probe against a weighting scale pan until the first maximum of interference is reached, and then keeps the deflection constant over the integration time of the scale. This process is repeated several times, at different maxima, which results in a weight versus displacement plot that directly allows calculation of the spring constant (Beekmans and Iannuzzi, 2015). This approach relies only on the measurements of wavelength and weights. The former is known with an accuracy of 10 pm. By using a scale with 100 ng

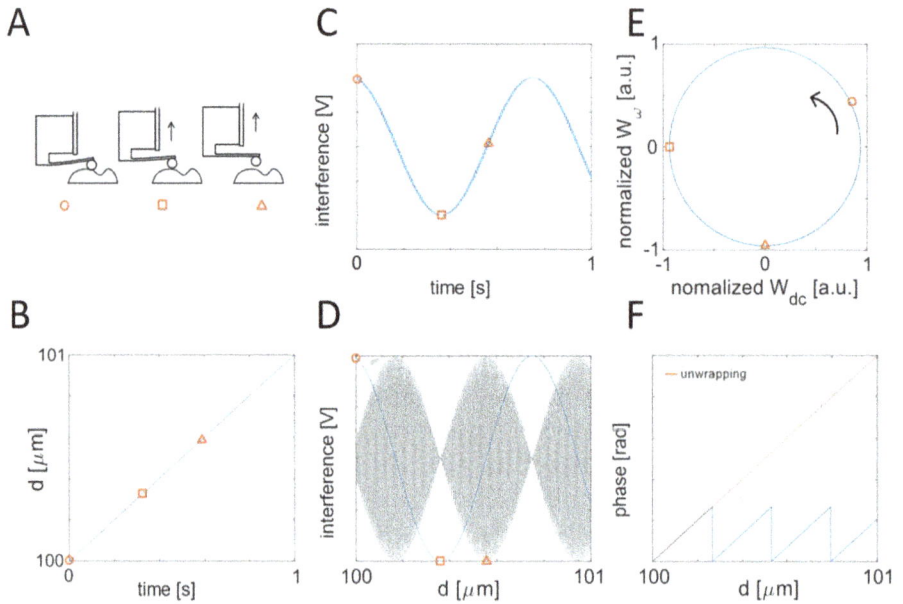

Figure 3.2.3: Simulation of the readout during the unloading phase of an indentation experiment, based on Equations 3.2.2–3.2.4. (A) Schematics of the unloading procedure. (B) Cavity variation over time. (C) Interferometric signal with wavelength modulation. (D) The low- and high-frequency components. (E) Rescaled W_{dc} and W_{ω} signals showing how the signal travels along the circle during unloading. (F) Phase variation during unloading and the unwrapped phase in red.

readability, the spring constant systematic error in calibration is as low as 100 µN/m (Beekmans and Iannuzzi, 2015).

Since the probe does not rely on any electronic components, it can be easily used in liquids. The materials that the probes are made of borosilicate glass (SiO_2) make them suitable for measuring in harsh conditions such as low and high temperatures (Chavan et al., 2011). The probes can be cleaned with common solvents such as isopropanol and ethanol over short periods (~30 s) and enzyme-based solutions such as Trypsin and Helizyme. The fiber and the cantilever support behave essentially as a monolithic body, which guarantees proper alignment and a consistent readout during any experiment, even for large cantilever bending. Furthermore, the probe, relying only on a cantilever on top of the fiber, can be very small, ultimately being part of the fiber itself (CUI et al., 2016, Rector et al., 2017, Tiribilli et al., 2011). Finally, this size and flexibility enable experimentation in conditions that are not achievable in conventional AFM, for example, it allows to test at the bottom of well plates or inside enclosed chambers, at the tip of needles (Beekmans and Iannuzzi, 2016), or in unconventional orientations.

This readout scheme and fabrication technology proved to give reliable results from indentation measurements, which quantitatively agree with macroscopic testing methods such as shear rheometry for homogeneous samples (Van Hoorn et al., 2016) and has been adapted in custom indentation setups combined with various microscopes to measure mechanical properties of soft tissues and cells such as the brain (Antonovaite et al., 2018), retina (Marrese et al., 2019b), embryos (Marrese et al., 2019a), and astrocytes (Antonovaite et al., 2020). Furthermore, the commercial nanoindenters have been featured in almost 200 publications; we will look over the main applications in Section 5.2.3.

3.2.2.2 Nanoindenter Setup Configurations

Three commercial fiber-optic indenter configurations are available, depending on the type of experiment and sample: bench-top Piuma Nanoindenter, microscope-compatible Chiaro Nanoindenter, and high-throughput mechanobiology platform Pavone.

The Piuma Nanoindenter consists of the piezo-transducer attached to a motorized closed-loop Z-stage, which can be moved vertically with the manual stage (see Figure 3.2.4B). The probe is clamped into the holder, which is attached to the piezo transducer, while the optical fiber of the probe is connected to the interferometer. The sample is placed on top of the motorized closed-loop stages, which can also be heated. The upright camera is aligned with the tip of the probe to image the sample from the top. Additionally, the inverted camera module can be mounted on top of the stages to image the transparent sample from underneath and position the tip of the probe in the region of interest (see Figure 3.2.4A dashed blue lines). The indenter can be placed on top of the antivibration system or just a simple table as the short mechanical loop ensures the low noise level of the system. The system has been employed to perform indentation mapping of various samples, both biological and not, at the μm–mm scale (see Figure 3.2.1A for the range of measurements).

The Chiaro Nanoindenter consists of a piezo-transducer mounted at the end of the manual sliding stage, which is connected to motorized XYZ-closed-loop stages (see Figure 3.2.4D). This whole part is placed on an L-bracket connected to the vertical post, which allows adjusting the height of the indentation head to fit the indentation probe in between the condenser and sample holder of the inverted microscope (see Figure 3.2.4C). The whole indentation arm is mounted on a breadboard with either a passive or active antivibration system. Alternatively, the indentation head can be mounted directly on the microscope stage to shorten the mechanical loop and, thus, the noise level. Moreover, a piezo-transducer can also be easily switched to a horizontal orientation to sense lateral forces. The system has been used to measure both the mechanical properties of single cells, microtissues, and cell-induced forces by precisely positioning the tip of the probe on top of the targeted region. Furthermore, more advanced mechanobiology experiments are possible by combining fluorescence and confocal imaging modalities.

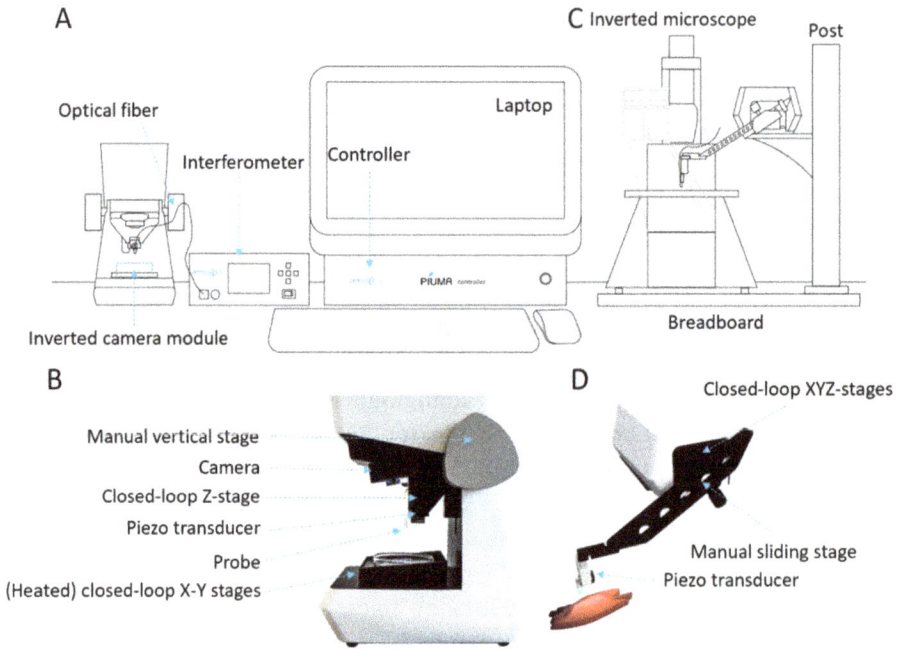

Figure 3.2.4: Components of (A) and (B) Piuma and (C) and (D) Chiaro Nanoindenters.

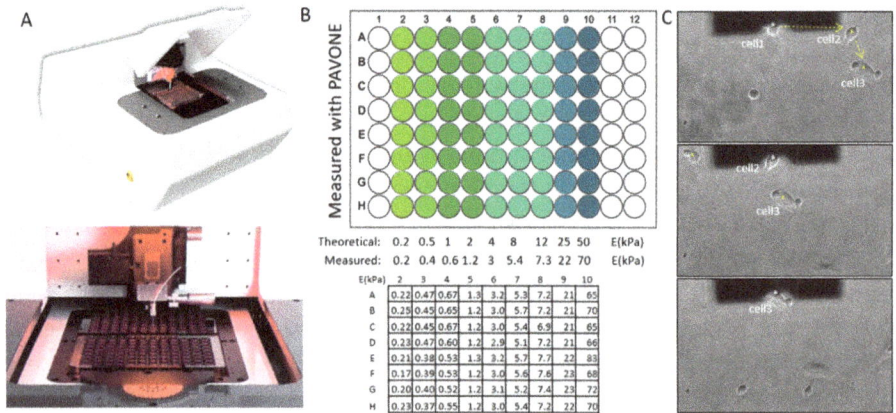

Figure 3.2.5: Pavone – a high-throughput mechanical testing platform. (A) The casing of the Pavone combines microscope, incubator, and indenter. Two well plates fit inside the sample holder area. (B) Example of results obtained with Pavone on a 96-well plate with various stiffness hydrogels (0.2–50 kPa). (C) Sequential single-cell mechanical testing with Pavone.

Many biomaterials are produced or cultured inside well plates, which enables high-throughput screening of biological activity such as viability, proliferation, toxicity, gene expression, epigenetics, and other omics. Mechanics can also be used as a biomarker to assess the cell culture state (Di Carlo, 2012) or as a quality control parameter for engineered materials (Eggert and Hutmacher, 2019). However, it was not possible to use mechanics for high-throughput screening due to manual exchange and localization of samples in conventional mechanical testing systems and incompatibility with the well-plate format, thereby the Pavone Nanoindentation platform was developed to integrate high-throughput mechanical testing inside the well plates together with imaging and incubation (Figure 3.2.5A). Two well plates filled over the whole well diameter with a material such as a biofilm, hydrogel, elastomer, or punch biopsy can be mechanically tested in a fully automatic manner with the highest well plate size of 384 (Figure 3.2.5B). For the samples that are scattered throughout the well diameter such as single cells, spheroids, organoids, or hydrogel spheres, a semiautomatic procedure can be used, where the user clicks to select on the imaging window multiple regions of interest, and the indenter measures these areas sequentially (Figure 3.2.5C). During the measurements, the environment can be regulated in terms of temperature, humidity, and CO_2 to provide incubator-like conditions while imaging of the sample can be performed at different time points during indentation. Depending on the study, the standard bright-field and phase-contrast imaging capabilities can be expanded with fluorescence, confocal, or multiphoton imaging to enable correlation studies between mechanical properties, structure, and function. Finally, the system can be made sterile to enable multiple-day mechanical testing experiments, such as following changes in mechanical properties during disease progression in 3D cell cultures.

3.2.2.3 Operation and Indentation Profiles

The operation of the Optics11 Life Nanoindenters consists of a quick calibration procedure, followed by the measurement process. Before starting the measurements, the user needs to perform the calibration of the probe inside the Petri dish filled with the liquid used in the experiment or air, depending on the sample condition. The calibration procedure consists of three steps: (1) a "wavelength scan," during which laser settings are automatically adjusted to have maximum visibility of the interference signal, which depends on the probe geometry and refractive index of the medium; (2) a "find surface" step, where the probe is automatically brought into contact with the glass; (3) a "calibration" step, where the piezo-transducer displaces the probe while in contact with a stiff surface to obtain geometrical correction factor due to mismatch between the tip and fiber positions and scaling factors for demodulation circle. This procedure takes 1–2 min and only needs to be repeated when the probe or medium is replaced.

Indentation measurements on samples start with the "find surface" step where the surface of the sample is found automatically and the probe is retracted to be out of contact from the sample. "Run experiment" step initiates the indentation measurement in either of the modes: indentation depth, load, or displacement where the profile is selected by the user. Experimental procedures such as "find surface," "matrix scan," "indentation," "take image," "move to well" can be set in sequences, and settings can be saved and reloaded the next time the same experiment needs to be performed.

Figure 3.2.6 shows three examples of such measurements. In the displacement control mode (Figure 3.2.6A), the profile of the piezo-transducer displacement is selected. Applied load and indentation depth will depend on the relation between the stiffness of the probe and sample such that the softer sample will be indented deeper and with faster indentation speed than the stiffer one. In indentation-depth-controlled mode (Figure 3.2.6C), the probe approaches the sample until the surface is found and then the feedback loop is initiated during which the piezo is adjusted to reach the selected indentation depth at the selected indentation speed. As a result, samples with varying stiffness can be measured at the same indentation depth, which is important when measuring nonlinear materials, and at the same indentation speed, which is important for viscoelastic materials. Similarly, measurements can be performed in a load-controlled mode. Furthermore, one can apply frequency or amplitude sweeps in either of the modes (Figure 3.2.6E) to obtain storage and loss moduli, and damping factor as a function of frequency or amplitude (Figure 3.2.6F). All raw data is given in a text file and can be analyzed with DataViewer software or custom code. The new online mechanical model sharing and analysis platform will be implemented in the coming year to standardize mechanical testing and analysis protocols.

3.2.3 Applications in Studying the Mechanics of Diseases

The Optics11 Life Nanoindenters have been used to characterize mechanical properties of various samples: hydrogels, biofilms, native and engineered tissues, ECM matrix, single cells and cell monolayers cultured on hard or soft substrates, spheroids, organoids, embryos, and oocytes, resulting in approximately 200 publications as of 2020. In many of them, both healthy and diseased conditions of samples were investigated. We overview how Optics11 Life Nanoindenters were used to study mechanical phenomena in diseases.

The eye is a pressurized vessel, where various ocular conditions such as glaucoma and myopia have been associated with the change in biomechanics of different ocular tissues. Therefore, mechanical testing of both intact and explanted tissues from the eye is used to understand their biomechanics and look for new treatment strategies. For example, Piuma Nanoindenter was used to measure three layers of cornea:

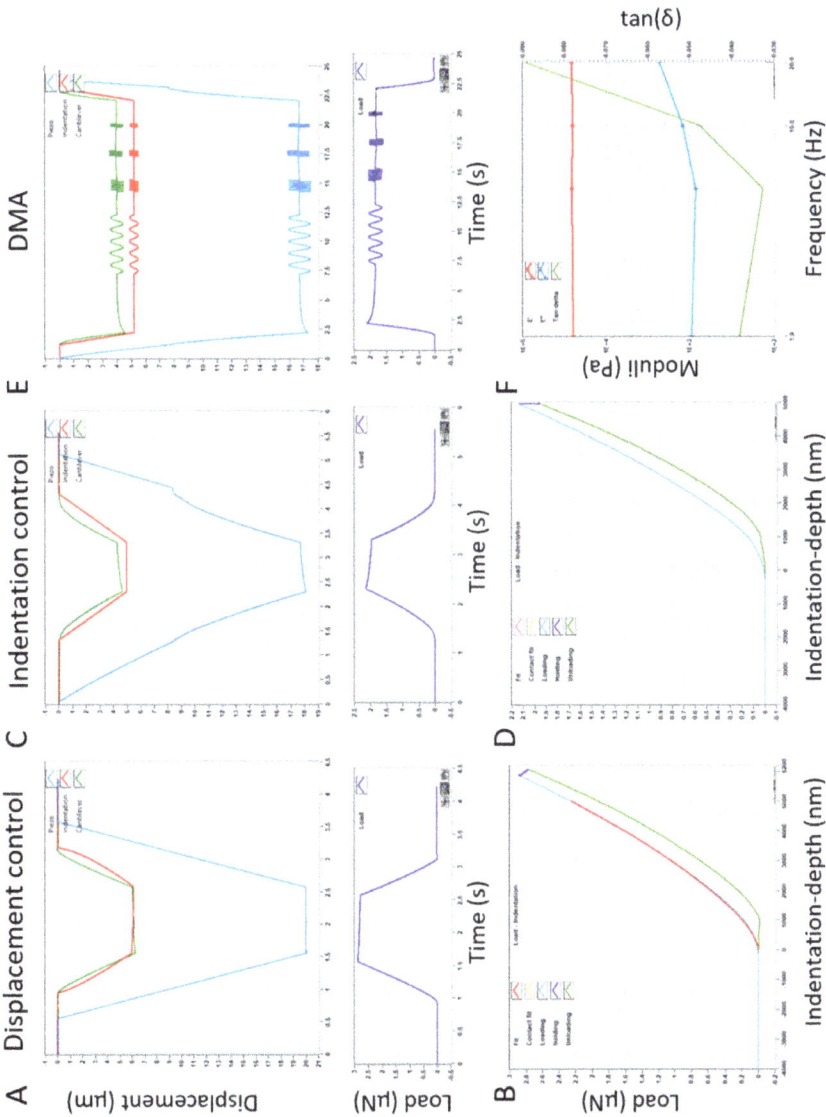

Figure 3.2.6: Indentation profiles. (A) Static indentation in the displacement-controlled mode with 20 μm displacement at 20 μm/s piezo-speed and 1 s holding time. Curves of piezo-displacement, cantilever deflection, indentation depth, and load are plotted as a function of time. (B) The corresponding load versus indentation depth curve with Hertz fit up to 5 μm indentation depth (in red). (C) Static indentation in the indentation-depth-controlled mode up to 5 μm indentation depth at 5 μm/s indentation speed and 1 s hold time. (D) The corresponding load versus indentation depth curve. Note that depth was constant during the hold period (dark blue). (E) Dynamic indentation profile, where 5 μm indentation depth is reached at 5 μm/s, then held at constant depth for 5 s, and then oscillations with an amplitude of 0.2 μm and frequencies between 1 and 20 Hz are applied. (F) Storage E′ and loss E″ moduli, and damping factor tan(δ) as a function of frequency.

epithelium, basement membrane, and stroma, while the eye was still intact (in situ) by immobilizing the head on the *XY*-stage and abrading the layers sequentially. As a result, Young's modulus of corneal epithelium was found to decrease when comparing the control healthy group with obese Type 2 prediabetes mice, which were related to the integrity of epithelium (Xu et al., 2020). Another study has been performed on the excised cornea by mounting it on a half dome to support the curved shape. As cornea was treated by irradiating it with a two-photon femtosecond laser to crosslink collagen, Young's moduli of irradiated regions significantly increased (Shavkuta et al., 2017). Moreover, both AFM and Piuma Nanoindenters were used to show how Young's modulus of human anterior lens capsules change with aging with similar data from both instruments (Efremov et al., 2020). Furthermore, a custom version of Chiaro Nanoindenter mounted on top of a multielectrode array has been used to induce electrical signals of retinal ganglion cells in explanted mice retinas by mechanically stimulating the photoreceptor layer (Marrese et al., 2019b). Therefore, multimode studies combining indentation with other biochemical or electrical analysis pave the way for understanding mechanobiological processes in ocular diseases.

In cardiovascular disease research, Optics11 Life Nanoindenters can be used in multiple ways: to test the mechanical properties of cardiac cells or their beating patterns, map mechanical properties of pathological cardiac tissues and vessels, and characterize substrate materials to mimic the mechanical environment of diseased tissues. For example, in the myocardial infarction research, various stiffness hydrogels were used as substrates to mimic the mechanical microenvironment of scarred and aged heart tissue for culturing cardiomyocytes (Acun et al., 2019, 2018, Nguyen et al., 2018). In these experiments, the contractile forces of single cells and their sheets in terms of beating force magnitude were measured with the Chiaro Nanoindenter through dwelling experiments where the displacement of the cantilever in the transverse direction is proportional to the contractile force. Also, mechanical testing of cardiac myocytes has been used to assess the changes in contractile function and stiffness by pharmacologically or genetically inducing rearrangements of the cytoskeletal network in the context of heart failure (Caporizzo et al., 2018, Chen et al., 2018, Coleman et al., 2020, Yingxian et al., 2020). In the research of diseases of vascular tissues, stiffness of live aorta was found to decrease in abdominal aortic aneurysm, which correlated with a decrease in the smooth muscle cell, collagen, and fibroblast quantitative fluorescence (Meekel et al., 2019). Another study on atherosclerotic arterial tissues found that calcified carotid plaques were 6 orders of magnitude stiffer than non-calcified, which agrees with similar studies using other indentation instruments (Cahalane and Walsh, 2020). At the cellular scale, mechanical properties of vascular smooth muscle cells from arteries were studied under different substrate stiffness conditions to understand the mechanical modulation of cell phenotype in relation to pathological arterial stiffening (Xie et al., 2018). These types of tests using Piuma or Chiaro Nanoindenters allowed the creation of novel disease models, study the relationships between mechanical properties, the function of cardiomyocytes, and composition at the cellular or tissue levels.

The biomechanics of cartilage depends on its extracellular matrix: type II collagen, sulfated glycosaminoglycan network and water. Changes in composition take place during osteoarthritis where mechanical implications have been vastly studied by indentation with Piuma (Gaumet et al., 2018, Grebenik et al., 2018, Lavet and Ammann, 2017, Moshtagh et al., 2016, Vindas Bolaños et al., 2017; Zhang et al., 2019). Fibrosis is also associated with changes in matrix stiffness due to increased deposition of fibrillar collagens. Fibrotic intestinal tissue regions from Crohn's disease patients were shown to be stiffer in comparison to non-fibrotic (Bokemeyer et al., 2019). Similarly, bioengineered cystic fibrosis connective airway tissue was found to be stiffer than normal (Mazio et al., 2020). Furthermore, mechanics is an important parameter when studying cancer progression. For example, the effects of tumor microenvironment stiffness have been investigated by culturing cancer cells on various stiffness substrates or within the ECM, characterized by Piuma Nanoindenter (De Gregorio et al., 2020, Reynolds et al., 2018; Zhang et al., 2020).

In tissue engineering and regenerative medicine research, mechanical properties represent one of the key parameters that guide the design and the fabrication of scaffolds. Not only do these have to behave similarly to their biological counterpart when tested macroscopically, but they also need to mimic microstructure and mechanical properties at the scale of cell interactions. On top of this, engineered materials need to promote cell migration and ECM formation and degrade at a predictable rate if meant for in vivo applications. Nanoindentation has been used extensively to tackle this problem. For example, Piuma has been used to assess the elastic properties of synthetic anisotropic composite scaffolds (Baklaushev et al., 2019; Bardakova et al., 2018; Chen et al., 2017; Reynolds et al., 2018; Tognato et al., 2019; Ye et al., 2017), and of bioderived, decellularized structures (Brancato et al., 2018; Gao et al., 2020; Ekaterina A Grebenik et al., 2019) used in cell culturing or surgical procedures. In other studies, nanoindentation was used to track the evolution of mechanical properties of scaffolds during degradation (Cao et al., 2020; Gao et al., 2020; Versteegden et al., 2019).

To summarize, Optics11 Life Nanoindenters allow capturing the mechanical alterations in cells and tissues that contribute to the understanding of mechanical implications on the onset, progression, and even treatment of diseases.

3.2.4 Conclusions and Future Perspectives

The complexity of the operation, the limited measurement scale, and constraints due to optical-triangulation-based readout of AFM indentation instruments have motivated Optics11 Life to develop a new type of indentation instruments. The key features of Piuma and Chiaro Nanoindenters are ease of use, flexibility, and expanded measurement scale, while Pavone indentation platform adds high-throughput, automation, and multimodality. These features are enabling researchers to obtain relevant data more efficiently. As a result, we have shown how Optics11 Life Nanoindenters have been used to study the mechanical phenomena in diseases. Experiments include single-cell mechanical characterization in various conditions, mechanical mapping of healthy and pathological tissues, and testing mechanical properties of hydrogels and other substrates when designing microenvironment of disease models. In this section, we look over the future perspectives of mechanical testing instrumentation.

With the increasing attention to the mechanics of biological materials in physiological and pathological conditions, new mechanical testing instrumentation is needed to fulfill the demands of growing research fields of three-dimensional cell culture models for tissue engineering and drug discovery, in both academia and industry. For example, the use of engineered tissues as disease models are opening new opportunities that were not possible with ex vivo studies. At the same time, they pose new challenges for mechanical testing due to a variety of sample geometries such as tissues on membrane inserts or inside microfluidic chambers and spherical or more complex shapes such as in the case of organoids, thereby the future indenters should have the functionality to measure these structures easily and reliably. Alternatively, new non-contact, image-based mechanical testing techniques such as Brillouin microscopy, acoustic force spectroscopy, magnetic resonance elastography, and optical coherence elastography might substitute contact mechanical testing.

Another important aspect is the development of automated and high-throughput mechanical testing methods. These are needed to assess the biological heterogeneity and increase the number of tested pathological conditions. Increasing the speed of mechanical testing would allow using mechanical properties for phenotyping of tissues and cell cultures routinely and even bring it to the clinical and drug screening applications. Additionally, measurements during disease development would provide more information about the progression of the disease than end-point measurements; thus, instruments should be able to do mechanical testing and imaging of cells and tissues under sterile and physiological conditions.

Further advancements should take place in combining mechanical testing with different imaging modalities and other measurement techniques such as spectroscopy and electrophysiology to better understand the structure–stiffness–function relationship such as coupling between mechanical and biochemical signaling (mechanotransduction), and how different tissue components modulate mechanical behavior. Therefore, the integration of different instruments to measure at the same time and location and correlative analysis between different outputs will be essential in the progress of the mechanobiology field. To fully capture the complexity of the mechanical behavior of biological tissues, there will be a need for both quasi-static and dynamic multiaxial testing methods. These should apply combinations of compression, tension, shear, and torque at each scale, from nano to macro. Finally, progress in the development of mechanical models should go along with experimental techniques to be able to extract relevant mechanical parameters in combination with structural information to create realistic material models.

Disclaimer: The authors of this chapter are affiliated with Optics11 Life, Amsterdam, *which produced the instruments described in this chapter.*

References

Acun, A., T. D. Nguyen and P. Zorlutuna (2018). "An aged human heart tissue model showing age-related molecular and functional deterioration resembling the native heart." BioRxiv 287334. https://doi.org/10.1101/287334.

Acun, A., T. D. Nguyen and P. Zorlutuna (2019). "In vitro aged, hiPSC-origin engineered heart tissue models with age-dependent functional deterioration to study myocardial infarction." Acta Biomaterialia **94**: 372–391. https://doi.org/10.1016/j.actbio.2019.05.064.

Antonovaite, N., S. V. Beekmans, E. M. Hol, W. J. Wadman and D. Iannuzzi (2018). "Regional variations in stiffness in live mouse brain tissue determined by depth-controlled indentation mapping." Scientific Reports **8**(1): 1–11. https://doi.org/10.1038/s41598-018-31035-y.

Antonovaite, N., T. A. van Wageningen, E. J. Paardekam, A. M. van Dam and D. Iannuzzi (2020). "Dynamic indentation reveals differential viscoelastic properties of white matter versus gray matter-derived astrocytes upon treatment with lipopolysaccharide." Journal of the Mechanical Behavior of Biomedical Materials **109**: 103783. https://doi.org/10.1016/j.jmbbm.2020.103783.

Baklaushev, V. P., V. G. Bogush, V. A. Kalsin, N. N. Sovetnikov, E. M. Samoilova, V. A. Revkova, K. V. Sidoruk, M. A. Konoplyannikov, P. S. Timashev, S. L. Kotova, K. B. Yushkov, A. V. Averyanov, A. V. Troitskiy, and J.-E. Ahlfors (2019). Tissue Engineered Neural Constructs Composed of Neural Precursor Cells, Recombinant Spidroin and PRP for Neural Tissue Regeneration. Scientific Reports **9**(1): 1–18. https://doi.org/10.1038/s41598-019-39341-9

Bardakova, K. N., E. A. Grebenik, E. V. Istranova, L. P. Istranov, Yu. V. Gerasimov, A. G. Grosheva, T. M. Zharikova, N. V. Minaev, B. S. Shavkuta, D. S. Dudova, S. V. Kostyuk, N. N. Vorob'eva, V. N. Bagratashvili, P. S. Timashev, and R. K. Chailakhyan (2018). Reinforced Hybrid Collagen Sponges for Tissue Engineering. Bulletin of Experimental Biology and Medicine **165**(1): 142–147. https://doi.org/10.1007/s10517-018-4116-8

Beekmans, S. V. and D. Iannuzzi (2015). "A metrological approach for the calibration of force transducers with interferometric readout." Surface Topography: Metrology and Properties **3**(2). https://doi.org/10.1088/2051-672X/3/2/025004.

Beekmans, S. V. and D. Iannuzzi (2016). "Characterizing tissue stiffness at the tip of a rigid needle using an opto-mechanical force sensor." Biomedical Microdevices **18**(1): 1–8. https://doi.org/10.1007/s10544-016-0039-1.

Bokemeyer, A., P. R. Tepasse, L. Quill, P. Lenz, E. Rijcken, M. Vieth, N. Ding, S. Ketelhut, F. Rieder, B. Kemper and D. Bettenworth (2019). "Quantitative phase imaging using digital holographic microscopy reliably assesses morphology and reflects elastic properties of fibrotic intestinal tissue." Scientific Reports **9**(1): 1–11. https://doi.org/10.1038/s41598-019-56045-2.

Brancato, V., M. Ventre, G. Imparato, F. Urciuolo, C. Meo, and P. A. Netti (2018). A straightforward method to produce decellularized dermis-based matrices for tumour cell cultures. Journal of Tissue Engineering and Regenerative Medicine **12**(1): e71–e81. https://doi.org/10.1002/term.2350

Cahalane, R. M. and M. T. Walsh (2020). "Nanoindentation of calcified and non-calcified components of atherosclerotic tissues." Experimental Mechanics. https://doi.org/10.1007/s11340-020-00635-z.

Cao, G., C. Wang, Y. Fan, and X. Li (2020). Biomimetic SIS-based biocomposites with improved biodegradability, antibacterial activity and angiogenesis for abdominal wall repair. Materials Science and Engineering: C **109**: 110538. https://doi.org/10.1016/j.msec.2019.110538

Caporizzo, M. A., C. Y. Chen, A. K. Salomon, K. B. Margulies and B. L. Prosser (2018). "Microtubules provide a viscoelastic resistance to myocyte motion." Biophysical Journal **115**(9): 1796–1807. https://doi.org/10.1016/j.bpj.2018.09.019.

Chavan, D., D. Andres and D. Iannuzzi (2011). "Note: Ferrule-top atomic force microscope. II. Imaging in tapping mode and at low temperature." Review of Scientific Instruments **82**(4): 129–131. https://doi.org/10.1063/1.3579496.

Chavan, D., T. C. Van De Watering, G. Gruca, J. H. Rector, K. Heeck, M. Slaman and D. Iannuzzi (2012). "Ferrule-top nanoindenter: An optomechanical fiber sensor for nanoindentation." Review of Scientific Instruments **83**(11). https://doi.org/10.1063/1.4766959.

Chen, C. Y., M. A. Caporizzo, K. Bedi, A. Vite, A. I. Bogush, P. Robison, J. G. Heffler, A. K. Salomon, N. A. Kelly, A. Babu, M. P. Morley, K. B. Margulies and B. L. Prosser (2018). "Suppression of detyrosinated microtubules improves cardiomyocyte function in human heart failure." Nature Medicine **24**(8): 1225. https://doi.org/10.1038/s41591-018-0046-2.

Chen, H., A. de B. F. B. Malheiro, C. van Blitterswijk, C. Mota, P. A. Wieringa, and L. Moroni (2017). Direct Writing Electrospinning of Scaffolds with Multidimensional Fiber Architecture for Hierarchical Tissue Engineering. ACS Applied Materials & Interfaces **9**(44): 38187–38200. https://doi.org/10.1021/acsami.7b07151

Coleman, A. K., H. C. Joca, G. Shi, W. J. Lederer and C. W. Ward (2020). Tubulin acetylation increases cytoskeletal stiffness to regulate mechanotransduction in striated muscle. BioRxiv. 2020.06.10.144931. https://doi.org/10.1101/2020.06.10.144931.

Cui, M., C. H. Van Hoorn and D. Iannuzzi (2016). "Miniaturized fibre-top cantilevers on etched fibres." Journal of Microscopy **264**(3): 370–374. https://doi.org/10.1111/jmi.12452.

De Gregorio, V., A. La Rocca, F. Urciuolo, C. Annunziata, M. L. Tornesello, F. M. Buonaguro, P. A. Netti and G. Imparato (2020). "Modeling the epithelial-mesenchymal transition process in a 3D organotypic cervical neoplasia." Acta Biomaterialia. https://doi.org/10.1016/j.actbio.2020.09.006.

Di Carlo, D. (2012). "A mechanical biomarker of cell state in medicine." Journal of Laboratory Automation **17**(1): 32–42. SAGE Publications: Los Angeles, CA. https://doi.org/10.1177/2211068211431630.

Dujardin, A., P. De Wolf, F. Lafont and V. Dupres (2019). "Automated multi-sample acquisition and analysis using atomic force microscopy for biomedical applications." PLOS One **14**(3): e0213853. https://doi.org/10.1371/journal.pone.0213853.

Efremov, Y., N. A. Bakhchieva, B. S. Shavkuta, A. A. Frolova, S. L. Kotova, I. A. Novikov, A. A. Akovantseva, K. S. Avetisov, S. E. Avetisov and P. S. Timashev (2020). "Mechanical properties of anterior lens capsule assessed with AFM and nanoindenter in relation to human aging, pseudoexfoliation syndrome, and trypan blue staining." Journal of the Mechanical Behavior of Biomedical Materials 104081. https://doi.org/10.1016/j.jmbbm.2020.104081.

Eggert, S. and D. W. Hutmacher (2019). "In vitro disease models 4.0 via automation and high-throughput processing." Biofabrication **11**(Issue 4): 043002. Institute of Physics Publishing. https://doi.org/10.1088/1758-5090/ab296f.

Gao, Y., Q. Liu, W. Kong, J. Wang, L. He, L. Guo, H. Lin, H. Fan, Y. Fan, and X. Zhang (2020). Activated hyaluronic acid/collagen composite hydrogel with tunable physical properties and improved biological properties. International Journal of Biological Macromolecules. https://doi.org/10.1016/j.ijbiomac.2020.07.319

Gaumet, M., I. Badoud and P. Ammann (2018). "Effect of hyaluronic acid-based viscosupplementation on cartilage material properties." Osteoarthritis and Cartilage **26**: S136. https://doi.org/10.1016/j.joca.2018.02.294.

Grebenik, Ekaterina A., L. P. Istranov, E. V. Istranova, S. N. Churbanov, B. S. Shavkuta, R. I. Dmitriev, N. N. Veryasova, S. L. Kotova, A. V. Kurkov, A. B. Shekhter, and P. S. Timashev (2019). Chemical cross-linking of xenopericardial biomeshes: A bottom-up study of structural and functional correlations. Xenotransplantation **26**(3): e12506. https://doi.org/10.1111/xen.12506

Grebenik, E. A., V. D. Grinchenko, S. N. Churbanov, N. V. Minaev, B. S. Shavkuta, P. A. Melnikov, D. V. Butnaru, Y. A. Rochev, V. N. Bagratashvili and P. S. Timashev (2018). "Osteoinducing scaffolds with multi-layered biointerface." Biomedical Materials **13**(5): 54103. https://doi.org/10.1088/1748-605X/aac4cb.

Lavet, C. and P. Ammann (2017). "Osteoarthritis like alteration of cartilage and subchondral bone induced by protein malnutrition is treated by nutritional essential amino acids supplements." Osteoarthritis and Cartilage **25**: S293. https://doi.org/10.1016/j.joca.2017.02.495.

Marrese, M., N. Antonovaite, B. K. A. Nelemans, T. H. Smit and D. Iannuzzi (2019a). "Micro-indentation and optical coherence tomography for the mechanical characterization of embryos: Experimental setup and measurements on chicken embryos." Acta Biomaterialia **97**: 524–534. https://doi.org/10.1016/j.actbio.2019.07.056.

Marrese, M., D. Lonardoni, F. Boi, H. van Hoorn, A. Maccione, S. Zordan, D. Iannuzzi and L. Berdondini (2019b). "Investigating the effects of mechanical stimulation on retinal ganglion cell spontaneous spiking activity." Frontiers in Neuroscience **13**. https://doi.org/10.3389/fnins.2019.01023.

Mazio, C., L. S. Scognamiglio, R. D. Cegli, L. J. V. Galietta, D. D. Bernardo, C. Casale, F. Urciuolo, G. Imparato and P. A. Netti (2020). "Intrinsic abnormalities of cystic fibrosis airway connective tissue revealed by an in vitro 3D stromal model." Cells **9**(6): 1371. https://doi.org/10.3390/cells9061371.

Meekel, J. P., G. Mattei, V. S. Costache, R. Balm, J. D. Blankensteijn and K. K. Yeung (2019). "A multilayer micromechanical elastic modulus measuring method in ex vivo human aneurysmal abdominal aortas." Acta Biomaterialia **96**: 345–353. https://doi.org/10.1016/j.actbio.2019.07.019.

Moshtagh, P. R., B. Pouran, N. Korthagen, A. Zadpoor and H. Weinans (2016). "Spatial variation of cartilage stiffness at micro and macro scale." Osteoarthritis and Cartilage **24**: S376. https://doi.org/10.1016/j.joca.2016.01.671.

Nguyen, D. T., N. Nagarajan and P. Zorlutuna (2018). "Effect of substrate stiffness on mechanical coupling and force propagation at the infarct boundary." Biophysical Journal **115**(10): 1966–1980. https://doi.org/10.1016/j.bpj.2018.08.050.

Offeddu, G. S., E. Axpe, B. A. C. Harley and M. L. Oyen (2018). "Relationship between permeability and diffusivity in polyethylene glycol hydrogels." AIP Advances **8**(10): 105006. https://doi.org/10.1063/1.5036999.

Rector, J. H., M. Slaman, R. Verdoold, D. Iannuzzi and S. V. Beekmans (2017). "Optimization of the batch production of silicon fiber-top MEMS devices." Journal of Micromechanics and Microengineering **27**(11). https://doi.org/10.1088/1361-6439/aa8c4e.

Reynolds, D. S., K. M. Bougher, J. H. Letendre, S. F. Fitzgerald, U. O. Gisladottir, M. W. Grinstaff and M. H. Zaman (2018). "Mechanical confinement via a PEG/Collagen interpenetrating network inhibits behavior characteristic of malignant cells in the triple negative breast cancer cell line MDA.MB.231." Acta Biomaterialia **77**: 85–95. https://doi.org/10.1016/j.actbio.2018.07.032.

Shavkuta, B. S., M. Y. Gerasimov, N. V. Minaev, D. S. Kuznetsova, V. V. Dudenkova, I. A. Mushkova, B. E. Malyugin, S. L. Kotova, P. S. Timashev, S. V. Kostenev, B. N. Chichkov and V. N. Bagratashvili (2017). "Highly effective 525\hspace0.167emnm femtosecond laser crosslinking of collagen and strengthening of a human donor cornea." Laser Physics Letters **15**(1): 15602. https://doi.org/10.1088/1612-202X/aa963b.

Tiribilli, B., G. Margheri, P. Baschieri, C. Menozzi, D. Chavan and D. Iannuzzi (2011). "Fibre-top atomic force microscope probe with optical near-field detection capabilities." Journal of Microscopy **242**(1): 10–14. https://doi.org/10.1111/j.1365-2818.2010.03476.x.

Tognato, R., A. R. Armiento, V. Bonfrate, R. Levato, J. Malda, M. Alini, D. Eglin, G. Giancane, and T. Serra (2019). A Stimuli-Responsive Nanocomposite for 3D Anisotropic Cell-Guidance and Magnetic Soft Robotics. Advanced Functional Materials **29**(9): 1804647. https://doi.org/10.1002/adfm.201804647

Van Hoorn, H., N. A. Kurniawan, G. H. Koenderink and D. Iannuzzi (2016). "Local dynamic mechanical analysis for heterogeneous soft matter using ferrule-top indentation." Soft Matter **12**(12): 3066–3073. https://doi.org/10.1039/c6sm00300a.

Versteegden, L. R., M. Sloff, H. R. Hoogenkamp, M. W. Pot, J. Pang, T. G. Hafmans, T. de Jong, T. H. Smit, S. C. Leeuwenburgh, E. Oosterwijk, W. F. Feitz, W. F. Daamen and T. H. van Kuppevelt (2019). A salt-based method to adapt stiffness and biodegradability of porous collagen scaffolds. RSC Advances **9**(63): 36742–36750. https://doi.org/10.1039/C9RA06651A

Vindas Bolaños, R. A., S. M. Cokelaere, J. M. Estrada Mcdermott, K. E. M. Benders, U. Gbureck, S. G. M. Plomp, H. Weinans, J. Groll, P. R. van Weeren and J. Malda (2017). "The use of a cartilage decellularized matrix scaffold for the repair of osteochondral defects: The importance of long-term studies in a large animal model." Osteoarthritis and Cartilage **25**(3): 413–420. https://doi.org/10.1016/j.joca.2016.08.005.

Xie, S. A., T. Zhang, J. Wang, F. Zhao, Y. P. Zhang, W. J. Yao, S. S. Hur, Y. T. Yeh, W. Pang, L. S. Zheng, Y. B. Fan, W. Kong, X. Wang, J. J. Chiu and J. Zhou (2018). "Matrix stiffness determines the phenotype of vascular smooth muscle cell in vitro and in vivo: Role of DNA methyltransferase 1." Biomaterials **155**: 203–216. https://doi.org/10.1016/j.biomaterials.2017.11.033.

Xu, P., A. Londregan, C. Rich and V. Trinkaus-Randall (2020). "Changes in epithelial and stromal corneal stiffness occur with age and obesity." Bioengineering **7**(1): 14. https://doi.org/10.3390/bioengineering7010014.

Ye, D., Q. Cheng, Q. Zhang, Y. Wang, C. Chang, L. Li, H. Peng, and L. Zhang (2017). Deformation Drives Alignment of Nanofibers in Framework for Inducing Anisotropic Cellulose Hydrogels with High Toughness. ACS Appl Mater Interfaces **9**(49): 43154–43162. https://doi.org/10.021/acsami.7b14900

Yingxian, C. C., A. K. Salomon, C. M. Alexander, C. Sam, N. A. Kelly, S. Curry, A. I. Bouush, K. Elisabeth, S. Saskia, J. Philip, M. Marie-Jo, C. Lucie, K. bedi and B. L. Prosser (2020). "Depletion of vasohibin 1 speeds contraction and relaxation in failing human cardiomyocytes." Circulation Research **0**(0). https://doi.org/10.1161/CIRCRESAHA.119.315947.

Zhang, H., F. Lin, J. Huang and C. Xiong (2020). "Anisotropic stiffness gradient-regulated mechanical guidance drives directional migration of cancer cells." Acta Biomaterialia. https://doi.org/10.1016/j.actbio.2020.02.004.

Zhang, X., D. Cai, F. Zhou, J. Yu, X. Wu, D. Yu, Y. Zou, Y. Hong, C. Yuan, Y. Chen, Z. Pan, V. Bunpetch, H. Sun, C. An, T. Yi-Chin, H. Ouyang and S. Zhang (2019). "Targeting downstream subcellular YAP activity as a function of matrix stiffness with Verteporfin-encapsulated chitosan microsphere attenuates osteoarthritis." Biomaterials 119724. https://doi.org/10.1016/j.biomaterials.2019.119724.

Pouria Tirgar, Allen J. Ehrlicher

3.3 Optical Tweezers and Force Spectrum Microscopy

3.3.1 Introduction

Microrheology is the study of material deformation and flow in response to applied forces at the microscale, and it has long been used as a tool to characterize biological samples and soft matter systems. The combination of externally controlled displacement and force measurement has given rise to three major techniques in active microrheology: atomic force microscopy (AFM), magnetic tweezers (MT), and optical tweezers (Liu et al., 2018, Rigato et al., 2017, Verdier et al., 2019, Rich et al., 2011). As AFM and MTs were previously discussed in this book, Chapters 3.1 and 3.4, respectively, here we focus solely on optical tweezers or optical traps (OTs), a family of instruments that employ focused light as their driving force. OTs are a no-contact technique for applying precisely calibrated forces to micron-scale objects in three dimensions using the focused beam of a laser (Molloy et al, 2002, Bui et al., 2018, Killian et al., 2018). Similar to beads in MTs and AFM cantilevers, the beads in OTs can be either used directly to probe the mechanics of a network (Neckernuss et al., 2015, Vos et al., 2017) or conjugated to biomolecules to detect their position or force within a trap (Bustamante et al., 2020, Wu et al., 2011).

OTs grew out of the idea of accelerating and trapping dielectric particles by radiation pressure, which was first introduced by Arthur Ashkin at Bell Labs in 1970 (Ashkin, 1970). This discovery evolved over the following 16 years into a platform capable of trapping particles using a single highly focused laser beam, a breakthrough that gave rise to the introduction of the first optical tweezers in 1986 (Ashkin and Dziedzic, 1987) and was later recognized by two Nobel prizes in 1997 and 2018. Since then, optical tweezers have come a long way and have proven to be an invaluable quantitative tool capable of manipulating objects from as small as atoms (Kim et al., 2016, Stuart, Kuhn, 2018, Samoylenko et al., 2020) to as big as 100 μm (Applegate et al., 2010, Jess et al., 2006) with forces spanning three orders of magnitude (0.1–100 pN) (Marton-falvi et al., 2017, He et al., 2019, Schwingel et al, 2013). OTs are inherently designed around microscope setups and hence benefit from advances in imaging systems. This allows for Angstrom-level displacement measurements at sub-millisecond resolutions

Acknowledgments: The authors gratefully acknowledge support from the Canada Research Chairs program (AE) and the McGill Engineering Doctoral Award (PT).

Pouria Tirgar, Allen J. Ehrlicher, Department of Bioengineering, McGill University, Montreal, Canada

https://doi.org/10.1515/9783110640632-009

(Zoldak et al, 2013, Fazal et al, 2011), making OTs a unique tool for quantitative force–displacement measurements.

With the increasingly widespread use of OTs, there has been growing interest among biologists and physicists in quantifying the forces at play in living organisms. For example, OTs have been extensively used to quantify the kinetics of motor proteins, for example, myosin (Liu et al., 2018, Gunther et al., 2020) and dynein (Ohashi et al., 2019, Ezber et al., 2020) or forces involved in the production and manipulation of genetic materials, like transcription (Ishibashi et al., 2014) and DNA stretching (Newton et al., 2019). When it comes to structural proteins, OTs are widely used to investigate forces involved in and generated by the cytoskeleton, like binding forces of actin and its cross-linking proteins (Francis et al., 2019) or intracellular stress caused by traction forces (Wei et al., 2020). All of these applications employ OTs to indirectly measure the force or displacement of a dynamic system. However, OTs can also be used to apply shear stress to a macromolecular network and study the microscale mechanics of soft materials (Chapman et al., 2014) and complex fluids (Paul et al., 2019), also known as microrheology.

Understanding the mechanical properties of soft matter systems has long been a topic of interest for researchers. From colloidal foods (e.g., ketchup; Juszcak et al., 2013) to the cell cytoplasm (Rigato et al., 2017), most of these materials have a nonlinear flow response to applied strain and do not fit the dichotomy of viscous fluids or elastic solids, but are a mixture of both. Depending on the nature of soft matter and the structure of its network, this viscoelastic behavior can drastically change at different size and timescales (Hu et al., 2017). For example, the cell cytoplasm is highly heterogeneous at the microscale and shows a viscoelastic response to small deformations (Berret, 2016). However, this heterogeneity is *"averaged"* and the system behaves like a viscous fluid when strained to higher extents (Puchkov, 2013), jeopardizing the full understanding of the network. These limitations illustrate the importance of systems like OTs in microrheology, which can probe microstructures by applying microscale deformations and detecting forces at different length scales.

Unlike passive microrheology, where thermally driven diffusive movements of microparticles are used to extract rheological data (Mason and Weiz, 1995), active microrheology techniques like OTs employ external forces to push microspheres through the material while tracing their position. This direct control over force–displacement enables OTs to measure the mechanics of soft matter out of thermal equilibrium (e.g., cells) and over a wider range of parameters (e.g., strain and frequency), well beyond those of diffusive forces. This makes OTs a truly unparalleled tool for unraveling the mechanics and forces in biologically complex systems at microscale.

This chapter aims to provide a general overview of OTs and their contribution to our understanding of the biomechanics of soft matter systems. Section 3.3.2 provides an overview of the physics behind OTs and the differences between different size regimes. This is followed by Section 3.3.3, which introduces some of the most common approaches to instrumentation and the use of OTs. Section 3.3.4 outlines the basics of microrheology with OTs, followed by a look at some of the main milestones in incorporating OTs in microrheology of cells in Section 3.3.5. In conclusion, Section 3.3.6 provides a perspective on how OT microrheology can contribute to our understanding of the structure of soft matter systems.

3.3.2 The Power of Light: The Working Principle of Optical Tweezers

The underlying principle of stable optical trapping of dielectric particles lies in the physics of the interaction of light with an object. These interactions can be divided into two main categories, reflection at the surface of the object and refraction, which occurs as the light passes through the object. Each of these interactions affects the energy and/or the direction of the light, thus changing its momentum, given that light carries momentum that is proportional to its speed and its direction of propagation. This momentum exchange is normally described by two distinct regimes, based on the diameter of the illuminated particle (d) and wavelength of light (λ). In one extreme, the dipole model is used when the particles are significantly smaller than the wavelength of light ($d \ll \lambda$), and is known as the Rayleigh regime. In the opposite scenario, when the object is significantly larger than the wavelength of the light ($d \gg \lambda$), the ray optics model is used (Ashkin, 1992, Harada and Asakura, 1996). When the object size is close to the wavelength of the light ($d \sim \lambda$), the more complex generalized Lorenz–Mie theory is required, which is beyond the scope of this text and described elsewhere (Lock, 2004, Lock and Gouesbet, 2009, Gouesbet, 2009).

3.3.2.1 Rayleigh Regime ($d \ll \lambda$)

In the dipole model, particles are usually on the nanometer scale or smaller. Because of this smaller size of the object, its impact on the light beam is very small, which allows ignoring the distinction of scattering components (refraction, reflection, and diffraction). So, the interactions of the object and the light can be treated as an induced dipole in an electromagnetic field.

In this regime close to a small molecule or an atom, the electric field of the light induces a dipole in the object. An induced dipole moment will be attracted toward the higher intensity of the light. This can be shown through the electrical energy applied to the object. The induced dipole moment for a Rayleigh scatterer with polarizability α in an electric field with E is

$$\mu = \alpha \cdot E \qquad\qquad (3.3.1)$$

This object will gain energy (U) in the electric field, that is,

$$U = -\frac{1}{2} \cdot \mu \cdot E = -\frac{1}{2} \cdot (\alpha \cdot E) \cdot E = -\frac{\alpha}{2} \cdot E^2 = -\frac{\alpha}{2} \cdot I \qquad (3.3.2)$$

where I is the intensity of the light if it is homogeneous. This equation also reveals an important fact: the energy of the dipole in the field (E) is proportional to the intensity of the light (I). Thus, if the electromagnetic field of the light is not homogeneous, the object will experience a force known as gradient force or Lorentz's force that pulls the object toward higher light intensities:

$$F_{\text{gradient}} = -\nabla \cdot E = -\nabla \cdot \left(-\frac{\alpha}{2} \cdot I\right) = \frac{\alpha}{2} \cdot \nabla I \qquad (3.3.3)$$

If a collimated and nonhomogeneous beam collides with an object much smaller than its wavelength, the object will experience a force toward the region with the highest intensity. Given the structure of a focused Gaussian beam, the intensity increases toward the center of the beam, and thus the direction of the intensity gradient (and that of the F_{gradient}) shown in eq. (3.3.3) will be toward the center of the OT. It is also noteworthy that in eq. (3.3.3) the force applied to the particle is a function of light intensity gradient, so the higher the intensity gradient the higher the trapping force. Thus, it is necessary to use the highest possible intensity gradient, forming the basis of using high numerical aperture (NA) in OTs.

The gradient force is not the only force applied to the object by light; the scattering of the light applies another force to the particle, referred to as recoil force. As shown in Figure 3.3.1, the momentum of the light before hitting the object is different than the one after the collision. As a result, the particle has to gain a momentum equal and opposite to this difference to conserve the total momentum of the system. This change of momentum generates a force that, according to Newton's second law, is equal to the rate of this change over time:

$$F = -\frac{dp}{dt} \qquad (3.3.4)$$

This allows one to calculate the scattering force applied to a small particle by an array of photons. In a beam of N photons, the momentum of each photon (p_{photon}) equals to

$$p_{\text{photon}} = \frac{h}{\lambda} \qquad (3.3.5)$$

where h is the Planck's constant and λ is the wavelength of the light. So, the momentum of the beam equals to

$$p = N \cdot p_{\text{photon}} = \frac{Nh}{\lambda} \qquad (3.3.6)$$

Equation (3.3.6) shows the total initial momentum of the beam. Depending on the specific condition of the collision, the change in the momentum of the beam after hitting the object can range from 0 (photons keep their average trajectory) to $2p$ (all photons are reflected). So, the secondary momentum of the beam can be written as

$$p = \varepsilon \cdot \frac{Nh}{\lambda} = \frac{\varepsilon Nh}{\lambda} \qquad (3.3.7)$$

where ε is the efficiency factor ranging from 0 to 1, considering the portion of the beams whose momentum is absorbed by the object. Combining eqs. (3.3.4) and (3.3.7), the scattering force applied to the object can be written as follows:

$$F_{\text{Scattering}} = -\frac{d}{dt}\left(\frac{\varepsilon Nh}{\lambda}\right) \qquad (3.3.8)$$

Replacing the wavelength of the light (λ) with the frequency (v) results in

$$F_{\text{Scattering}} = -\frac{d}{dt}\left(\frac{\varepsilon Nhv}{c}n_m\right) = -\frac{d}{dt}\left(\varepsilon \frac{Nhv}{c}n_m\right) = -\varepsilon \cdot \frac{n_m}{c} \cdot \frac{d}{dt}(Nhv) \qquad (3.3.9)$$

where c is the speed of the light and n_m is the refractive index of the medium. Moreover, the value of $(d/dt)(Nhv)$ in eq. (3.3.9) is the power of the light beam (w). So, the force applied to the object through the scattering of the light is

$$F_{\text{Scattering}} = -\varepsilon \cdot w \cdot \frac{n_m}{c} = -\frac{\varepsilon w n_m}{c} \qquad (3.3.10)$$

As shown in Figure 3.3.1A, if the laser beam is not tightly focused, the component of the gradient force toward the objective is negligible. As a result, while the gradient force is still toward the highest intensity (i.e., the axis), the bead will be pushed away by the scattering force in the direction of light propagation. This can be resolved by either using two counter-propagating beams to balance axial scattering forces or by using a more tightly focused laser beam as shown in Figure 5.1B. A highly focused beam has a higher curvature beam that generates a large enough force component toward the objective to cancel the scattering force and restore the particle in a three-dimensional trap.

3.3.2.2 Mie or Ray Optics Regime ($d \gg \lambda$)

When the size of the particle is much larger than the wavelength of the light, a single beam can be tracked as it interacts with and passes through the object. In general, the refractive index of the object should be larger than its surrounding medium for successful trapping, commonly with a ratio of ~1.1 to 1.3. Trapping might still be possible if the ratio is smaller than 1 but large enough that allows for diffraction effects

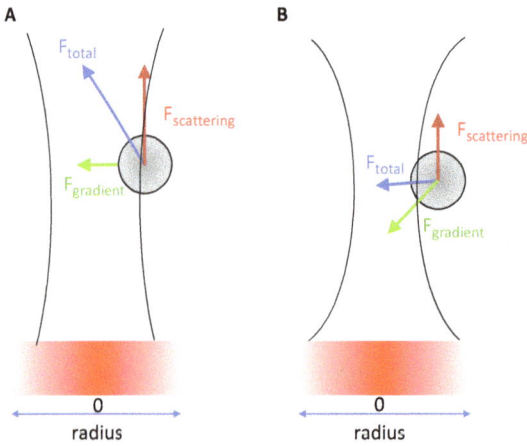

Figure 3.3.1: The force profile in (A) slightly diverging and (B) tightly focused laser beams when the object is smaller than the wavelength of the light (Rayleigh regime).

to be neglected (Huf et al, 2015, Ambrosio et al, 2010), which is assumed for the rest of this discussion.

As shown in Figure 3.3.2a, when a beam of light passes through an object and refracts, it bends and changes its direction, which in turn changes its momentum.

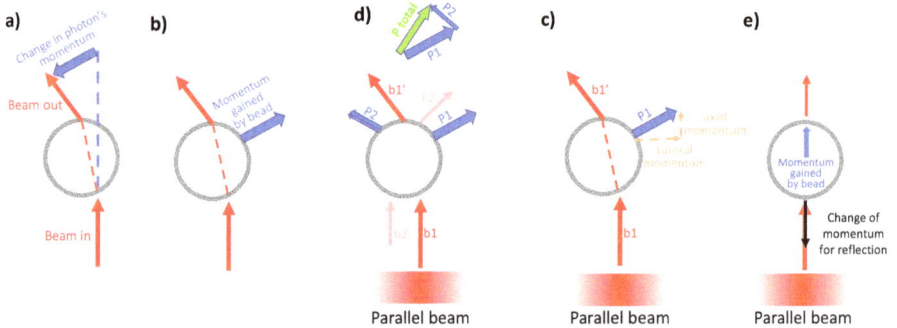

Figure 3.3.2: The schematic of force–momentum exchanges in an optical trap in the ray regime, where objects are bigger than the wavelength of the light.

This change of momentum requires the bead to undergo an equal but opposite momentum change for the system to conserve the total momentum according to Newton's third law (Figure 3.3.2b). This momentum transfer applies a reacting force to the object that the beam traveled through, providing a means to push an object by the sole act of shining a beam of the laser through it. However, to create a stable point, there should be forces applied to the bead from both sides and a net-zero force at the "trap" point.

This can be achieved by using a gradient intensity profile instead of a single beam, with the highest intensity in the center. As shown in Figure 3.3.2c, when the object locates away from the center, though the low-intensity beams from the nearest edge push it further away from the center, the high-intensity beams from the center apply a higher center-oriented momentum and pull the bead to the trap. However, the force applied to the bead has two components: a lateral component and an upward axial component (Figure 3.3.2d). While the lateral component allows for precise positioning of the object in the *XY*-plane, the axial force keeps pushing the bead away from the light source. Even when the bead is centered and the beam passes without any refraction (Figure 3.3.2e), it still loses some momentum due to reflecting on the surface of the bead, and hence the bead will still experience an upward force due to the lost momentum.

This axial movement was the basis of the introduction of dual-beam traps by Ashkin (1970), where two beams from opposite directions applied axial forces in opposite directions and create a stable trap. However, this almost doubled the optical components used in the system and made the system significantly more complicated. So, in 1986, Ashkin et al. (1986) introduced a newer version of the OT, which is schematically shown in Figure 3.3.3. Here, the light is highly focused using a high NA objective. This structure divides the axial space into two sections: before and after the focal point. If the object is below the focal point, it will experience a similar condition to parallel beams and will be pushed toward the focal point (Figure 3.3.3a). However, if it goes above the focal point (Figure 3.3.3b), it will be pulled back toward the objective because of the gradient in *Z*. This results in a stable trap position at the focal point of the objective in 3D space and necessitates the use of high NA objectives in optical tweezers.

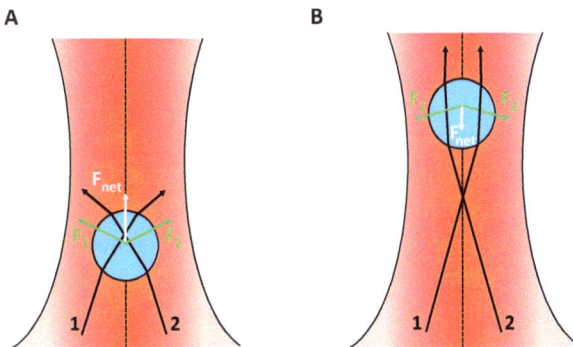

Figure 3.3.3: Lateral and axial forces on a trapped particle in a tightly focused beam (A) before and (B) after the point of focus. Before the focus point, the object is pushed away by refracted rays, while refracted rays after the focus create a backward gradient force toward the focus.

3.3.3 OT Instrumentation, Components, and Variations

Given the wide range of applications, there are many variations of OTs with different configurations and hence capabilities. This section is an overview of OTs setup/embodiments, main components, and their common variations regarding their use in microrheology. For further information on the basics of building and calibrating OTs, readers are referred to excellent previously published works (Molloy and Padgett, 2002, Neuman et al, 2008, Pesce et al., 2015).

Despite the increase in commercially available options, optical tweezers are generally very expensive, custom-built instruments. However, for many applications, OTs can be built around a commercial optical microscope as the basic backbone with extra components and modifications (Lee et al., 2007). This has made inverted microscopes the most common platform for OTs. Figure 3.3.4 shows a basic schematic of an OT setup built around an inverted microscope. In the most basic structure, a collimated beam of laser is added to the microscope that creates the trap in the image plane as a diffraction-limited focused spot. As mentioned before, the need for a steep intensity gradient necessitates the use of a high NA objective lens that can be used for both creating the OT and imaging the sample. The laser beam should be expanded to fill the back focal plane of this objective, which can be done with setups like the Keplerian telescope shown in Figure 3.3.4 (Hernandez et al, 2013). Most light microscopes can accommodate these modifications without compromising any of their imaging capacity, which enables OTs to be combined with most traditional microscopy modes from bright and dark fields to confocal and super-resolution imaging (Ma et al., 2019, Heller et al., 2013). The rest of this section provides an overview of the variations in these components and the considerations for their use in OTs.

3.3.3.1 Optical Components

As described in Section 3.3.2, the laser source is the key element of an OT system and it should meet at least three main criteria to be used for quantitative measurements. Firstly, the Gaussian intensity profile of the laser allows for generation of strong OTs in small, focused spots. Secondly, the laser source used in OTs should have superior power amplitude stability with minimal fluctuations. Power, and hence intensity, fluctuations directly translate to force fluctuation and can be a strong source of noise in measurements. And lastly, the laser source should provide high pointing stability, as if the position of the beam remains stable to prevent any unwanted trap movement and noise in displacement data. Besides these physical requirements, there are also concerns of structural damage when OTs are used with delicate or biological samples that are extremely wavelength sensitive.

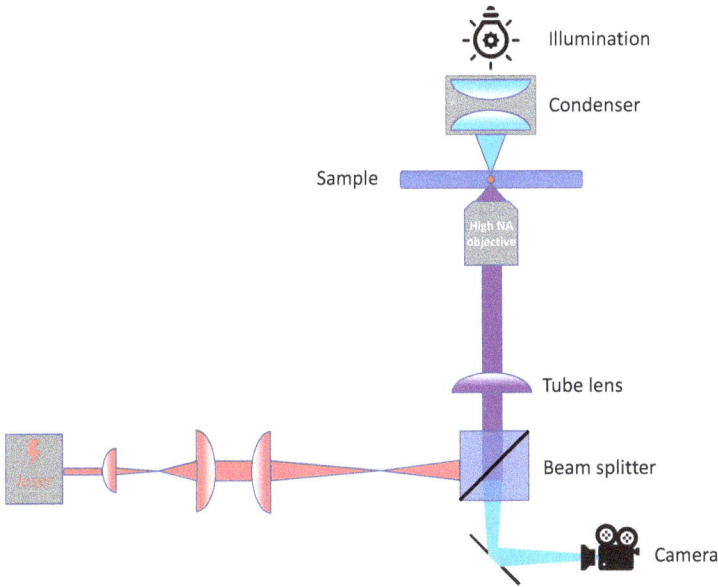

Figure 3.3.4: Schematic design of a simple optical tweezer system installed around an inverted microscope.

The light source used in the first designs by Ashkin was an argon laser at 514.5 nm. However, when OT applications were expanded to handling biological samples, it became evident that the blue light damages the trapped sample. This was due to the high absorption ratio of water, and proteins, in the lower end of the visible spectrum. It was later shown that the absorption ratio of biological samples decreases as the wavelength increases from the ultraviolet regime, showing a minimum absorption in the near-infrared (IR) region around 800–1,100 nm and again increases for higher wavelengths (Blázquez-Castro et al., 2019). This prompted Ashkin et al. (1987) to employ IR light sources in 1987, making diode-pumped neodymium yttrium aluminum garnet (Nd:YAG) emitting at 1,064 nm the dominant laser source for biological applications for the following decades. These lasers are highly efficient, which reduces the amount of heat transmitted to laser housing and optics and reduces thermal drift. Other more recent options include diode lasers that can still provide suitable power at relatively lower costs and with a longer lifetime (Nasim and Jamil, 2014).

Another critical component of the optical path in OTs is the objective. As discussed before, generating a stable OT requires a steep focus of the laser on the image plane. This necessitates the use of high NA (1.2 or higher) objectives. It is noteworthy that the performance of the objective also depends on its immersion solution (water or oil) and its transmission at the trapping wavelength. However, most high NA objectives are optimized for visible light applications. So, the transmission properties of objectives should be checked before using them with near-IR laser sources.

3.3.3.2 Light-Steering Mechanisms to Control the Trap Position

The fundamental need of any rheology, micro or bulk, is the ability to apply controlled stress or strain to a sample. In the case of OTs, this translates to moving the trapped bead in the medium. The most basic way to displace the trap is to move the stage and keep the objective fixed. Of course, the simplicity of this approach comes at the cost of precision. The backlash of gears in most motorized stages challenges their displacement reproducibility. This problem can be addressed by piezoelectric stages that are fast and more accurate. However, piezo-controlled stages are mostly limited in their range of movement and cannot be used when the trap needs to move across the field of view. Additionally, in many cases moving the whole sample is not ideal and might lead to disturbing or changing the properties of the sample. So, OTs often use optical approaches to spatially control the trap position.

The most common optical approach to move the trap position across the field of view is to change the angle of incidence of the laser beam in the back focal plane of the objective lens. The simplest approach to do so is to use an adjustable mirror mount in the light path. Tilting the mirror will change the beam angle on the back aperture and hence laterally moves the trap position (Lee et al., 2007). The mirror can be controlled manually or automatically using motors or piezo-actuators. Despite its simplicity, when there is a need for more than one trap it should be either through the addition of extra lasers or splitting and recombining a single beam (Fällman and Axner, 2017). Hence, systems with more than two traps mostly use other light control mechanisms like galvanometers.

Galvanometers are among the fastest mechanical beam scanners and allow for the beam to be rapidly scanned over the field of view. This enables the system to operate multiple traps in a time-sharing approach, so if the travel time is less than the time needed by objects to diffuse out of the trap, multiple objects can be trapped at the same time (Sasaki et al., 1991). However, the scanning speed of galvanometers is limited due to the mass of the scanning mirror and mounts. This limits this technique to relatively large particles and high viscosity medium, giving rise to nonmechanical beam scanners for multitrap systems (Mio et al., 2000). The two most common nonmechanical approaches are scanning the laser beam between different positions (time-multiplexing) using acousto-optic deflectors (AODs) or using dynamically controlled diffractive elements like liquid crystal spatial light modulators (LC-SLMs).

Based on a single solid-state scanner, AODs offer stable and fast beam scanning without the drawbacks of mechanical scanners like the loss of alignment because of bearing wear (Friedman et al., 2000). AODs are commonly made of an acousto-optic crystal e.g., paratellurite (Antonov, 2019) or TeO_2 (Wallin et al., 2011) connected to a piezoelectric transducer on one end and attached to an acoustic absorber on the other end. An RF source drives the piezo-transducer and induces sound waves that propagate through the crystal, which in turn creates regions with different refractive indices on the crystal. This diffraction grating can deflect an incident laser beam to different degrees based on the frequency of the wave, allowing the laser beam to be moved sequentially between hundreds of traps with time multiplexing (Enders et al, 2016). In this case, the trap visits multiple positions frequently enough for trapped objects not to have time to diffuse away. The main drawback of this configuration is that the traps are not permanent in time and particles experience fluctuations in trapping force. Moreover, since each AOD can create a single-axis diffraction grating, two orthogonally aligned AODs are commonly used to create traps in the 2D plane (Neuman and Block, 2004).

Unlike AODs, SLMs are pixelated devices based on LC where the refractive index of each pixel can be independently controlled with the voltage applied to it. When a laser beam is illuminated on the SLM, the phase front of the reflected beam is shifted based on the refractive index of each pixel of the SLM (Gibson, 2016). This allows SLMs to split a single beam and create multiple traps simultaneously. Unlike mirror and AODs, which use time-sharing to create pseudo-simultaneous multiple traps, SLMs create temporally constant traps by illuminating all traps at the same time instead of moving the beam between the traps (Grier, 2003). Another significant advantage of SLMs over other methods is their ability to individually control and move each trap in the 3D space. This allows a whole new level of configurations like creating traps with arbitrary 3D configurations (Liesener et al., 2000) or measuring multiple samples at the same time (Mejean et al., 2009), which were impossible using previously discussed methods and gave rise to a new approach called holographic optical tweezers (HOTs).

First introduced in 1998 (Dufresne and Grier, 1998), HOTs use computer-generated holograms to create multiple tightly focused spots in 3D space from a single beam of laser. Soon after their invention, it was shown that multiple holograms can be combined to create multiple simultaneous traps (Grier, 2003). This unique advantage of HOTs turned them into a widely used technique for multitrap systems for the past three decades. However, despite their merits, HOTs mostly have high computational demands, which increases for applications requiring a large array of traps or closed-loop control (Montes-Usagetui et al, 2006). One of the main sources of calculation demand in HOTs is ghost traps (Hesseling et al., 2011). Since SLMs only control the phase of the beam and not its amplitude, the light intensity might be higher than expected in some parts of the hologram and create unwanted "ghost" traps. While they were inevitable in many cases in the early years, recent advances in hologram and trap modification can significantly minimize and eliminate these traps (Spalding et al., 2008). Such improvements in spatial and temporal control over independent traps paved the way for combining OTs with other techniques and configurations. While the details of such instrumentations are beyond the scope of this work, readers are referred to published works on optical fiber tweezers (Zhao et al., 2020), optical stretchers (Guck et al., 2001), fleezers (Whitley et al., 2017), optical clocks (Madjarov et al., 2019), and vortex-based OTs (Liang et al., 2018, Cheng, Tao, 2016).

3.3.4 Microrheology with optical tweezers

Given the wide range of displacement and frequency in microrheological experiments, there is a wealth of approaches for analysis and interpretation of the force–displacement data collected from OTs that are discussed previously. However, the basis of most methods is using analogous definitions with bulk rheology for easier understanding and also the integration of experiments at micro- and macro-scales. The simplest microrheology experiment is based on the same experimental foundation as the bulk rheology, applying small-amplitude deformations x by oscillating a trapped bead at different frequencies ω. In a calibrated OT system, the force F required for each deformation can be measured and used to generate a force–displacement response of the material at each frequency. Assuming the displacements are small enough to fall in the linear viscoelastic regime, the viscoelastic moduli of a network surrounding a microsphere with radius R can be calculated based on its maximum displacement (X_{max}), its associated force (F_{max}), and the phase shift between the two sine curves (Robertson-Anderson, 2018):

$$G'(\omega) = \frac{F_{max}}{6\pi R x_{max}} \cdot \cos(\Delta\Phi) \tag{3.3.11}$$

$$G''(\omega) = \frac{F_{max}}{6\pi R x_{max}} \cdot \sin(\Delta\Phi) \tag{3.3.12}$$

$$\eta^*(\omega) = \frac{(G'(\omega)^2 + G''(\omega)^2)^{\frac{1}{2}}}{\omega} \tag{3.3.13}$$

This approach has been widely used to study the heterogeneous and anisotropic mechanics of subcellular structures like DNA, microtubules, actin, and intermediate filaments at the microscale (Neckernuss et al., 2015, Guo et al., 2013, Laan et al., 2008, Footer et al., 2007, Gross et al., 2011).

Another advantage of active microrheology with OTs is the ability to study the response to mesoscale displacements that are larger than passive diffusion and smaller than bulk deformations. In this measurement regime, force F is measured upon bead movement of x with a constant velocity of v. For a spherical bead with radius R, these values can be translated to standard rheological terms of stress (σ), strain (γ), and strain rate ($\dot{\gamma}$):

$$\sigma = \frac{F}{\pi R^2} \tag{3.3.14}$$

$$y = \frac{x}{2R} \tag{3.3.15}$$

$$\dot{y} = \frac{3v}{\sqrt{2R}} \tag{3.3.16}$$

One main result of the force–displacement data is the effective differential modulus that can be determined in each direction of X and Y as a measure of elasticity:

$$K(x) = \frac{dF(x)}{dx} \tag{3.3.17}$$

$$K(y) = \frac{dF(y)}{dy} \tag{3.3.18}$$

The effective differential moduli are useful parameters to assess the stiffness of a solid-like structure or the steady-state viscosity of materials that exhibit strain-independent forces at large deformations (Chapman, Robertson-Anderson, 2014). However, it should be noted that the displacement speed can affect these values and complicate the analysis given the time-dependent relaxation seen in many biological systems (Rigato et al., 2017, de Sousa et al., 2020, Nishizawa et al., 2017). One widely investigated and interesting case of such behavior is cell cytoplasm. It is shown that at shorter timescales (<0.1 s), cytoplasmic movements are at equilibrium and governed by thermal fluctuations, while at longer timescales active biological forces are dominant (Gupta et al, 2017). Interestingly, the mechanics of the cytoplasm is not isotropic and is shown using OTs to be a function of morphology (Gupta et al., 2019). Such experiments helped evolving models of cytoplasm that mostly consider it as an elastic meshwork consisting of mostly the cytoskeleton immersed in the viscous cytosol fluid (Moeendarbary et al., 2013). While these models can successfully describe the complex mechanical behavior of the cytoplasm, they contradicted observations that required further investigation of how life interacts with mechanics in the cytoplasm.

3.3.5 Combining Optical Tweezers with Cell-Driven Particle Motion

Optical tweezers offer elusive measurements of the cytoplasm moduli (Guo et al., 2013), a complex viscoelastic yet predominantly solid-like material. Considering that the cytoplasm is solid-like, it is puzzling how organelles larger than the mesh size of the cytoplasmic network appear to diffuse through it. Depleting cells of the intermediate filament increases the mobility of cytoplasmic particles (Guo et al., 2013) but does not explain the driving forces of the particles or organelles in transport. Due to the apparent random diffusive transport of intracellular particles, many previous works have assumed that forces driving movement arise from thermal fluctuations as in classical Brownian motion. In the case where a material is in thermal equilibrium, this concept of solely thermal forces driving movement holds and enables the application of *"passive microrheology"* (Mason et al., 1997). Based on visual similarities, researchers have thus applied these concepts to cells; however, this approach fails as the cells are inherently far from equilibrium. Indeed, this random transport appears to be an active process, as reducing the myosin and ATP activity reduces bead transport in reconstituted systems (Mizuno et al., 2007).

To understand the diverse nonequilibrium forces acting in the cytoplasm, one must infer them from mechanics and displacement data. Force spectrum microscopy (FSM) bridges this gap by combining mechanical measurements of the cytoplasm with quantitative particle transport (Guo et al., 2014). FSM to date has revealed several important aspects of ensemble cellular forces: firstly, the active components of the cellular force spectrum overwhelmingly dominate thermal contributions at timescales longer than approximately 1 s. This clarifies that the transport of objects larger than the ~50 nm cytoplasmic mesh is considered to be exclusively driven by active processes. Similarly to reconstituted systems, when cellular myosin activity is reduced, this transport is also attenuated, and when ATP is depleted, transport ceases. This reveals that the cell is mechanically predominantly an elastic solid; however, it displays internal transport behavior like a fluid and represents a remarkable natural materials' innovation that captures ideal aspects of both phases. Secondly, these forces not only drive large mesh-bound objects but even nanometer-scale objects such as small proteins and molecules which can diffuse through the cytoplasm are accelerated by active processes, yielding approximately a twofold increase in diffusivity due to ATP activity. Thirdly, Guo et al. showed that the force spectrum is an order of magnitude larger in metastatic cells as compared to benign cells, consistent with the perspective of increased metabolic activity during cancer progression (Guo et al., 2014).

Since development, the methodology and concepts of FSM have had a broad-reaching impact in diverse areas of cellular biophysics, from measuring the metabolic changes during metastasis (Yubera et al, 2020), to the fundamental nature of cellular mechanosensing (Wei et al., 2020). Researchers have shown that these active random forces not only increase diffusive transport but also selectively move larger organelles to regions of higher cytoskeletal density (Wolgemuth et al, 2020). These studies demonstrate that these previously neglected stochastic active forces play a central role in cellular mechanobiology.

3.3.6 The Perspective of Optical Tweezers in Mechanobiology

In retrospect over the last decades, few quantitative approaches have had a more transformative impact on our knowledge of biophysics than optical tweezers: from the first single-beam traps to multibeam stretchers that deform cells, to polarized traps that can manipulate not only the position, but the torsion applied to cells and proteins in 3D space. Continued quantum leaps in laser efficiency, cost, and size will continue to open new doors in optical tweezers' design and implementation, enabling measurements that were previously impossible. In particular, combinations of existing approaches, for example, FSM's hybridization of active microrheology and particle tracking, offer the capability to examine active processes in cell mechanics that are otherwise inaccessible. Nevertheless, at least as important as technical advances are the ongoing efforts to theoretically integrate these data into nonequilibrium and reactive models of cellular biophysics.

References

Ambrosio, L. A. and H. E. Hernández-Figueroa (2010). "Fundamentals of negative refractive index optical trapping: Forces and radiation pressures exerted by focused Gaussian beams using the generalized Lorenz-Mie theory." Biomedical Optics Express **1**(5): 1284–1301.

Antonov, S. N. (2019). "Paratellurite-based acoustooptical deflectors. Methods for increasing their efficiency and widening the scanning angle." Instruments and Experimental Techniques **62**(3): 386–392.

Applegate, R. W., D. W. M. Marr, J. Squier and S. W. Graves (2010) "Particle size limits of optical trapping and deflection for sorting using diode laser bars". CLEO/QELS: 2010 Laser Science to Photonic Applications, San Jose, CA, USA, 2010, 1–2. doi: 10.1364/CLEO.2010.CTuJJ7

Ashkin, A., J. M. Dziedzic, J. E. Bjorkholm and S. Chu (1986). "Observation of a single-beam gradient force optical trap for dielectric particles." Opt Angular Momentum **11**(5): 288–290.

Ashkin, A., J. M. Dziedzic and T. Yamane (1987). "Optical trapping and manipulation of single cells using infrared laser beams." Nature **330**. 769–771.

Ashkin, A. and J. M. Dziedzic (1987). "Optical trapping and manipulation of viruses and bacteria." Science **235**(4795): 1517–1520.

Ashkin, A. (1970). "Acceleration and trapping of particles by radiation pressure." Physical Review Letters **24**(4): 156–159.

Ashkin, A. (1992). "Forces of a single-beam gradient laser trap on a dielectric sphere in the ray optics regime." Biophysical Journal **61**(2): 569–582.

Berret, J. F. (2016). "Local viscoelasticity of living cells measured by rotational magnetic spectroscopy." Nature Communications **7**. 10134.

Blázquez-Castro, A. (2019). "Optical tweezers: Phototoxicity and thermal stress in cells and biomolecules." Micromachines **10**(8): 507.

Bui, A. A. M., A. V. Kashchuk, M. A. Balanant, T. A. Nieminen, H. Rubinsztein-Dunlop and A. B. Stilgoe (2018). "Calibration of force detection for arbitrarily shaped particles in optical tweezers." Scientific Reports **8**. 10798.

Bustamante, C., C. Bustamante, L. Alexander, K. Macluba and C. M. Kaiser (2020). "Single-molecule studies of protein folding with optical tweezers." Annual Review of Biochemistry **89**. 443–470.

Chapman, C. D., K. Lee, D. Henze, D. E. Smith and R. M. Robertson-Anderson (2014). "Onset of non-continuum effects in microrheology of entangled polymer solutions." Macromolecules **47**(3): 1181–1186.

Chapman, C. D. and R. M. Robertson-Anderson (2014). "Nonlinear microrheology reveals entanglement-driven molecular-level viscoelasticity of concentrated DNA." Physical Review Letters **113**(9): 098303.

Cheng, S. and S. Tao (2016). "Vortex-based line beam optical tweezers." Journal of Optics **18**(10): 105603.

de Sousa, J. S., R. S. Freire, F. D. Sousa, M. Radmacher, A. F. B. Silva, M. V. Ramos, et al. (2020). "Double power-law viscoelastic relaxation of living cells encodes motility trends." Scientific Reports **10**(1): 4749.

Dufresne, E. R. and D. G. Grier (1998). "Optical tweezer arrays and optical substrates created with diffractive optics." The Review of Scientific Instruments **69**(5): 1974–1977.

Endres, M., H. Bernien, A. Keesling, H. Levine, E. R. Anschuetz, A. Krajenbrink, et al. (2016). "Atom-by-atom assembly of defect-free one-dimensional cold atom arrays." Science **354**(6315): 1024–1027.

Ezber, Y., V. Belyy, S. Can and A. Yildiz (2020). "Dynein harnesses active fluctuations of microtubules for faster movement." Nature Physics **16**(3): 312–316.

Fällman, E. and O. Axner (2017). "Design for fully steerable dual-trap optical tweezers." Applied Optics **36**(10): 2107–2113.

Fazal, F. M. and S. M. Block (2011). "Optical tweezers study life under tension." Nature Photonics **5**(6): 318–321.

Footer, M. J., J. W. J. Kerssemakers, J. A. Theriot and M. Dogterom (2007). "Direct measurement of force generation by actin filament polymerization using an optical trap." Proceedings of the National Academy of Sciences of the United States of America **104**(7): 2181–2186.

Francis, M. L., S. N. Ricketts, L. Farhadi, M. J. Rust, M. Das, J. L. Ross, et al. (2019). "Non-monotonic dependence of stiffness on actin crosslinking in cytoskeleton composites." Soft Matter **15**(44): 9056–9065.

Friedman, N., A. Kaplan and N. Davidson (2000). "Acousto-optic scanning system with very fast nonlinear scans." Optics Letters **25**(24): 1762–1764.

Gibson, G. M. (2016). "Optical tweezers configurations." In Microrheology with optical tweezers: Principles and applications, vol. 4, Jenny Stanford Publishing, 103–133.

Grier, D. G. (2003). "A revolution in optical manipulation." Nature **424**(6950): 810–816.

Gross, P., N. Laurens, L. B. Oddershede, U. Bockelmann, E. J. G. Peterman and G. J. L. Wuite (2011). "Quantifying how DNA stretches, melts and changes twist under tension." Nature Physics **7**(9): 731–736.

Guck, J., R. Ananthakrishnan, H. Mahmood, T. J. Moon, C. C. Cunningham and J. Käs (2001). "The optical stretcher: A novel laser tool to micromanipulate cells." Biophysical Journal **81**(2): 767–784.

Guo, M., A. J. Ehrlicher, S. Mahammad, H. Fabich, M. H. Jensen, J. R. Moore, et al. (2013). "The role of vimentin intermediate filaments in cortical and cytoplasmic mechanics." Biophysical Journal **105**(7): 1562–1568.

Guo, M., A. J. Ehrlicher, M. H. Jensen, M. Renz, J. R. Moore, R. D. Goldman, et al. (2014). "Probing the stochastic, motor-driven properties of the cytoplasm using force spectrum microscopy." Cell **158**(4): 822–832.

Gupta, S. K. and M. Guo (2017). "Equilibrium and out-of-equilibrium mechanics of living mammalian cytoplasm." Journal of the Mechanics and Physics of Solids **107**. 284–293.

Gupta, S. K., Y. Li and M. Guo (2019). "Anisotropic mechanics and dynamics of a living mammalian cytoplasm." Soft Matter **15**(2): 190–199.

Gouesbet, G. (2009). "Generalized Lorenz-Mie theories, the third decade: A perspective." Journal of Quantitative Spectroscopy & Radiative Transfer **110**(14–16): 1223–1238.

Gunther, L. K., J. A. Rohde, W. Tang, J. A. Cirilo, C. P. Marang, B. D. Scott, et al. (2020). "FRET and optical trapping reveal mechanisms of actin-activation of the power stroke and phosphate-release in myosin V." The Journal of Biological Chemistry **295**(51): 17383–17397.

Harada, Y. and T. Asakura (1996). "Radiation forces on a dielectric sphere in the Rayleigh scattering regime." Optics Communications **124**(5–6): 529–541.

He, C., S. Li, X. Gao, A. Xiao, C. Hu, X. Hu, et al. (2019). "Direct observation of the fast and robust folding of a slipknotted protein by optical tweezers." Nanoscale **11**(9): 4101–4107.

Heller, I., G. Sitters, O. D. Broekmans, G. Farge, C. Menges, W. Wende, et al. (2013). "STED nanoscopy combined with optical tweezers reveals protein dynamics on densely covered DNA." Nature Methods **10**(9): 910–916.

Hernández Candia, C. N., S. Tafoya Martínez and B. Gutiérrez-Medina (2013). "A minimal optical trapping and imaging microscopy system." PLoS One **8**(2): e57383.

Hesseling, C., M. Woerdemann, A. Hermerschmidt and C. Denz (2011). "Controlling ghost traps in holographic optical tweezers." Optics Letters **36**(18): 3657–3659.

Hu, J., S. Jafari, Y. Han, A. J. Grodzinsky, S. Cai and M. Guo (2017). "Size- and speed-dependent mechanical behavior in living mammalian cytoplasm." Proceedings of the National Academy of Sciences of the United States of America **114**(36): 9529–9534.

Huff, A., C. N. Melton, L. S. Hirst and J. E. Sharping (2015). "Stability and instability for low refractive-index-contrast particle trapping in a dual-beam optical trap." Biomedical Optics Express **6**(10): 3812.

Ishibashi, T., M. Dangkulwanich, Y. Coello, T. A. Lionberger, L. Lubkowska, A. S. Ponticelli, et al. (2014). "Transcription factors IIS and IIF enhance transcription efficiency by differentially modifying RNA polymerase pausing dynamics." Proceedings of the National Academy of Sciences of the United States of America **111**(9): 3419–3424.

Jess, P. R. T., V. Garcés-Chávez, D. Smith, M. Mazilu, L. Paterson, A. Riches, et al. (2006). "Dual beam fibre trap for Raman micro-spectroscopy of single cells." Optics Express **14**(12): 5779–5791.

Juszczak, L., Z. Oczadły and D. Gałkowska (2013). "Effect of modified starches on rheological properties of ketchup." Food and Bioprocess Technology **6**(5): 1251–1260.

Killian, J. L., F. Ye and M. D. Wang (2018). "Optical tweezers: A force to be reckoned with." Cell **175**(6): 1445–1448.

Kim, H., W. Lee, H. G. Lee, H. Jo, Y. Song and J. Ahn (2016). "In situ single-atom array synthesis using dynamic holographic optical tweezers." Nature Communications **7**. 13317.

Laan, L., J. Husson, E. L. Munteanu, J. W. J. Kerssemakers and M. Dogterom (2008). "Force-generation and dynamic instability of microtubule bundles." Proceedings of the National Academy of Sciences of the United States of America **105**(26): 8920–8925.

Lee, W. M., P. J. Reece, R. F. Marchington, N. K. Metzger and K. Dholakia (2007). "Construction and calibration of an optical trap on a fluorescence optical microscope." Nature Protocols **2**(12): 3226–3238.

Lee, H., J. M. Ferrer, F. Nakamura, M. J. Lang and R. D. Kamm (2010). "Passive and active microrheology for cross-linked F-actin networks in vitro." Acta Biomaterialia **6**(4): 1207–1218.

Liang, Y., Y. Cai, Z. Wang, M. Lei, Z. Cao, Y. Wang, et al. (2018). "Aberration correction in holographic optical tweezers using a high-order optical vortex: Publisher's note." Applied Optics **57**(13): 3618–3623.

Liesener, J., M. Reicherter, T. Haist and H. J. Tiziani (2000). "Multi-functional optical tweezers using computer-generated holograms." Optics Communications **185**(1–3): 77–82.

Liu, C., M. Kawana, D. Song, K. M. Ruppel and J. A. Spudich (2018). "Controlling load-dependent kinetics of β-cardiac myosin at the single-molecule level." Nature Structural & Molecular Biology **25**(6): 505–514.

Liu, W. and C. Wu (2018). "Rheological study of soft matters: A review of microrheology and microrheometers." Macromolecular Chemistry and Physics **219**. 1700307.

Lock, J. A. (2004). "Calculation of the radiation trapping force for laser tweezers by use of generalized Lorenz-Mie theory. II. On-axis trapping force." Applied Optics **43**(12): 2545–2554.

Lock, J. A. and G. Gouesbet (2009). "Generalized Lorenz-Mie theory and applications." Journal of Quantitative Spectroscopy & Radiative Transfer **110**(11): 800–807.

Ma, G., C. Hu, S. Li, X. Gao, H. Li and X. Hu (2019). "Simultaneous, hybrid single-molecule method by optical tweezers and fluorescence." Nami Jishu Yu Jingmi Gongcheng/Nanotechnology Precision Engineering **2**(4): 145–156.

Madjarov, I. S., A. Cooper, A. L. Shaw, J. P. Covey, V. Schkolnik, T. H. Yoon, et al. (2019). "An atomic-array optical clock with single-atom readout." Physical Review X **91**(4): 41052.

Martonfalvi, Z., P. Bianco, K. Naftz, G. G. Ferenczy and M. Kellermayer (2017). "Force generation by titin folding." Protein Science **26**(7): 1380–1390.

Mason, T. G. and D. A. Weitz (1995). "Optical measurements of frequency-dependent linear viscoelastic moduli of complex fluids." Physical Review Letters **74**(7): 1250–1253.

Mason, T. G., K. Ganesan, J. H. Van Zanten, D. Wirtz and S. C. Kuo (1997). "Particle tracking microrheology of complex fluids." Physical Review Letters **79**(17): 3282–3285.

Mejean, C. O., A. W. Schaefer, E. A. Millman, P. Forscher and E. R. Dufresne (2009). "Multiplexed force measurements on live cells with holographic optical tweezers." Optics Express **17**(8): 6209–6217.

Mio, C., T. Gong, A. Terray and D. W. M. Marr (2000). "Design of a scanning laser optical trap for multiparticle manipulation." The Review of Scientific Instruments **71**(5): 2196–2200.

Mizuno, D., C. Tardin, C. F. Schmidt and F. C. MacKintosh (2007). "Nonequilibrium mechanics of active cytoskeletal networks." Science **315**(5810): 370–373.

Moeendarbary, E., L. Valon, M. Fritzsche, A. R. Harris, D. A. Moulding, A. J. Thrasher, et al. (2013). "The cytoplasm of living cells behaves as a poroelastic material." Nature Materials **12**(3): 253–261.

Molloy, J. E. and M. J. Padgett (2002). "Lights, action: Optical tweezers." Contemporary Physics **43**(4): 241–258.

Montes-Usategui, M., E. Pleguezuelos, J. Andilla, E. Martín-Badosa and I. Juvells (2006) "Algorithm for computing holographic optical tweezers at video rates". Proc. SPIE 6326, Optical Trapping and Optical Micromanipulation III, 63262X.

Nasim, H. and Y. Jamil (2014). "Diode lasers: From laboratory to industry." Optics and Laser Technology **56**. 211–222.

Neckernuss, T., L. K. Mertens, I. Martin, T. Paust, M. Beil and O. Marti (2015). "Active microrheology with optical tweezers: A versatile tool to investigate anisotropies in intermediate filament networks." Journal of Physics D: Applied Physics **49**(4): 045401.

Neuman, K. C. and S. M. Block (2004). "Optical trapping." The Review of Scientific Instruments **75**(9): 2787–2809.

Neuman, K. C. and A. Nagy (2008). "Single-molecule force spectroscopy: Optical tweezers, magnetic tweezers and atomic force microscopy." Nature Methods **5**(6): 491–505.

Newton, M. D., B. J. Taylor, R. P. C. Driessen, L. Roos, N. Cvetesic, S. Allyjaun, et al. (2019). "DNA stretching induces Cas9 off-target activity." Nature Structural & Molecular Biology **26**(3): 185–192.

Nishizawa, K., M. Bremerich, H. Ayade, C. F. Schmidt, T. Ariga and D. Mizuno (2017). "Feedback-tracking microrheology in living cells." Science Advances **3**(9): e1700318.

Ohashi, K. G., L. Han, B. Mentley, J. Wang, J. Fricks and W. O. Hancock (2019). "Load-dependent detachment kinetics plays a key role in bidirectional cargo transport by kinesin and dynein." Traffic **20**(4): 284–294.

Paul, S., A. Kundu and A. Banerjee (2019). "Single-shot phase-sensitive wideband active microrheology of viscoelastic fluids using pulse-scanned optical tweezers." Journal of Physics: Condensed Matter **31**. 504001.

Pesce, G., G. Volpe, O. M. Maragó, P. H. Jones, S. Gigan, A. Sasso, et al. (2015). "Step-by-step guide to the realization of advanced optical tweezers." Journal of the Optical Society of America B **32**(5): B84.

Puchkov, E. O. (2013). "Intracellular viscosity: Methods of measurement and role in metabolism." Biochemistry Supplement Series A: Membrane and Cell Biology **7**(4): 270–279.

Rich, J. P., J. Lammerding, G. H. McKinley and P. S. Doyle (2011). "Nonlinear microrheology of an aging, yield stress fluid using magnetic tweezers." Soft Matter **7**(21): 9933–9943.

Rigato, A., A. Miyagi, S. Scheuring and F. Rico (2017). "High-frequency microrheology reveals cytoskeleton dynamics in living cells." Nature Physics **13**(8): 771–775.

Robertson-Anderson, R. M. (2018). "Optical tweezers microrheology: From the basics to advanced techniques and applications." ACS Macro Letters **7**(8): 968–975.

Samoylenko, S. R., A. V. Lisitsin, D. Schepanovich, I. B. Bobrov, S. S. Straupe and S. P. Kulik (2020). "Single atom movement with dynamic holographic optical tweezer." Laser Physics Letters **17**. 025203.

Sasaki, K., M. Koshioka, H. Misawa, N. Kitamura and H. Masuhara (1991). "Laser-scanning micromanipulation and spatial patterning of fine particles." Japanese Journal of Applied Physics **30**(5): L907–L909.

Schwingel, M. and M. Bastmeyer (2013). "Force mapping during the formation and maturation of cell adhesion sites with multiple optical tweezers." PLoS One **8**(1): e54850.

Spalding, G. C., J. Courtial and R. Di Leonardo (2008). "Holographic optical tweezers." In Structured light and its applications. Cambridge, Academic Press, chapter 6, 139–168.

Stuart, D. and A. Kuhn (2018). "Single-atom trapping and transport in DMD-controlled optical tweezers." New Journal of Physics **20**. 023013.

Wallin, A. E., H. Ojala, G. Ziedaite and E. Hggstrm (2011). "Dual-trap optical tweezers with real-time force clamp control." The Review of Scientific Instruments **82**. 083102.

Wei, M. T., S. S. Jedlicka and H. D. Ou-Yang (2020). "Intracellular nonequilibrium fluctuating stresses indicate how nonlinear cellular mechanical properties adapt to microenvironmental rigidity." Scientific Reports **10**(1): 5902.

Whitley, K. D., M. J. Comstock and Y. R. Chemla (2017). "High-resolution "fleezers": Dual-trap optical tweezers combined with single-molecule fluorescence detection." Methods in Molecular Biology **1486**. 183–256.

Wolgemuth, C. W. and S. X. Sun (2020). "Active random forces can drive differential cellular positioning and enhance motor-driven transport." Molecular Biology of the Cell **31**(20): 2283–2288.

Wu, Y., D. Sun and W. Huang (2011). "Mechanical force characterization in manipulating live cells with optical tweezers." Journal of Biomechanics **44**(4): 741–746.

Verdier, C., Y. Abidine, V. Laurent, A. Duperray, C. Verdier, Y. Abidine, et al. (2019). "Cancer cell microrheology using AFM." Computer Methods in Biomechanics and Biomedical Engineering **22**(sup1): S260–S261.

Vos, B. E., L. C. Liebrand, M. Vahabi, A. Biebricher, G. J. L. Wuite, E. J. G. Peterman, et al. (2017). "Programming the mechanics of cohesive fiber networks by compression." Soft Matter **13**(47): 8886–8893.

Yubero, M. L., P. M. Kosaka, Á. San Paulo, M. Malumbres, M. Calleja and J. Tamayo (2020). "Effects of energy metabolism on the mechanical properties of breast cancer cells." Communications Biology **3**(1): 590.

Zhao, X., N. Zhao, Y. Shi, H. Xin and B. Li (2020). "Optical fiber tweezers: A versatile tool for optical trapping and manipulation." Micromachines **11**(2): 114.

Žoldák, G., J. Stigler, B. Pelz, H. Li and M. Rief (2013). "Ultrafast folding kinetics and cooperativity of villin headpiece in single-molecule force spectroscopy." Proceedings of the National Academy of Sciences of the United States of America **110**(45): 18156–18161.

Peng Li, Anita Wdowicz, Conor Fields, Gil Lee

3.4 Magnetic Tweezers: From Molecules to Cells

3.4.1 Introduction

Magnetic tweezers (MT) is a technique similar to atomic force microscopy (AFM) or optical tweezers, where a force is transduced to a micron-sized object, while its position is detected with an optical sensor. In MT, force is applied to a superparamagnetic (SPM) or ferromagnetic microparticle with a magnetic field created by one or more electro- or permanent magnets.[1] Unlike AFM, the magnetic particle is not trapped in a 3D space by the magnetic field. As a result, it cannot be manipulated arbitrarily. Despite this limitation, MT remains a powerful tool that has played a key role in our understanding of single-molecule mechanics (Koster et al., 2005, Strick et al., 2000) and mechanobiology (Ingber et al., 2014, Roca-Cusachs et al., 2013, Del Rio et al., 2009, Wang et al., 2009, Dichtl and Sackmann, 2002, Wang et al., 1993a). We believe that MTs will continue to play an important role in our understanding of the biomechanics of disease, due to its proven capacity to rapidly screen cellular responses to pico-Newton-scale forces (Kilinc et al., 2015, 2014, Shang and Lee, 2007, Lee et al., 2000) as well as its potential application in vivo.

The ability to use magnetic particles to study the biomechanical properties of cells was initially recognized almost a century ago (Heilbornn, 1922). Concurrent advances in magnetic materials, alongside microscopy, have permitted control over the force, and detection of the position of a microscopic magnetic particle with a precision exceeding 0.01 pN and 1 nm, respectively (Lionnet et al., 2012b, c, Mosconi et al., 2011, Neuman and Nagy, 2008, Kilinc and Lee, 2014). Nowadays, pico-Newton-scale forces can easily be applied to micron-sized SPM particles bound to a surface in MT. Moreover, several groups have developed magnetic systems also capable of applying torque. The precision with which the force can be applied to these particles is defined by the form of the magnetic field, and modeled biomolecular properties are being used as internal controls (Kilinc et al., 2016). One limitation of the MT approach is its restriction to a maximum force of approximately 100 pN for a micron-sized SPM particle (see further). Due to their strong optical signature, optical microscopy has been used predominantly to track microparticles in

Acknowledgment: This work was supported by the Science Foundation of Ireland Grants 15/AI/3127, 08/RP1/B1376, and 08/IN1/B207.

Peng Li, Anita Wdowicz, Conor Fields, Gil Lee, School of Chemistry and Conway Institute for Biomolecular and Biomedical Science, University College Dublin, Republic of Ireland

https://doi.org/10.1515/9783110640632-010

MT. Using optical interference or fluorescence microscopy, the position of these particles relative to a glass surface can be determined with sub-nanometer precision (Kilinc and Lee, 2014, Lionnet et al., 2012c). Ultimately, the high force and displacement precision of MT makes it a powerful tool for understanding the biomechanical properties of single molecules and cells.

MT has played a crucial role in pioneering biomechanic measurements, from single molecules to cellular systems. Weak protein–protein intermolecular interactions may be studied using MT using the force clamp mode, where the lifetime of a bond is measured under a constant force by tracking the position of the magnetic microparticles (Shang and Lee, 2007, Kilinc et al., 2012). Furthermore, MTs have been used to define the mechanical properties of double-stranded DNA and protein–DNA interactions, as both tensile and torsional forces can be applied to macromolecules assembled between bead and surface (Koster et al., 2005, Strick et al., 2000). Single-molecule interactions have been studied on the surface of cells to reveal the biomechanics of cellular adhesion (Wang et al., 1993b, Roca-Cusachs et al., 2013). MT has also been used to define the viscoelastic properties of the cytoskeleton of cells using oscillating forces (Dichtl and Sackmann, 2002). Moreover, optically active biosensors have made it possible to characterize the way force produced by MT is distributed within a cell and link these forces to biochemical signaling pathways (Kilinc et al., 2015). Additionally, these studies have provided insight into how cells interact and alter extracellular matrix ligands, producing mechanical signals that allow continuous plasticity and reshaping to take place. Several MT studies relevant to the biomechanical nature of disease have been reviewed in Section 3.4.5 to provide an understanding of the applications and limitations of the MT technique.

The goal of this chapter is to present biomedical scientists with an introduction to the principles of MT as they apply to the biophysical characterization of molecular interactions. Section 3.4.2 introduces the principles of creating controlled 3D magnetic forces for MT. Section 3.4.3 provides an overview of the physical and chemical properties of magnetic microparticles that are frequently used for MT. Section 3.4.4 reviews the techniques that have been used to detect the position of the particles using either light or magnetic fields. This chapter is concluded with remarks on future trends likely to develop using MTs to screen the in vitro and in vivo biomechanical response of cells.

3.4.2 Magnetic Systems for Magnetic Tweezers

MT studies are often performed with a magnet system designed for a specific biomechanical measurement. The design problem is defined by the orientation, magnitude, and range of the tweezer's forces, which may be oriented normal to (F_z), or in the plane (F_θ), or perpendicular to the plane (F_ϕ) of a microscope slide. Other design criteria are the area over which we wish to apply these forces and the temporal resolution of variations in these forces. The design parameters that are available to create these forces are the configuration of the magnetic system, that is, number of magnets, orientation of magnets, the source of the magnetic field and the means for changing its magnitude, as well as the properties of the magnetic particle, that is, size, shape, and magnetic material properties. This section introduces a simple electromagnet and permanent magnet system, with the goal of providing insight into the scaling of the magnetic fields and the temporal response of each system.

Let us start by considering the interaction of a 3D magnetic field with a microscopic magnetic particle, which for the purpose of this discussion can be modeled as a simple dipole. In the most general case, the magnetic force $\boldsymbol{F}$ applied to the magnetic dipole is related to the magnitude and orientation of the magnetic field

$$\boldsymbol{F} = (\boldsymbol{m} \, \nabla)\boldsymbol{B} \qquad\qquad (3.4.1)$$

where $\nabla \boldsymbol{B}$ is the gradient of the magnetic flux density and $\boldsymbol{m}$ is the effective magnetic moment of the particle. In the case of an SPM microparticle, the magnetic moment will be aligned with the magnetic field

$$m = VM = V\chi H = V\chi B/\mu_o$$

where V is the volume of particle, M is the magnetization of particle, χ is the susceptibility of particle, H is the external magnetic field, and μ_o is the vacuum permeability. Thus, in the special case of an SPM microparticle, a magnetic force will be oriented in the direction of the $\boldsymbol{B}$ and can be deduced from eq. (3.4.1) to be

$$F = V\chi\nabla\frac{B^2}{2\mu_o}$$

In the special case where the $\boldsymbol{B}$ is oriented normal to the axis of a microscope slide on which the molecule or cells of interest are immobilized, this becomes a 1D problem

$$F = V\chi\frac{\partial B}{\partial z} \qquad\qquad (3.4.2)$$

where $\partial B/\partial z$ is the gradient of the magnetic field at the point of the microparticle dipole. Note that off-axis components of the magnetic field may cause the microparticles to contact the coverslip, resulting in a torque on the particle (Lee et al., 2000). Thus, the magnetic force on an SPM microparticle is directly proportional to the magnetic field gradient for the special case when $\boldsymbol{B}$ is oriented normal to a coverslip and the absolute magnitude of the magnetic field exceeds approximately 0.1 T (i.e., the magnetization of an SPM microbead is a function of magnetic field, as described in Section 3.4.3).

The force generated on a ferromagnetic particle differs significantly from an SPM particle. A torque is exerted on a ferromagnetic bead due to its permanent magnetic moment. This torque, τ, is defined by the orientation of the magnetic dipole associated with the magnetic particle and the orientation of the magnetic field:

$$\tau = m \times B \tag{3.4.3}$$

The magnetization of ferromagnetic materials, such as iron or platinum–iron alloy, is typically oriented along its long axis.

Magnet systems used in MT are typically assembled from electromagnets and/or rare-earth permanent magnets. Electromagnets possess dynamic magnetic fields with a wide temporal bandwidth, while permanent magnetic systems require a mechanical manipulator to change the magnitude of the field. An electromagnet has a metal core, which is usually precisely shaped at one end, encased in a solenoid coil. This core is constructed from magnetic alloys with high saturation and low remanence. The direction and magnitude of the magnetic force is precisely manipulated using a programmed supply current, with more complex fields being generated using multiple electromagnets (Li et al., 2013). Although an electromagnet can produce 0.5 T flux densities near its tip, this requires a high level of current and produces large amounts of heat in the solenoid coil. Thus, the coils must be cooled if high fields are used over periods of time exceeding 1 min. By contrast, high-graded, millimeter-sized rare-earth magnets, for example, neodymium boron iron (NdBFe), produce magnetic flux densities over 1 T. Usually, a pair of permanent magnets is assembled in a yoke to produce high magnetic field gradients and/or torque on magnetic microparticles.

Figure 3.4.1a presents a schematic of an electromagnetic tweezer system integrated with an inverted optical microscope that we use in our laboratory. The electromagnet is composed of a solenoid with 1,500 copper wire turns (the outer diameter × length × inner diameter of the solenoid are 20 × 50 × 7 mm) that surround the soft magnetic alloy core (the diameter × length × tip diameter of the core are 5 × 60 × 0.8 mm). The core is a soft-magnetic alloy (49% cobalt, 2% vanadium, and balance iron) that has a high magnetic saturation, that is, it has a saturation induction of 2.4 T, maximum permeability of 12,000, and coercive force of 72 A/m. The electromagnet is driven with a DC power supply (Kikusui PWR800L, Japan) with programmable current controller. The transient response of this system is approximately 1.5 ms. The solenoid is cooled with a heat exchanger that is based on a water-cooling jacket. This cooling system is able to maintain the temperature of the electromagnet at 30 °C for 1 A of supplied current at an ambient temperature of 19 °C.

High magnetic field gradients are produced by machining the electromagnet core into a point. This point is then positioned within 1 mm of the microfabricated sample chamber, as shown in Figure 3.4.1b. The position of the tip of the electromagnet is controlled with a programmable motorized *xyz*-stage, that is, Zaber Tech T-LSM050B, Canada, equipped with a customized mounting arm. This stage is used to position the tip of the electromagnet above the microfluidics cell with a precision of 25 μm and repeatability exceeding 6 μm. A nonmagnetic stage, mounting arms, and microscopic objectives are used (Zeiss observer Z1, Germany) to minimize possible interference from ferromagnetic materials.

The electromagnetic system's performance was characterized using finite element modeling and local magnetic field measurements. Finite element modeling indicates that B decays rapidly away from the tip in all directions, as shown in Figure 3.4.1c, thereby generating a large field gradient in its immediate vicinity. As a result of these properties, this electromagnet can produce forces up to 100 pN on a 1 μm SPM bead of $\chi = 0.3$ at a distance of 1 mm from the tip using only 1 A of current supply, as shown in Figure 3.4.1c. However, as the magnetic force on the bead decays rapidly away from the tip of electromagnet, the tip must be positioned accurately within a few millimeters of the sample.

Figure 3.4.2 presents a schematic of a permanent magnetic system that is used in our laboratory for MT measurements. Two NdBFe magnets sized 12.7 × 12.7 × 6.4 mm are assembled with their poles facing each other across a 1 mm air gap. The NdFeB magnets are N42 grade and have a remanence of approximately 1.3 T. To precisely position the magnetic system relative to the magnetic particles, it is mounted on a motorized xyz-manipulator (Eppendorf InjectMan NI 2, Germany) in the same inverted optical microscope mentioned earlier. The force applied to the SPM beads is controlled by moving the permanent magnet assembly relative to the microfluidic chamber, and the motorized system has a response time on the order of seconds, that is, the slew rate of the motor is 10 mm/s. Figure 3.4.2a presents the results of a finite element computation of the magnet flux produced by the magnetic system along the axis of symmetry between the two magnets. The color map presents the magnetic flux density in the plane and the red arrows indicate the direction of B (in the plane of symmetry between the magnets). The maximum value of B is approximately 0.9 T in the center of the gap between the magnets. Figure 3.4.2b presents the measured B and calculated F as a function of distance from the front of the magnet. Like the electromagnet, the B of the permanent magnet decays rapidly away from the magnets but the rate of decay is significantly slower. The magnetic force present on the right axis of Figure 3.4.2b was calculated for a 1 μm diameter spherical SPM of $\chi = 0.3$. The permanent magnet produces weaker magnetic force than the electromagnet for distances less than 5 mm. The advantage of the permanent magnet system, however, is that it produces stable pN-scale tweezer forces over extended areas and periods of time. We have successfully used this permanent magnetic system to simultaneously execute MT measurements on hundreds of cells in a single microfluidic cell.

We have found that a microfluidic system, such as the one shown in Figure 3.4.1b, is an important functional component of an MT instrument. These lab-on-a-chip (LOC) systems are readily constructed from one or more layers of polydimethylsiloxane bonded to a glass coverslip. The LOC system provides a gas-permeable cell culture chamber that can be used to study cultured cells over several weeks. The magnetic microparticles, cell culture media, and specific chemical compounds are introduced to the chip using one of three channels connected through a series of small openings

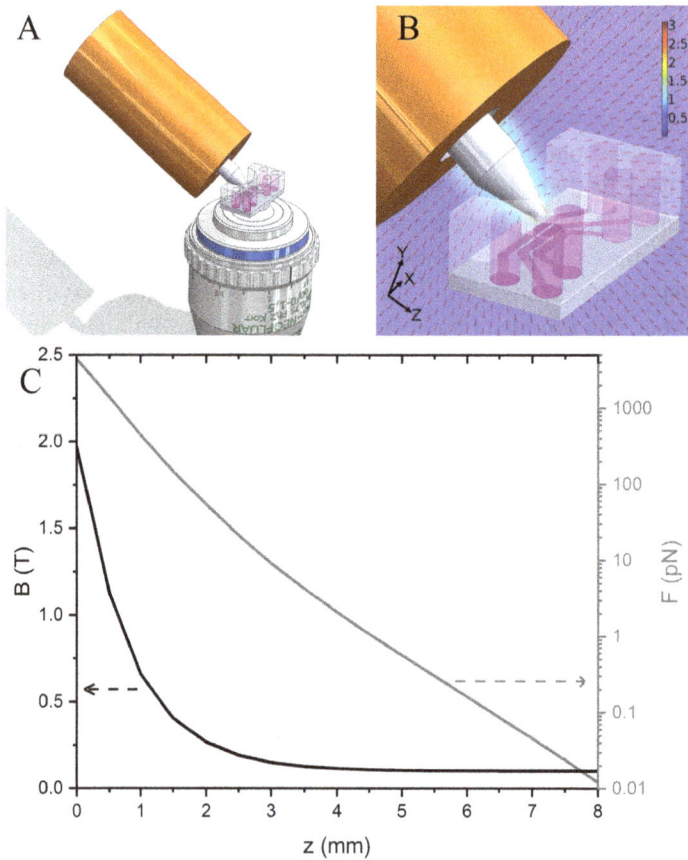

Figure 3.4.1: Design of an electromagnetic system with integrated microfluidic cell for MT. (A) Magnetic field is produced by an electromagnet integrated with a microfabricated cell culture system and inverted optical microscope. (B) Results of a finite element model calculation of the field generated by the electromagnet for 1 A of current. The colored map presents the magnetic flux density in a cross section through the axis of the core and the arrows indicate the direction of B (unit of color legend is Tesla). (C) The scaling behavior of B and F for the electromagnet is presented on the left and right axes, respectively. The magnitude of B has been measured in the direction along the axis of the core as a function of the distance from the tip using a magnetometer. Magnetic force in the direction along the axis of the alloy core has been calculated for a 1 μm diameter spherical SPM particle of $\chi = 0.3$ as a function of the distance from the tip of alloy core.

(Lesniak et al., 2014). These openings between the channels permit the introduction of specific biochemicals/biochemical gradients across the cell culture. Furthermore, this LOC setup makes it possible to accurately determine the position of the magnet relative to the magnetic particles with sub-millimeter precision. This is done by (1) centering the tip of the electromagnet at the focal point of the objective, (2) moving it to a position several centimeters away from the optical path while inserting the

A

B

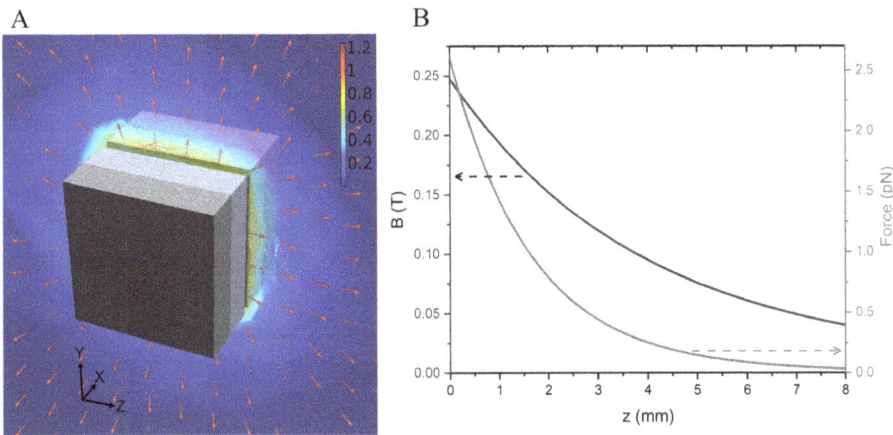

Figure 3.4.2: Design of a permanent magnet system for MT. (A) Finite element model of magnetic fields are produced by two NdBFe permanent magnet of size of 12.7 × 12.7 × 6.4 mm. (B) Magnetic flux density, B (left axis), and magnetic force, F (right axis), produced by the permanent magnet system as a function of the distance along the central plane in the gap of two magnets, that is, along the x-axis of the magnet assembly. The zero point has been designated as the front of the magnet. The force has been calculated for a 1 μm diameter SPM bead with a χ of 0.3.

microfabricated cell, and (3) placing the tip on the surface of the microfabricated cell to define its relative position. By utilizing the motorized microscope stage and magnet system described here, hundreds of MT magnetic microparticle measurements can be made simultaneously.

3.4.3 Chemical and Physical Properties of Superparamagnetic Microparticles

SPM is a state of matter where the direction of the magnetic moment of the single-domain nanoparticle can freely move between two stable orientations on the easy axis due to thermal energy (Fields et al., 2016). Without an external magnetic field, the magnetization of the nanoparticles averages zero, due to flipping of the orientation of the magnetic field in the nanoparticle. Particles exhibiting SPM should not interact with each other in the absence of a strong magnetic field, which would lead to the formation of unwanted aggregates. Crucially, this property allows re-dispersion of the magnetic particles after removal of an external magnetic field. Ferromagnetic and ferrimagnetic materials consist of magnetic domains – regions in which the direction of magnetization is primarily uniform. Frequently cited SPM materials include iron, nickel, cobalt, gadolinium oxide, and iron oxide. Magnetite (Fe_3O_4) and maghemite (γ-Fe_2O_3) are the most widely used SPM nanoparticles for a

number of reasons: low toxicity, well-established synthesis routes, and high saturation magnetization.

Iron oxide nanoparticles of the size of 10 nm are suitable for many applications
such as magnetic resonance imaging (MRI) contrast agents, drug delivery, cell separation, and water treatment (Gupta and Gupta, 2005, Oberteuffer et al., 1975, Franzreb
et al., 2006). However, for MT applications, due to the smaller size of nanoparticles,
stronger magnetic field gradients (approximately 10^4 T/m) are required to produce the
desired levels of force. The incorporation of large amounts of iron oxide nanoparticles
into micron and sub-micron-sized structures produces an additive effect of the magnetic moments of the individual nanoparticles, while still retaining the SPM properties
of the nanoparticles. It is therefore more commonplace in MT applications to use
500–5,000 nm assemblies of SPM nanoparticles.

As previously described by our group (Fields et al., 2016), Figure 3.4.3 presents
three strategies for the assembly of SPM iron oxide nanoparticles into microparticles.
Nanoparticles that are distributed in a polymer microparticle matrix have been described (Liu et al., 2003, Ugelstad et al., 1983, Levison et al., 1998, Yang et al., 2008).
Alternatively, nanoparticle self-assembly to form a spherical SPM has also been reported. In this case, the microparticle is coated with a polymer layer to provide available chemical groups to which biomolecules such as DNA and proteins can be
appended (Uhlen, 1989, Muzard et al., 2012). This type of polymer-coated magnetic
microparticle is referred to as a core–shell structure. Thirdly, microparticles have
been synthesized by precipitating maghemite nanoparticles inside the pores of polymer microparticles previously formed by emulsion polymerization (Häfeli, 1997,
Ugelstad et al., 1983). A fourth category, where SPM nanoparticles are physically or
chemically adsorbed onto a micron-sized polymer particle, has also been reported.
However, this technique results in microparticles with low magnetic loading due to
the polymer core comprising the majority of the particle volume (Bizdoaca et al.,
2002). The synthesis of polymer magnetite composites is typically accomplished
through suspension, or mini-emulsion-templated polymerization, where nanoparticles
are suspended in monomer-containing droplets which are polymerized through the
addition of a polymerization initiator, encapsulating the nanoparticles (Hai et al.,
2009, Ramírez and Landfester, 2003, Csetneki et al., 2004, Ma et al., 2005). While the
polymer component of these particles is a useful support for coupling functional
groups for different applications, the magnetic loadings are generally low, in the order
of 30–40 wt% (Hai et al., 2009, Ramírez and Landfester, 2003, Zheng et al., 2005).

Magnetic beads must be chemically coated to enable conjugation of biomolecules. Typically, organic polymers and surfactants, or inorganic layers, such as silica
are chosen. Due to its ease of use and different functional groups available in the
form of commercial silanes, silica represents a popular choice for inorganic coating.
However, other examples such as alumina, zirconium manganese ferrites, gold, and
silver have also been reported for the generation of inorganic magnetic sorbents
(Fields et al., 2016). Organic polymer coatings can be divided into natural polymer

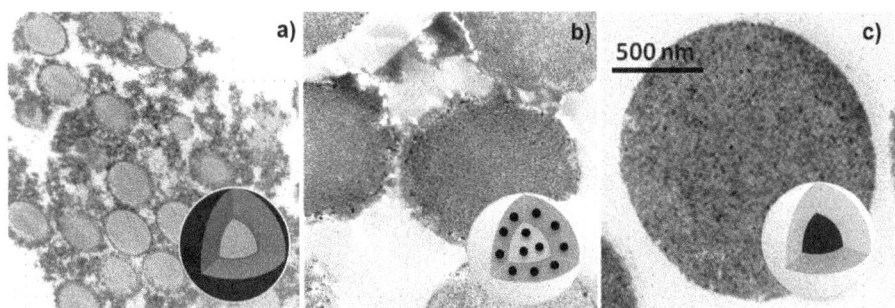

Figure 3.4.3: Schematic and TEM images of commonly used magnetic bead constructs. (a) A core–shell particle structure where a dense magnetic shell is coated onto a polymer core. The magnetic content in these particles ranges from 10 to 20 wt%. TEM image depicts a cross section through beads embedded in a wax matrix. (b) A distributed particle structure in which the nanoparticles have precipitated within a polymer matrix. Magnetic loadings in this case can be up to 32 wt%. (c) A solid core particle structure where a dense magnetic core is coated with an ultra-thick polymer layer. Magnetic loading in core–shell particles is typically >70 wt% wt (Fields et al, 2016).

coatings, including biocompatible polysaccharides such as dextran, starch, heparin, pullulan, chitosan, and alginate. Nonnatural polymer coatings including polyvinyl-pyrrolidone (Guowei et al., 2007) provide increased steric repulsion between particles due to their long hydrophobic chains. Other examples of nonnatural polymer coatings include polyvinyl alcohol (Liu et al., 2008), polyethylimine, polymethylmetha-crylate (Gass et al., 2006), poly-2-methacryloyloxyethyl phosphorylcholine, and polyamidoamine (Strable et al., 2001).

Controlling polymer thickness and grafting density is highly desirable for the design of functionalized magnetic adsorbents. This is made possible using surface-initiated polymerization techniques, such as atom transfer radical polymerization (ATRP) chemistry. Bioconjugation of a variety of biological entities, such as biotin (Kang et al., 2009), antibody fragments (Iwata et al., 2008), and peptides (Glinel et al., 2008), has been made possible using the ATRP approach. After synthesis and coating, the resulting particle has the appearance of an SPM core and an outer layer consisting of a polymer, which often provides specific functional groups for bioconjugation to affinity ligands (see Figure 3.4.4; Fields et al., 2016).

For conjugation of protein-based ligands to the particle surface, there are four functional group options available – primary amines (NH_2), carboxyls (COOH), sulfhydryls (SH), and carbonyls (CHO). Commonly, peptides and proteins are immobilized through free amine groups using EDC/NHS cross-linking chemistry (Bartczak and Kanaras 2011). Coupling to carbonyl groups is commonly performed using hydrazine cross-linkers (Moghimi and Moghimi 2008). Thiol-reactive groups for coupling to Cys-containing affinity ligands include maleimides, iodoacetamides, and disulfides and may be conjugated to magnetic matrices via the use of sulfo-SMCC cross-linking

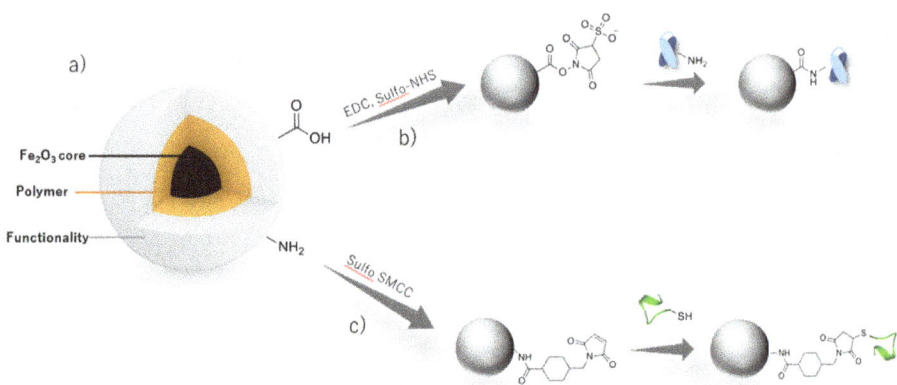

Figure 3.4.4: Schematic of multilayered bead–ligand bioconjugation with common amine and carboxyl functionalities (Fields et al., 2016). (a) Image depicting the layered particle structure, showing the Fe_2O_3 superparamagnetic core, the surrounding polymer shell, and the resulting chemical functionality conferred to the particle. (b) Schematic of conjugation of antibody fragments to a carboxyl-derivatized particle surface, using EDC and sulfo-NHS chemical cross-linkers (Muzard et al., 2012). (c) Schematic of conjugation of a thiolated synthetic peptide to an amine-derivatized bead using sulfo-SMCC cross-linking chemistry (Fields et al., 2012). In each case, the immobilized ligand is irreversibly and covalently tethered to the magnetic support (Fields et al., 2016).

agents (Kalia and Raines 2010), as depicted in Figure 3.4.4. The versatility of bifunctional cross-linkers makes them extremely useful for enabling covalent immobilization between two previously incompatible functional groups and should be considered in the design of affinity ligand-functionalized biomaterials.

Also noteworthy is the use of biotin–avidin technology, which is regularly used in magnetic separation. Nucleic acids are routinely recovered from crude samples by immobilizing biotinylated, complimentary strands on streptavidin-coated magnetic particles (Mojsin et al., 2006). While it is accepted that orientating the ligand in a site-specific manner is beneficial in affinity-ligand adsorbent design, the role of other variables is less understood. Spacer arms become necessary when the nonbinding domain of the ligand does not offer substantial spacing between the structure of the adsorbent and the ligand-binding domain to permit the ligand to interact freely with the target. The significance of spacer arm length has been previously communicated in the literature (Fuentes et al., 2006, Hubbuch and Thomas 2002b). Computational analysis and simulations on the effect of linker length also state that longer spacer arms decrease negative steric effects (Ghaghada et al., 2005, Jeppesen et al., 2001).

Table 3.4.1 (Fields et al., 2016) presents a list of commercially available SPM beads and their physical properties. Magnetic microparticles in the size range of 1–2 μm are used in MT applications, due to their high magnetic moment and the relatively low magnetic field gradients needed to generate pN forces. The magnetic moment of the

Table 3.4.1: Comparison of commercially available magnetic constructs (n.a refers to data that was not available).

Manufacturer	Product	Diameter (μm)	Surface functionalities	Composition	Density (g/cm^3)	Magnetism	Magnetic content
Thermo-Fischer	Dynabeads (MyOne, M270, M-450)	1, 2.8, 4.4	Carboxylic acid, Streptavidin, antibodies, protein, antigens, DNA/RNA	Iron oxide nanoparticle core with polystyrene shell	1.4–1.7	Superparamagnetic	37%
Magnostics	HiMag	0.5, 0.75, 1	Silanol, polymer, antibodies	Magnetite nanoparticle closed packed	2.25	Superparamagnetic	90%
Spherotech Inc.	SPHREO magnetic particles	1–120	Amino, carboxyl, diethylamino, dimethylamino, hydroxyethyl	Magnetite nanoparticles coated onto a polystyrene core	n.a.	Superparamagnetic or paramagnetic	~15%
Merck-Millipore	Pure proteome magnetic beads	0.3, 1, 2.5, 10	Carboxyl, Streptavidin, protein, N-hydroxy-succinimide (NHS)	Magnetic nanoparticles in a silica or polystyrene matrix	n.a.	Paramagnetic	n.a.
GE Lifescience	Sera-mag magnetic beads	1	Carboxyl, Streptavidin, Neutravidin, oligo, amine, protein	Magnetite nanoparticles and polystyrene core	1.5, 2.0	Superparamagnetic	~40%, ~60%
Polysciences Inc.	Biomag	~1.5	Carboxyl, Streptavidin, amine, oligo (dT), antigen, antibody	Silanized iron oxide matrix, irregular shape	2.5	Superparamagnetic	n.a.

particles is largely dependent on their loading of inorganic magnetic material, which is presented in the final column of Table 3.4.1. Magnetic polystyrene particles have been synthesized with 55 wt% magnetite (Zheng et al., 2005), and polymer–magnetite composite particles have been synthesized with 30–40 wt% magnetite (Omi et al., 2001). Dynabeads contain between approximately 17 and 32 wt% maghemite depending on the size of the microparticle (Fonnum et al., 2005). Higher magnetic loadings are desirable when a large number of measurements need to be performed rapidly. Size distributions and uniformity of magnetic loading must be considered, as variation in these parameters can lead to loss of variation in MT force.

3.4.4 Magnetic Particle Tracking Using Optical Microscopy and Magnetometry

The position of the MT magnetic particles is made possible by the use of optical or magnetic sensors. Typically, optical microscopy is the most common particle tracking technique used in MT publications due to its high resolution and the availability of complementary fluorescence imaging techniques. Coupling these techniques makes it possible to simultaneously determine the position of a bead and the macromolecular properties of a cell (Kilinc and Lee, 2014, Kilinc et al., 2015). Alternatively, magnetometry can be used to track the position of SPM magnetic particles at surfaces. Microfabricated giant magnetoresistance and Hall effect sensor arrays have been developed and are also capable of tracking individual SPM particles with nanometer resolution (Baselt et al., 1998, Sinha et al., 2014).

Optical microscopy can easily determine the lateral position (i.e., xy-position) of a 1 μm diameter SPM particle with micron resolution using transmitted light illumination and a high-quality 100×, high-NA oil immersion objective. Nanometer-scale resolution is achievable with a microscope equipped with a high-quality CCD camera, assuming two conditions are satisfied: (1) a centrosymmetric particle (e.g., spherical particle) is used and (2) the particle only moves a few pixels from one frame to the next (Gosse & V., 2002). Digital images that are acquired at 60 Hz frame rates are then analyzed using customized software. This software first fits a 1D Gaussian intensity profile, $I(x)$, to a bead over several y-pixels, respectively, for each frame. The correlation function is then calculated for the bead, $C(x_o)$, using a fast Fourier transform algorithm. The maximum of $C(x_o)$ occurs at a position δx, which is twice the shift of the bead's position from the profile center. A pixel of the CCD camera typically corresponds to about 100 nm, so precise measurement of x requires subpixel resolution, which is achieved by polynomial interpolation and low-pass filtering. The same procedure is also performed to find the center in the y-direction.

The position of the magnetic microparticle may be determined with nanometer precision relative to the surface of a coverslip using reflected interference contrast

(Lionnet et al., 2012a) or total internal reflection fluorescence (Kilinc and Lee, 2014). Due to its ease of use and broad application, we describe the reflection interference contrast technique here. High-resolution diffraction images of centrosymmetric beads are formed using parallel illumination of the beads. Parallel illumination is produced using a slightly focused illumination beam, generated by a monochromatic light source, such as a light-emitting diode. Under these illuminating conditions, the bead image produces a series of diffraction rings, and the shapes of which depend on the relative distance between the bead and the focal plane. When the bead is in focus, these rings disappear; however, as the bead moves out of focus, their diameter increases. By precisely stepping the focal plane through a series of positions, for example, by moving the objective with a piezo-electric device, it is possible to form a stack of calibration images that record the shape of the diffraction rings versus the distance from the focal plane. Customized software is then used to determine the out-of-focus distance for a new bead image by comparing its diffraction pattern to the calibration stack (Lionnet et al., 2012a).

3.4.5 Selected Examples of Magnetic Tweezer Studies of Cellular Mechanics

MTs have been used to study mechanotransduction and the effect of force on cellular stiffness, viscosity, deformation, and intercellular events associated with the applied strain. These studies form the basis of our current understanding of the complex in vivo biological interactions, whereby cells and tissues are exposed to shear stress and tension as part of their complex 3D physiological environment, for example, blood flow, interstitial flow, and compression from muscle tissue activity (Mitchell and King, 2013, Jansen et al., 2017). As pioneers in cellular biomechanics, Wang and Ingber used magnetic twisting to demonstrate that the ECM receptor integrin β1 was a mechanoreceptor (Wang et al., 1993a). This was the first published identification of an adhesion receptor as a transducer of external mechanical stimulus causing a change in the cell's cytoskeleton. This methodology was superior to previous methods at the time which depended on deformation of large areas of the cell in a nonspecific fashion (Petersen et al., 1982, Wang and Ingber, 1994). Adherent endothelial cells can be bound to spherical ferromagnetic microbeads coated with an integrin recognition peptide Arg–Gly–Asp (RGD). Precoating the beads with RGD mimicked cell reactions to external forces exerted by the ECM and its components. The orientation and magnitude of force were controlled with a complex external magnetic field, that is, a strong external magnetic field (1,000 G) was used to orient the particle and a weaker magnet (0–25 G) was used to apply torque. The change in cell surface-bound particle rotation (angular strain) was measured using a magnetometer. After application of the magnetic field in both directions, cells bound to the RGD bead displayed increased stiffness and resistance to mechanical deformation. This effect was specific to the RGD peptide but not to a control peptide (Gly-Arg-Gly-Glu-Ser-Pro) or when a competitor peptide Gly-Arg-Gly-Asp-Ser-Pro (GRGDSP) was introduced to the culture medium. As previously mentioned, extracellular mechanical forces not only exert an effect on cell rigidity but activate series of molecular events leading to restructuring of the cytoskeletal network. In this elegant series of experiments, the authors showed that cell stiffening correlated with focal adhesion formation through recruitment of talin, vinculin, and α-actinin to the side of the RGD bead but not a control-acetylated low-density-lipoprotein-coated bead.

An important finding of cellular biomechanics was the determination that the endogenous properties of the cytoskeleton depend on the interaction with the ECM. Wang demonstrated that both integrin activation and changes in mechanical force are sufficient to affect the structure of the entire cytoskeleton (Petersen et al., 1982, Wang and Ingber, 1994). The specificity of the observed ECM–cell interaction was elegantly described in an experiment comparing cells with intact membranes and membrane-permeabilized (saponin-treated) cells. In both cases, application of mechanical force with MTs resulted in the same trend of increased cell stiffness in response to mechanical stimulation. As a result, the possibility of intracellular osmotic or hydrostatic pressure affecting changes in cellular stiffening, viscosity, and deformation were excluded. These events can be attributed to intracellular changes in the cytoskeletal structure in response to biomechanical stimulation. Research into the cytoskeletal properties using MT has made it possible to model, and further understand, the biomechanics of living cells. In particular, the studies described above support the tensegrity model – now recognized as an underlying principle in mechanobiology, giving cells and tissues their structural resilience. This was a revolutionary step toward nullifying the simplistic view of cells as a 3D isotropic model, uniform in all directions.

Additionally, MT has been used to study cellular biomechanics and apoptosis using immunohistochemistry to identify chemical signaling (Kilinc et al., 2015). In this study, Fe–Au nanorods were used to specifically target and kill cancer cells in conjugation with mechanical stimulation. The benefits of target-specific magnetic particles in the context of cancer diagnosis is a widely explored topic; however, the exact targeting mechanism of particles and the molecular events that follow are not completely understood. Previously, our group has used magnetic nanorods bound to MCF7 breast cancer cells and manipulated the forces in an effort to understand the biomechanical and biomolecular responses of particle-bound cells to the mechanical stimulus. Nanorod particles typically exhibit a larger magnetic moment but a smaller surface area, allowing them to target receptors on specific regions of cells. Magnetic nanorods were functionalized with heregulin (HRG), a ligand of ErbB family of tyrosine kinase receptors, known to be overexpressed on cancer cells. Cyclic stretching forces at predefined magnitudes were applied to individual nanorod–cell complexes over fixed time. Mechanical stimulation of cell–bead complexes resulted in additional phosphorylation of extracellular signal-regulated kinase 1/2 (ERK) using an optically activated force-sensing protein expressed in the cells. The molecular activity of ERK was observed for applied mechanical forces ranging between 4.1 and 12.7 pN. ERK activity was observed not only at the nanorod-binding site but also in the entire cell. Therefore, stimulation of a small number of receptors at the tip referred to as an "active zone" provides a sufficient stimulus leading to activation of downstream signal transduction pathways such as ERK signaling. This finding led us to further investigate combined approaches for effective cell death of MCF-7 cells. Here, mechanical stimulation in conjunction with pharmaceutical stimulation of ERK pathways with a B-Raf inhibitor resulted in global ERK hyperactivation and cancer cell apoptosis. This study involving MT allowed us to further understand the molecular responses of "tagged tumor cells" and equipped us with additional insight into how cancer cells respond, specifically to mechanical stimulation. It is our hope that these types of studies involving MT will allow the development of new therapeutic approaches to cancer treatment.

3.4.6 The Future of Magnetic Tweezers in Mechanobiology and Medicine

In this chapter, we have introduced you to the principles of the MT technique and highlighted examples where it has been used effectively to study biomechanics at the molecular scale. We believe that this technique is complementary to AFM for the study of the biomechanical aspects of disease. The two advantages of this technique that demonstrate the most promise for medicine are (1) its application in the screening of therapeutics targeting the biomechanical properties of cells and (2) the study of the forces in vivo associated with biomechanical aspects of disease.

A useful feature of MT is that it can apply uniform forces to magnetic microparticles across areas containing 1,000 cells or more. In the past, our group has used the permanent magnet system and a three-chamber LOC device to rapidly screen the responses of hundreds of mouse neuron growth cones to various applied forces (Kilinc et al., 2014). The three chambers of this microfluidics device provided us with the means to establish biochemical gradients on a surface, or in the culture media, in which the axons were isolated. Here, MT was used to apply force to the receptors on the surface of the growth cones, but MT measurements have also been performed within cells, by inserting the magnetic microparticles into the cells by microinjection or electroporation. This approach allowed us to determine the response of the growth cones to simultaneous biomechanical and biochemical signaling, which is critical to nervous system regeneration. Thus, MT provides a means for applying local, asymmetric forces to many cells in the presence of gradients of short- or long-range biochemical stimuli.

Also noteworthy is the emerging application of MT for the study of the biomechanical aspects of disease in 3D studies within living systems. Magnetic fields are routinely applied in humans for MRI and a number of reports have emerged detailing the manipulation and imaging of magnetic microparticles directly in vivo for therapeutic applications (Dobson, 2008, Du et al., 2017). One such imaging tool that has been described is magnetic particle imaging (MPI), a noninvasive tomographic technique capable of detecting SPM nanoparticles with very high resolution (Gleich and Weizenecker, 2005). To date, several commercial MPI scanners have become available.

In conclusion, MT has played an important role in pioneering biomechanic measurements, from single molecules to cellular systems. We anticipate MT technology will continue to grow and develop, due largely, to its application in high-throughput screening and in vivo measurements. [1] Note on magnetic material properties: Ferromagnetic materials exhibit spontaneous magnetization and retain their magnetic properties in absence of external magnetic field. Paramagnetic materials become magnetized and form internal induced magnetic field in the direction of an external field, but the induced field disappears when the external field is removed. On the contrary, the diamagnetic materials form internal induced magnetic field in the direction opposite to external field. SPM is a form of magnetism that occurs in ferromagnetic or ferrimagnetic nanoparticles. In an external magnetic field, the SPM materials can be magnetized and exhibit paramagnetism but with relative larger susceptibility. In the absence of an external magnetic field, the magnetization of SPM materials appears to be in average zero.

References

Bartczak, D. and A. G. Kanaras. (2011). "Preparation of peptide-functionalized gold nanoparticles using one pot EDC/sulfo-NHS coupling." Langmuir **27**(16): 10119–10123.

Baselt, D. R., G. U. Lee, M. Natesan, S. W. Metzger, P. E. Sheehan, and R. J. Colton (1998). "A Biosensor based on Magnetoresistance Technology." Biosensors and Bioelectronics **13**: 731–739.

Bizdoaca, E., M. Spasova, M. Farle, M. Hilgendorff, and F. Caruso (2002). "Magnetically directed self-assembly of submicron spheres with a Fe3O4 nanoparticle shell." Journal of Magnetism and Magnetic Materials **240**: 44–46.

Csetneki, I., M. K. Faix, A. Szilágyi, A. L. Kovács, Z. Németh and M. Zrinyi (2004). "Preparation of magnetic polystyrene latex via the miniemulsion polymerization technique." Journal of Polymer Science Part A: Polymer Chemistry **42**: 4802–4808.

del Rio, A., R. Perez-Jimenez, R. Liu, P. Roca-Cusachs, J. M. Fernandez and M. P. Sheetz. (2009). "Stretching single talin rod molecules activates vinculin binding." Science **323**: 638–641.

Dichtl, M. A. and E. Sackmann (2002). "Microrheometry of semiflexible actin networks through enforced single-filament reptation: Frictional coupling and heterogeneities in entangled networks." Proceedings of the National Academy of Sciences of the United States of America **99**: 6533–6538.

Dobson, J. (2008). "Remote control of cellular behaviour with magnetic nanoparticles." Nature Nanotechnology **3**: 139–143.

Du, V., N. Luciani, S. Richard, G. Mary, G. Gay, G. Mazue, M. Reffay, P. Menasché, O. Agbulut and C. Wilhelm (2017). Magnetic nanoparticle-mediated heating for biomedical applications. Nature Communications **8**: 400.

Fields, C., P. Mallee, J. Muzard, and G. U. Lee. (2012). "Isolation of Bowman-Birk-Inhibitor from soybean extracts using novel peptide probes and high gradient magnetic separation." Food Chemistry **134**(4): 1831–1838.

Fields, C., P. Li, J. J. O'Mahony and G. U. Lee. (2015). "Advances in affinity ligand-functionalized nanomaterials for biomagnetic separation." Biotechnology & Bioengineering **113**(1): 11–25.

Fonnum, G., C. Johansson, A. Molteberg, S. Morup, and E. Aksnes (2005). "Characterisation of Dynabeads® by magnetization measurements and Mössbauer spectroscopy." Journal of Magnetism and Magnetic Materials **293**: 41–47.

Franzreb, M., M. Siemann-Herzberg, T. J. Hobley and O. R. Thomas (2006). "Protein purification using magnetic adsorbent particles." Applied Microbiology and Biotechnology **70**: 505–516.

Fuentes, M., C. Mateo, R. Fernández-Lafuente, and J. M. Guisán (2006). "Detection of polyclonal antibody against any area of the Protein-Antigen using immobilized Protein-Antigens: The critical role of the immobilization protocol." Biomacromolecules **7**(2): 540–544.

Gass, J., P. Poddar, J. Almand, S. Srinath, and H. Srikanth (2006). "Superparamagnetic polymer nanocomposites with uniform Fe3O4 nanoparticle dispersions." Advanced Functional Materials **16**(1): 71–75.

Ghaghada, K. B., J. Saul, J. V. Natarajan, R. V. Bellamkonda and A. V. Annapragada (2005). "Folate targeting of drug carriers: a mathematical model." Journal of Controlled Release **104**(1): 113–128.

Gleich, B. and J. Weizenecker (2005). "Tomographic imaging using the nonlinear response of magnetic particles." Nature **435**: 1214.

Glinel, K., A. M. Jonas, T. Jouenne, J. Leprince, L. Galas, and W. T. Huck (2008). "Antibacterial and antifouling polymer brushes incorporating antimicrobial peptide." Bioconjugate Chemistry **20**(1): 71–77.

Gosse, C. and V. Croquette (2002). "Magnetic tweezers: micromanipulation and force measurement at the molecular level." Biophysical Journal **82**: 3314–3329.

Guowei, D., K. Adriane, X. Chen, C. Jie, and L. Yinfeng (2007). "PVP magnetic nanospheres: Biocompatibility, in vitro and in vivo bleomycin release." International Journal of Pharmaceutics **328**(1): 78–85.

Gupta, A. K. and M. Gupta (2005). "Synthesis and surface engineering of iron oxide nanoparticles for biomedical applications." Biomaterials **26**: 3995–4021.

Häfeli, U. (1997). Scientific and clinical applications of magnetic carriers. Springer.

Hai, N. H., N. H. Luong, N. Chau and N. Q. Tai (2009). "Preparation of magnetic nanoparticles embedded in polystyrene microspheres." Journal of Physics: Conference Series. IOP Publishing. p.012009.

Heilbornn, A. (1922). "Eine neue Methode zur Bestimmung der Viskosität lebender Protoplasten." Jahrbücher für Wissenschaftliche Botanik **61**: 284–338.

Hubbuch, J. J. and O. R. Thomas (2002b). "High-gradient magnetic affinity separation of trypsin from porcine pancreatin." Biotechnology and Bioengineering **79**(3): 301–313.

Ingber, D. E., N. Wang and D. Stamenovic (2014). "Tensegrity, cellular biophysics, and the mechanics of living systems." Reports on Progress in Physics **77**: 046603.

Iwata, R., R. Satoh, Y. Iwasaki, and K. Akiyoshi (2008). "Covalent immobilization of antibody fragments on well-defined polymer brushes via site-directed method." Colloids and Surfaces B: Biointerfaces **62**(2): 288–298.

Jansen, K. A., P. Atherton and C. Ballestrem (2017). "Mechanotransduction at the cell-matrix interface." Seminars in Cell and Developmental Biology **71**: 75–83.

Jeppesen, C., J. Y. Wong, T. L. Kuhl, J. N. Israelachvili, N. Mullah, S. Zalipsky, and C. M. Marques (2001). "Impact of polymer tether length on multiple ligand-receptor bond formation." Science **293**(5529): 465–468.

Kalia, J. and R. T. Raines (2010). "Advances in Bioconjugation." Current Organic Chemistry **14**(2): 138–147.

Kang, S. M., I. S. Choi, K-B. Lee and Y. Kim (2009). "Bioconjugation of poly (poly (ethylene glycol) methacrylate)-coated iron oxide magnetic nanoparticles for magnetic capture of target proteins." Macromolecular Research **17**(4): 259–264.

Kilinc, D., A. Blasiak, J. J. O'Mahony and G. U. Lee (2014). "Low Piconewton Towing of CNS Axons against Diffusing and Surface-Bound Repellents Requires the Inhibition of Motor Protein-Associated Pathways." Scientific Reports **4**: 7128.

Kilinc, D., A. Blasiak, J. J. O'Mahony, D. M. Suter and G. U. Lee. (2012). "Magnetic Tweezers-Based Force Clamp Reveals Mechanically Distinct apCAM Domain Interactions." Biophysical Journal **103**: 1120–1129.

Kilinc, D., C. L. Dennis and G. U. Lee (2016). "Bio-nano-magnetic materials for localized mechanochemical stimulation of cell growth and death." Advanced Materials **28**: 5672–5680.

Kilinc, D. and G. U. Lee (2014). "Advances in magnetic tweezers for single molecule and cell biophysics." Integrative Biology (Camb) **6**: 27–34.

Kilinc, D., A. Lesniak, S. A. Rashdan, D. Gandhi, A. Blasiak, P. C. Fannin, A. von Kriegsheim, W. Kolch and G. U. Lee (2015). "Mechanochemical Stimulation of MCF7 Cells with Rod-Shaped Fe–Au Janus Particles Induces Cell Death Through Paradoxical Hyperactivation of ERK." Advanced Healthcare Materials **4**: 395–404.

Koster, D. A., Croquette, V., Dekker, C., Shuman, S. & Dekker, N. H., 2005, Friction and torque govern the relaxation of DNA supercoils by eukaryotic topoisomerase IB. Nature **434**: 671–674.

Lee, G. U., Metzger, S., M. Natesan, C. Yanavich and Y. F. Dufrene (2000). "Implementation of force differentiation in the immunoassay." Analytical Biochemistry **287**: 261–271.

Lesniak, A., D. Kilinc, S. A. Rashdan, A. von Kriegsheim, B. Ashall, D. Zerulla, W. Kolch and G. U. Lee (2014). "In vitro study of the interaction of heregulin-functionalized magnetic–optical nanorods with MCF7 and MDA-MB-231 cells." Faraday Discuss **175**: 189–201.

Levison, P. R., S. E. Badger, J. Dennis, P. Hathi, M. J. Davies, I. J. Bruce and D. Schimkat (1998). "Recent developments of magnetic beads for use in nucleic acid purification." Journal of Chromatography A **816**: 107–111.

Li, P., D. Kilinc, Y. F. Ran and G. U. Lee (2013). "Flow enhanced non-linear magnetophoretic separation of beads based on magnetic susceptibility." Lab on a Chip **13**: 4400–4408.

Lionnet, T., J.-F. Allemand, A. Revyakin, T. R. Strick, O. A. Saleh, D. Bensimon and V. Croquette (2012a). Cold Spring Harb Protoc. 34–49.

Lionnet, T., J. F. Allemand, A. Revyakin, T. R. Strick, O. A. Saleh, D. Bensimon and V. Croquette (2012b). Cold Spring Harbor Protocols **2012**: 34–49.

Lionnet, T., J. F. Allemand, A. Revyakin, T. R. Strick, O. A. Saleh, D. Bensimon and V. Croquette (2012c). Cold Spring Harbor Protocols **2012**: 133–138.

Liu, T-Y., S-H. Hu, K-H. Liu, D-M. Liu, and S-Y. Chen (2008). "Study on controlled drug permeation of magnetic-sensitive ferrogels: Effect of Fe3O4 and PVA." Journal of Controlled Release **126**(3): 228–236.

Liu, X., H. Liu, J. Xing, Y. Guan, Z. Ma, G. Shan and C. Yang (2003). China Particuology **1**: 76–79.

Ma, Z.-Y., Y.-P. Guan, X.-Q. Liu and H.-Z. Liu (2005). Journal of Applied Polymer Science **96**: 2174–2180.

Mitchell, M. J. and M. R. King (2013). Frontiers in Oncology **3**: 44.

Moghimi, M. and S. M. Moghimi (2008). "Lymphatic targeting of immuno-PEG-liposomes: evaluation of antibody-coupling procedures on lymph node macrophage uptake." Journal of Drug Targeting **16**(7): 586–590.

Mojsin, M., J. S. Đurović, I. Petrović, A. Krstić, D. Drakulić, T. Savić, and M. Stevanović (2006). "Rapid detection and purification of sequence specific DNA binding proteins using magnetic separation." Journal of the Serbian Chemical Society **71**(2): 135–141.

Mosconi, F., J. F. Allemand and V. Croquette (2011). The Review of Scientific Instruments **82**: 034302.

Muzard, J., C. Fields, J. J. O'Mahony and G. U. Lee (2012). Journal of Agricultural and Food Chemistry **60**: 6164–6172.

Neuman, K. C. and A. Nagy (2008). Nature Methods **5**: 491–505.

Oberteuffer, J., I. Wechsler, P. G. Marston and M. McNallan (1975). Magnetics, IEEE Transactions On **11**: 1591–1593.

Omi, A. K., Y. Shimamori, A. Supsakulchai, M. Nagai and G.-H. Ma S (2001). Journal of Microencapsulation **18**: 749–765.

Petersen, N. O., W. B. McConnaughey and E. L. Elson (1982). Proceedings of the National Academy of Sciences **79**: 5327–5331.

Ramírez, L. P. and K. Landfester (2003). Macromolecular Chemistry and Physics **204**: 22–31.

Roca-Cusachs, P., A. Del Rio, E. Puklin-Faucher, N. C. Gauthier, N. Biais and M. P. Sheetz (2013). Proceedings of the National Academy of Sciences of the United States of America **110**: E1361–1370.

Shang, H. and G. U. Lee (2007). Journal of the American Chemical Society **129**: 6640–6646.

Sinha, B., T. S. Ramulu, K. W. Kim, R. Venu, J. J. Lee and C. G. Kim (2014). Biosensors & Bioelectronics **59**: 140–144.

Strable, E., J. W. Bulte, B. Moskowitz, K. Vivekanandan, M. Allen, and T. Douglas (2001). "Synthesis and characterization of soluble iron oxide-dendrimer composites." Chemistry of Materials **13**(6): 2201–2209.

Strick, T. R., V. Croquette and D. Bensimon (2000). Nature **404**: 901–904.

Ugelstad, J., L. Soderberg, A. Berge and J. Bergstrom (1983). Nature **303**: 95–96.

Uhlen, M. (1989). Nature **340**: 733–734.

Wang, N., J. Butler and D. Ingber (1993a). Science **260**: 1124–1127.

Wang, N., J. P. Butler and D. E. Ingber (1993b). Science **260**: 1124–1127.

Wang, N. and D. E. Ingber (1994). Biophysical Journal **66**: 2181–2189.

Wang, N., J. D. Tytell and D. E. Ingber (2009). Nature Reviews. Molecular Cell Biology **10**: 75–82.

Yang, S., H. Liu and Z. Zhang (2008). Journal of Polymer Science. Part A, Polymer Chemistry **46**: 3900–3910.

Zheng, W., F. Gao and H. Gu (2005). Journal of Magnetism and Magnetic Materials **288**: 403–410.

Jorge Otero, Daniel Navajas

3.5 Cell and Tissue Stretcher

Physical forces have a considerable impact on cells, tissue function, and development in vivo, since both cells and tissues are mechanosensitive (Na et al., 2008). Indeed, when cells are subjected to external forces, they respond at a biological level (Davies, 1995), for instance, by modulating ion channels (Blumenthal et al., 2014) or activating transmembrane proteins such as integrins (Giancotti and Ruoslahti, 1999). Therefore, it is important that in vitro cell culture settings mimic the natural cell microenvironment in vivo, as much as possible. However, physiological strain patterns are complex, they depend on the specific organ, and range from periodical and multidirectional stimuli such as in lungs (Farré et al., 2018b) to asynchronous and unidirectional strains in skeletal muscles (Collinsworth et al., 2000). Therefore, it is of major importance that in vitro settings reproduce the specific mechanical strains, which depend on the organ where the cells and tissues are placed in vivo. Fortunately, there are several commercially available stretching devices (e.g., Flexcell from Flexcell Int. Corp., USA, and STREX from STREX Inc., Japan) that allow seeding and culture of cells on top of a flexible membrane, which is stretched by pneumatic actuation, thereby applying strain to the cultured cells.

Nevertheless, commercially available stretching devices present certain limitations such as low-throughput and nonlinear strain profiles. Therefore, these devices cannot universally mimic the variety of strains to which cells and tissues are subjected to in the different organs. To overcome these problems, considerable research effort to develop custom-made devices has been carried out by several groups, mainly by using micro-electro-mechanical systems (MEMS), microfluidics, and lab-on-a-chip technologies (Davis et al., 2015, Friedrich et al., 2019). Although specific device design should be closely adapted to the biological question being investigated, some methodological aspects are common to a majority of devices and experiments. This chapter describes experimental settings suitable to measure cell and tissue mechanics under different levels of stretch and highlights advanced approaches incorporating optical

Acknowledgments: This book chapter was funded in part by the Spanish Ministry of Sciences, Innovation and Universities (PGC2018-097323-A-I00, DPI2017-83721-P), and by the Marie Sklodowska-Curie Action, Innovative Training Networks 2018, EU grant agreement no. 812772.

Jorge Otero, Unitat de Biofísica i Bioenginyeria, Facultat de Medicina i Ciències de la Salut, Universitat de Barcelona, Barcelona, Spain, CIBER de Enfermedades Respiratorias, Madrid, Spain
Daniel Navajas, Unitat de Biofísica i Bioenginyeria, Facultat de Medicina i Ciències de la Salut, Universitat de Barcelona, Institute for Bioengineering of Catalonia, Barcelona, Spain.
dnavajas@ub.edu

https://doi.org/10.1515/9783110640632-011

microscopy and combining other biophysical techniques such as AFM, traction microscopy, or optical trapping.

3.5.1 Stretch Methods

Macroscopic stretching is mainly used to measure passive mechanics of tissues or skeletal muscle single fibers. The method mainly consists of clamping the sample and stretching it by a controlled actuator, while measuring the force exerted by using a transducer (Mohammadkhah et al., 2016). If the geometry of the sample and its unstretched length are known, the stress–strain curve can be computed from the measurements. These measurements of the passive response of tissues can be correlated with diseases related to extracellular matrix remodeling (Farré et al., 2018a).

As clamping is not feasible for cell stretching at the microscopic scale, the most common approach is to culture the cells on top of a flexible membrane, which is subjected to the desired stretch. If the cells perfectly adhere to the substrate, its stretch is entirely transmitted to cells through membrane deformation. There are different modes to apply cell strain, in vitro, to mimic the diverse stretches that cells experience, in vivo. The simplest configuration is using uniaxial testing, where the mechanical stimulus is applied to the sample over just one direction (Clark et al., 2001) (Figure 3.5.1a). The actually applied strain can be simply calculated as the ratio between the length change and the unstretched length of the membrane. Uniaxial stretch devices have the advantage of simple design, but they often produce a heterogeneous strain field (Matsumoto et al., 2007). Multiaxial stretching devices are more complex settings implemented on the basis of biaxial (where the strain is applied in two perpendicular directions, Figure 3.5.1b) or multiaxial (where isotropic strain is applied to the sample, Figure 3.5.1c) settings. These multiaxial devices present a more homogenous strain field than uniaxial setting, especially at the center of the membrane. However, they are more difficult to fabricate, drive, and control, and their calibration is more complex. Regardless of whether the deformation challenge applied is uniaxial or biaxial, achieving in-plane stretch of the membrane is important for cell imaging with optical microscopes. Indeed, if the sample is stretched out of plane, there is loss of focus originated by the vertical displacement of the sample (Figure 3.5.1d). To minimize this problem, in-plane designs have been developed by using a post to avoid the vertical displacement of the membrane, while applying a negative pressure (Figure 3.5.1e). Usually, the use of lubrication between the membrane and the post is required to prevent excessive friction, which could yield to sample heating or device damage (Jorba et al., 2019).

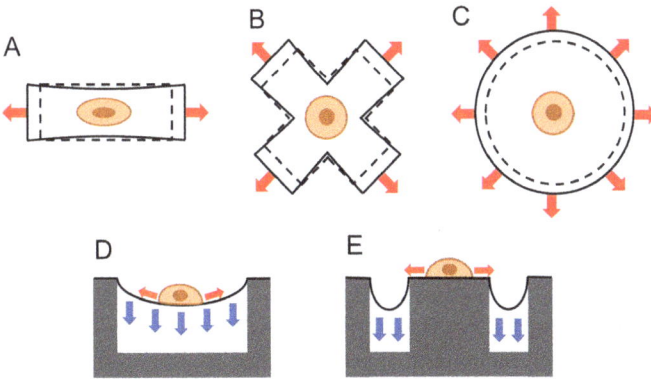

Figure 3.5.1: Different stretch techniques. Uniaxial (A). Multiaxial stretch methods, where only two axes are stretched (biaxial, B); or the whole membrane is isometrically stretched (equiaxial, C). Out-of-plane (D) and in-plane (E) stretch.

3.5.2 Instrumentation

The optimal design of the stretching device depends on the desired readout for the experiments to be conducted. For fundamental studies in mechanobiology, where the target is usually to determine how each mechanical stimulus modulates the cell response, only one method is commonly used to apply strain. In contrast, for tissue engineering applications, multiple stimuli are applied simultaneously to better mimic the in vivo environment. Thus, devices that are able to apply different combined mechanical tests have been developed for this purpose (Hu et al., 2013). Another important aspect for device design is whether the cells should be analyzed after strain application (e.g., cell staining or supernatant analysis) (McAdams et al., 2006) or live cell imaging during the experiments is required. In the latter case, as mentioned above, the device should apply the strain in-plane (Huang et al., 2010). Hence, there are two major design components as regards the stretching device: the type of membrane and the mechanical actuation principle (Kamble et al., 2016).

The vast majority of membranes used in the devices for cell stretching are fabricated on poly-dimethylsiloxane (PDMS) (Friedrich et al., 2017). PDMS is elastic, biocompatible, and optically transparent, and its mechanical properties can be tuned during the preparation process. Also, PDMS is one of the gold standards for the fabrication of polymer-based lab-on-a-chips, so it is easy to integrate in the fabrication process of the devices (de Jong et al., 2006). However, PDMS is highly hydrophobic, and, therefore, its surface should be functionalized before seeding the cells on top of the membrane. Noteworthy, a drawback of PDMS is that it may absorb small hydrophobic molecules, a process that can interfere in certain studies (Wang et al., 2012). As an alternative, polyurethane thin, flexible membranes have been recently developed by spin coating, and tested as potential PDMS substitutes (Arefin et al., 2017, Ergene et al., 2019).

To effectively stretch the membrane and to apply strain to the cells, different types of actuators can be used. Pneumatic settings are the most commonly implemented because they are easier to control. Moreover, pneumatic actuators present the advantage of avoiding any direct contact with the sample. By applying positive or negative air pressure below the membrane, the strain is applied to the cells cultured on top of it. The simplest design is the balloon approach, which presents the problem of out-of-plane displacement (Shimizu et al., 2011). More sophisticated designs apply pressure in lateral chambers, while the Z movement of the membrane is constricted by a post. In this way, in-plane stretching of the membrane is accomplished (Huang and Nguyen, 2013). Piezoelectric (Deguchi et al., 2015) and electromagnetic (Tang et al., 2010) actuators appear as promising alternatives to pneumatic ones, especially in devices fabricated using MEMS technologies instead of polymer soft lithography. These actuators present higher precision in the control of the applied strain and higher dynamic response and are, therefore, well-suited for oscillatory stretch experiments where rheological properties need to be measured at higher frequencies (Sander et al., 2017). Interestingly, MEMS-based solutions offer high-throughput measurements and the possibility to perform the experiments at the single cell level (Scuor et al., 2006).

3.5.3 Sample Preparation

The first choice for sample preparation is to select the mechanical properties of the membrane, as it is well known that this factor (mainly stiffness) considerably influences the way cells behave (Smith et al., 2018). As mentioned before, the elastic modulus of PDMS can be tuned in the preparation process. When very soft membranes are needed, a common solution is to use polyacrylamide gels (Sun et al., 2012) or PDMS blends (Palchesko et al., 2012). To improve cell adhesion, PDMS surface can be treated by oxygen plasma to improve its wettability (Duffy et al., 1998). Nevertheless, problems with the recovery of the hydrophobicity appear in course of time and under certain conditions (Bodas and Khan-Malek, 2007, Amerian, Amerian et al, 2019). To overcome this problem, a common solution is to silanize the surface of the membrane for the formation of a self-assembled monolayer and then to bioactivate the silanized surface with adherent proteins (Chuah et al., 2015). The choice for the functionalization procedure directly depends on the cells to be cultured and stretched. There is no universal strategy, and hence, every new experiment setting will usually require the tuning of an adequate protocol for effective cell adhesion to the membrane. Type I collagen (Qian et al., 2018), RGD peptides (conjugated with sulfo-SANPAH) (Li et al., 2006), polydopamine (Fu et al., 2017), or fibronectin (Liao et al., 2019) are just examples of the wide range of proteins that have been successfully used to improve the adhesion of different cell types to the PDMS surface. In the case of polyurethane membranes, similar problems arise regarding the physical and chemical methods suitable for the bioactivation of the membrane surfaces fabricated with this material (Bax et al., 2014).

3.5.4 Calibration

The stretch and strain applied to the membrane can be analytically calculated by knowing the geometries and the elastic modulus of the materials. For example, in isotropic systems, a 10% increase in membrane area is obtained by a 5% radial displacement of the actuator. Nevertheless, in practice, due to defects in the fabrication, leaks and other potential problems, it is common to calibrate the effective displacement of the membrane for a given driving pressure applied (or voltage, in the case of piezoelectric or electromagnetic actuators). The calibration should be done specifically for each combination of cells and coating and for each device fabrication.

The most common solution to calibrate the displacement is to use an optical microscope. One possible method earlier described required plotting nine dots on the membrane and calculating the average displacement between them, when the membrane is subjected to different deformation signals (Schaffer et al., 1994). A more sophisticated solution is to incorporate microbeads in the preparation of the membrane or in the sample surface (Trepat et al., 2004) and to measure the average displacement between them (Schürmann et al., 2016) (Figure 3.5.2a). In the case of out-of-plane displacement membranes, if the optical microscope has a motorized Z stage, an easy and reliable way to calibrate the central displacement of the membrane is to focus the top of the membrane and to measure the change of focus with the actuation in the membrane (Campillo et al., 2016). This method is quite easy and produces results comparable to the microbeads' calibration method (Figure 3.5.2b). Differences in the stretch experienced at different directions of the membrane in biaxial or equiaxial devices strongly depend on the device design. In case the design is optimal, the interdirection differences are so low that only one of the directions of the membrane needs to be calibrated (Figure 3.5.2c). In certain experiments, however, the membrane needs to be calibrated once the cells are seeded, and no microbeads can be placed in the sample. The common solution in this case is to directly measure the deformation of the sample (cells or tissue) and to compute the effective stretch as the change in the area of the sample by using image processing techniques (Jorba et al., 2019).

If the membrane is purely elastic, the calculation of stresses is straightforward, in case the elastic modulus is known or measured. Nevertheless, for local stress assessment, sensors should be integrated within the devices to measure the forces on the cell. The common solution is to combine mechanical stretch with other techniques such as atomic force microscopy or optical/magnetic tweezers.

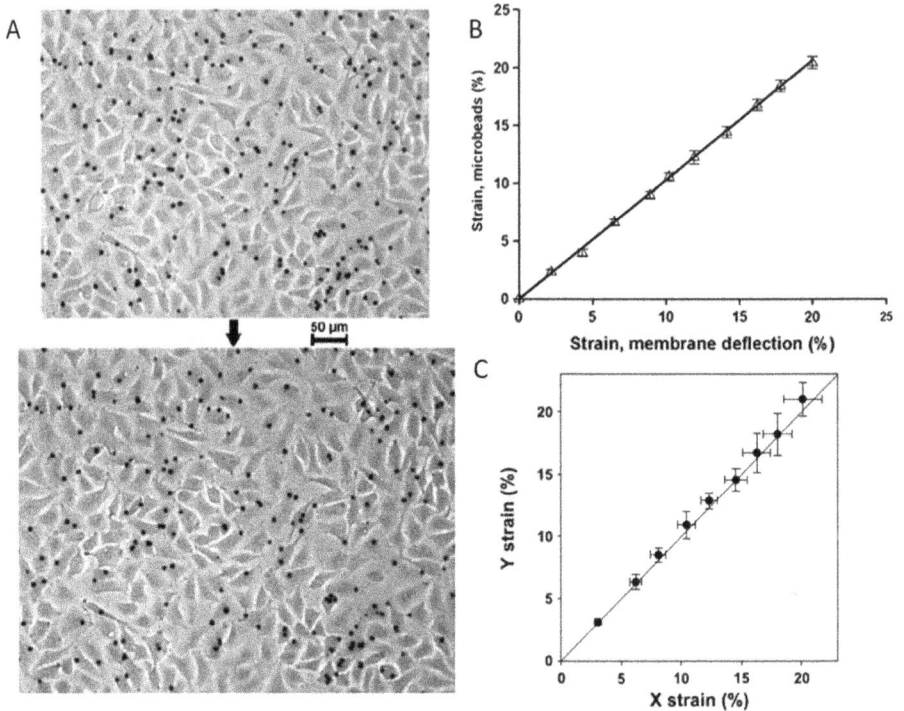

Figure 3.5.2: Calibration. Unstretched and stretched membrane of an epithelial cell monolayer with magnetic microbeads from Trepat et al. (2004) (A). Comparison of the microbeads position quantification and change in the Z focus of the optical microscope methods for calibration of an out-of-plane membrane from Campillo et al. (2016) (B). Comparison between the X- and Y-directions' stretch in an equibiaxial device from Trepat et al. (2004) (C).

3.5.5 Single-Cell Stretching

An important evolution in the field of cell mechanics is the development of techniques for single-cell stretching. They are commonly based on optical or microfluidic devices, or a combination of both. Optical stretching (Guck et al., 2001) is based on the use of two opposed laser beams to apply a force on the cell. The total force acting on the cell is zero, but the additive forces in the surface stretch the cell in the direction of the laser beams (Figure 3.5.3a). Microfluidic stretching is accomplished by circulating the cells in restricted-dimension channels and measuring the deformation induced by the fluidic drag force and the constriction of the microchannel (Iragorri et al., 2018) (Figure 3.5.3b). Interestingly, the evolution of both techniques allows the use of optofluidic devices, which combine microfluidic channels with laser beams: the microfluidic channel is used to position the cells in the desired position, while the laser radiation traps the cells by applying a contactless force on the cell surface (Yang et al., 2016). In combination with the fluidic drag force and the quantitative analysis of the images of cells trapped and recovered, differences in the mechanical properties of cells can be assessed (Figure 3.5.3c). Recently, a device capable of analyzing the mechanics of one cell per second has been reported, showing a high sensitivity in the differentiation between healthy and stiffened red blood cells (Yao et al., 2020).

3.5.6 Related Techniques and Future Outlook

The availability of commercial stretching equipment for biological samples is speeding up research in the field. However, customized developments will probably continue being the most commonly used solution, because the need to answer new biological questions usually requires advanced experimental approaches. However, there are marked differences between devices aimed to study basic mechanobiology, where usually only one variable should be investigated, and devices for tissue engineering, where several variables should be tuned to apply the optimum cell culture microenvironment (Nonaka et al., 2019).

One of the most interesting applications of cell stretching devices is their combination with other biophysical measurement techniques. Devices for such purposes need to be designed in a way that allows the combined application of the stretching stimulus and the conduction of the experiment with the complimentary technique. The incorporation of magnetic probes, by using microbeads and coils (optical magnetic twisting cytometry), allows the measurement of the viscoelastic properties of cells subjected to stretch (Trepat et al., 2004). It is a complex technique, and the device should be accurately designed to simultaneously apply the membrane stretch and the magnetic field to the sample (Figure 3.5.4a). By combining mechanical stretch with traction force microscopy (Bashirzadeh et al., 2019), the sheet tension in epithelial islands have been measured, opening the doors for studying the mechanobiology of collective cells subjected to mechanical stretch (Figure 3.5.4b). Moreover, the strain stiffening commonly observed in biopolymer networks (Janmey et al., 1991) has been effectively observed and quantified in living cells and in decellularized extracellular matrix from the lung by combined application of stretching and atomic force microscopy (Ahrens et al., 2019, Jorba et al., 2019). These measurements of the micromechanics of the lung extracellular matrix when subjected to different strains has revealed that alveolar epithelial cells sense different surrounding stiffness during the different cycles of breathing (Figure 3.5.4c).

The field of cell and tissue stretching is evolving rapidly. The technological advances and the integration of complementary techniques has shown that stretch is an important stimulus to be taken into account in developing relevant in vitro models in mechanobiology and tissue engineering. Nevertheless, the low throughput

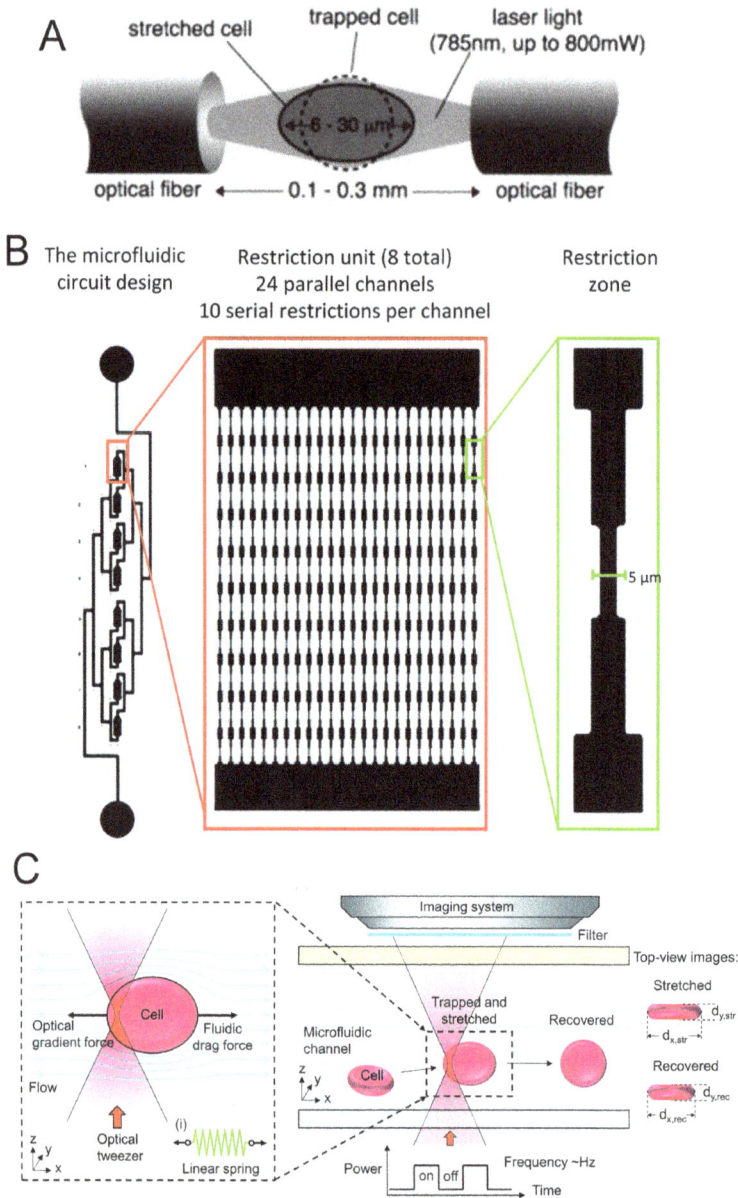

Figure 3.5.3: Single-cell stretching. Principle of operation of optical stretch (Guck et al., 2001) (A). Microfluidic chip for cell stretching (Iragorri et al., 2018) (B). Schematic of the principle of optical stretch technique in combination with microfluidics (Yao et al., 2020) (C).

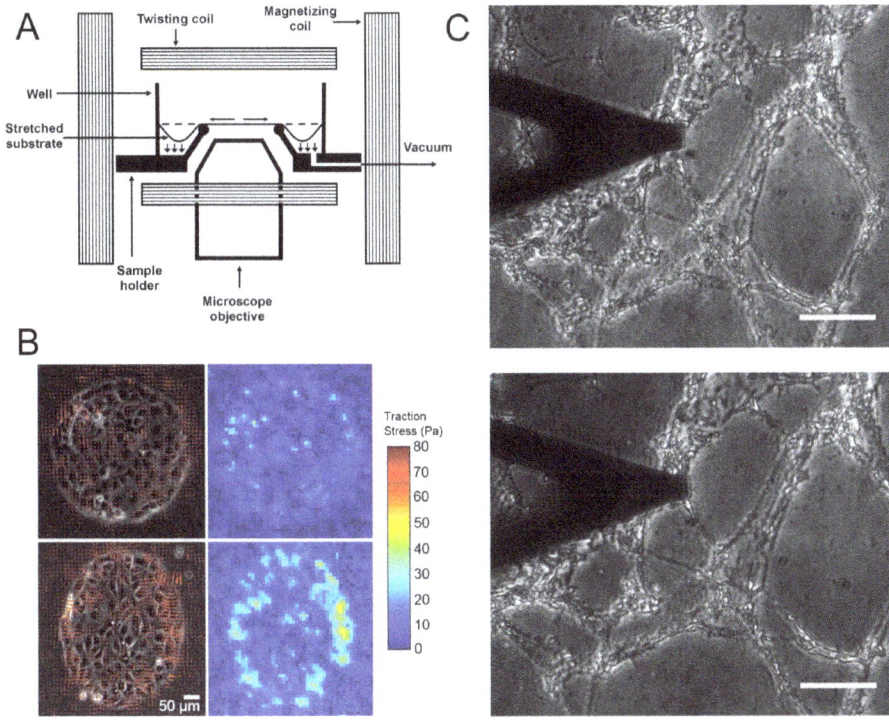

Figure 3.5.4: Combined techniques. Device design implemented for simultaneous cell stretching and magnetic twisting cytometry experiments (Trepat et al., 2004) (A). Traction forces evaluated in a uniaxially stretched epithelial cell island (Bashirzadeh et al., 2019) (B). Simultaneous application of strain and AFM measurement of lung extracellular matrix (Jorba et al., 2019) (C).

and high cost of fabrication of current devices is a limitation that should be overcome, mainly by using techniques such as additive manufacturing. Recently, 3D printed lab-on-a-chip devices have been reported, exhibiting a clear advance when compared with the fast prototyping of new stretch devices (Ho et al., 2015). It is also important to realize that the majority of current devices are designed for 2D cell culture. With the recent increase of research and progress in 3D cell culture, a new generation of devices should be developed to apply stretch into 3D scaffolds mimicking the microenvironment of the extracellular matrix, in vivo (Saldin et al., 2017). In this way, in vitro models will be much more realistic and, hence, improve experimental setting to better mimic the physiological conditions that cells experience within native tissues and organs.

References

Ahrens, D., W. Rubner, R. Springer, N. Hampe, J. Gehlen, T. M. Magin, B. Hoffmann and R. Merkel (2019). "A combined AFM and lateral stretch device enables microindentation analyses of living cells at high strains." Methods and Protocols 2(2): 43.

Amerian, M., M. Amerian, M. Sameti and E. Seyedjafari (2019). "Improvement of PDMS surface biocompatibility is limited by the duration of oxygen plasma treatment." Journal of Biomedical Materials Research. Part A 107(12): 2806–2813.

Arefin, A., J. Huang, D. Platts, V. Hypes, J. Harris, R. Iyer and P. Nath (2017). "Fabrication of flexible thin polyurethane membrane for tissue engineering applications." Biomedical Microdevices 19(4): 98.

Bashirzadeh, Y., S. Dumbali, S. Qian and V. Maruthamuthu (2019). "Mechanical response of an epithelial island subject to uniaxial stretch on a hybrid silicone substrate." Cellular and Molecular Bioengineering 12(1): 33–40.

Bax, D. V., A. Kondyurin, A. Waterhouse, D. R. McKenzie, A. S. Weiss and M. M. Bilek (2014). "Surface plasma modification and tropoelastin coating of a polyurethane co-polymer for enhanced cell attachment and reduced thrombogenicity." Biomaterials 35(25): 6797–6809.

Blumenthal, N. R., O. Hermanson, B. Heimrich and V. P. Shastri (2014). "Stochastic nano roughness modulates neuron-astrocyte interactions and function via mechanosensing cation channels." Proceedings of the National Academy of Sciences of the United States of America 111(45): 16124–16129.

Bodas, D. and C. Khan-Malek (2007). "Hydrophilization and hydrophobic recovery of PDMS by oxygen plasma and chemical treatment – An SEM investigation." Sensors and Actuators. B, Chemical 123(1): 368–373.

Campillo, N., I. Jorba, L. Schaedel, B. Casals, D. Gozal, R. Farré, I. Almendros and D. Navajas (2016). "A novel chip for cyclic stretch and intermittent hypoxia cell exposures mimicking obstructive sleep apnea." Frontiers in Physiology 7. 319.

Chuah, Y. J., S. Kuddannaya, M. H. Lee, Y. Zhang and Y. Kang (2015). "The effects of poly (dimethylsiloxane) surface silanization on the mesenchymal stem cell fate." Biomater Science 3(2): 383–390.

Clark, C. B., T. J. Burkholder and J. A. Frangos (2001). "Uniaxial strain system to investigate strain rate regulation in vitro." Review of Scientific Instruments 72(5): 2415–2422.

Collinsworth, A. M., C. E. Torgan, S. N. Nagda, R. J. Rajalingam, W. E. Kraus and G. A. Truskey (2000). "Orientation and length of mammalian skeletal myocytes in response to a unidirectional stretch." Cell and Tissue Research 302(2): 243–251.

Davies, P. F. (1995). "Flow-mediated endothelial mechanotransduction." Physiological Reviews 75(3): 519–560.

Davis, C. A., S. Zambrano, P. Anumolu, A. C. Allen, L. Sonoqui and M. R. Moreno (2015). "Device-based in vitro techniques for mechanical stimulation of vascular cells: A review." Journal of Biomechanical Engineering 137(4): 040801.

de Jong, J., R. G. Lammertink and M. Wessling (2006). "Membranes and microfluidics: A review." Lab on a Chip 6(9): 1125–1139.

Deguchi, S., S. Kudo, T. S. Matsui, W. Huang and M. Sato (2015). "Piezoelectric actuator-based cell microstretch device with real-time imaging capability." AIP Advances 5(6): 067110.

Duffy, D. C., J. C. McDonald, O. J. Schueller and G. M. Whitesides (1998). "Rapid prototyping of microfluidic systems in poly(dimethylsiloxane)." Analytical Chemistry 70(23): 4974–4984.

Ergene, E., B. S. Yagci, S. Gokyer, A. Eyidogan, E. A. Aksoy and P. Y. Huri (2019). "A novel polyurethane-based biodegradable elastomer as a promising material for skeletal muscle tissue engineering." Biomedical Materials 14(2): 025014.

Farré, N., J. Otero, B. Falcones, M. Torres, I. Jorba, D. Gozal, I. Almendros, R. Farré and D. Navajas (2018a). "Intermittent hypoxia mimicking sleep apnea increases passive stiffness of myocardial extracellular matrix. A multiscale study." Frontiers in Physiology 9. 1143.

Farré, R., J. Otero, I. Almendros and D. Navajas (2018b). "Bioengineered lungs: A challenge and an opportunity." Archivos de bronconeumologia 54(1): 31–38.

Friedrich, O., M. Haug, B. Reischl, G. Prölß, L. Kiriaev, S. I. Head and M. B. Reid (2019). "Single muscle fiber biomechanics and biomechatronics–The challenges, the pitfalls and the future." The International Journal of Biochemistry & Cell Biology 114. 105563.

Friedrich, O., D. Schneidereit, Y. A. Nikolaev, V. Nikolova-Krstevski, S. Schürmann, A. Wirth-Hücking, A. Merten, D. Fatkin and B. Martinac (2017). "Adding dimension to cellular mechanotransduction: Advances in biomedical engineering of multiaxial cell-stretch systems and their application to cardiovascular biomechanics and mechano-signaling." Progress in Biophysics and Molecular Biology 130. 170–191.

Fu, J., Y. J. Chuah, W. T. Ang, N. Zheng and D. A. Wang (2017). "Optimization of a polydopamine (PD)-based coating method and polydimethylsiloxane (PDMS) substrates for improved mouse embryonic stem cell (ESC) pluripotency maintenance and cardiac differentiation." Biomater Science 5(6): 1156–1173.

Giancotti, F. G. and E. Ruoslahti (1999). "Integrin signaling." Science 285(5430): 1028–1032.

Guck, J., R. Ananthakrishnan, H. Mahmood, T. J. Moon, C. C. Cunningham and J. Käs (2001). "The optical stretcher: A novel laser tool to micromanipulate cells." Biophysical Journal 81(2): 767–784.

Ho, C. M., S. H. Ng, K. H. Li and Y. J. Yoon (2015). "3D printed microfluidics for biological applications." Lab on a Chip 15(18): 3627–3637.

Hu, J. J., Y. C. Liu, G. W. Chen, M. X. Wang and P. Y. Lee (2013). "Development of fibroblast-seeded collagen gels under planar biaxial mechanical constraints: A biomechanical study." Biomechanics and Modeling in Mechanobiology 12(5): 849–868.

Huang, L., P. S. Mathieu and B. P. Helmke (2010). "A stretching device for high-resolution live-cell imaging." Annals of Biomedical Engineering 38(5): 1728–1740.

Huang, Y. and N. T. Nguyen (2013). "A polymeric cell stretching device for real-time imaging with optical microscopy." Biomedical Microdevices 15(6): 1043–1054.

Iragorri, M. A. L., S. El Hoss, V. Brousse, S. D. Lefevre, M. Dussiot, T. Xu, A. R. Ferreira, Y. Lamarre, A. C. S. Pinto and S. Kashima (2018). "A microfluidic approach to study the effect of mechanical stress on erythrocytes in sickle cell disease." Lab on a Chip 18(19): 2975–2984.

Janmey, P. A., U. Euteneuer, P. Traub and M. Schliwa (1991). "Viscoelastic properties of vimentin compared with other filamentous biopolymer networks." The Journal of Cell Biology 113(1): 155–160.

Jorba, I., G. Beltrán, B. Falcones, B. Suki, R. Farré, J. M. García-Aznar and D. Navajas (2019). "Nonlinear elasticity of the lung extracellular microenvironment is regulated by macroscale tissue strain." Acta biomaterialia 92. 265–276.

Kamble, H., M. J. Barton, M. Jun, S. Park and N. T. Nguyen (2016). "Cell stretching devices as research tools: Engineering and biological considerations." Lab on a Chip 16(17): 3193–3203.

Li, B., J. Chen and J. H. Wang (2006). "RGD peptide-conjugated poly(dimethylsiloxane) promotes adhesion, proliferation, and collagen secretion of human fibroblasts." Journal of Biomedical Materials Research. Part A 79(4): 989–998.

Liao, W., Y. Hashimoto, Y. Honda, P. Li, Y. Yao, Z. Zhao and N. Matsumoto (2019). "Accelerated construction of an." PeerJ 7. e7036.

Matsumoto, T., Y. C. Yung, C. Fischbach, H. J. Kong, R. Nakaoka and D. J. Mooney (2007). "Mechanical strain regulates endothelial cell patterning in vitro." Tissue Engineering 13(1): 207–217.

McAdams, R. M., S. B. Mustafa, J. S. Shenberger, P. S. Dixon, B. M. Henson and R. J. DiGeronimo (2006). "Cyclic stretch attenuates effects of hyperoxia on cell proliferation and viability in human alveolar epithelial cells." American Journal of Physiology. Lung Cellular and Molecular Physiology 291(2): L166–174.

Mohammadkhah, M., P. Murphy and C. K. Simms (2016). "The in vitro passive elastic response of chicken pectoralis muscle to applied tensile and compressive deformation." Journal of the Mechanical Behavior of Biomedical Materials **62**. 468–480.

Na, S., O. Collin, F. Chowdhury, B. Tay, M. Ouyang, Y. Wang and N. Wang (2008). "Rapid signal transduction in living cells is a unique feature of mechanotransduction." Proceedings of the National Academy of Sciences of the United States of America **105**(18): 6626–6631.

Nonaka, P. N., B. Falcones, R. Farre, A. Artigas, I. Almendros and D. Navajas (2019). "Biophysically preconditioning mesenchymal stem cells improves treatment of ventilator-induced lung injury." Archivos de bronconeumologia **56**(3): 179–181.

Palchesko, R. N., L. Zhang, Y. Sun and A. W. Feinberg (2012). "Development of polydimethylsiloxane substrates with tunable elastic modulus to study cell mechanobiology in muscle and nerve." PLoS One **7**(12): e51499.

Qian, Z., D. Ross, W. Jia, Q. Xing and F. Zhao (2018). "Bioactive polydimethylsiloxane surface for optimal human mesenchymal stem cell sheet culture." Bioact Mater **3**(2): 167–173.

Saldin, L. T., M. C. Cramer, S. S. Velankar, L. J. White and S. F. Badylak (2017). "Extracellular matrix hydrogels from decellularized tissues: Structure and function." Acta Biomater **49**. 1–15.

Sander, M., H. Dobicki and A. Ott (2017). "Large amplitude oscillatory shear rheology of living fibroblasts: Path-dependent steady states." Biophysical Journal **113**(7): 1561–1573.

Schaffer, J. L., M. Rizen, G. J. L'Italien, A. Benbrahim, J. Megerman, L. C. Gerstenfeld and M. L. Gray (1994). "Device for the application of a dynamic biaxially uniform and isotropic strain to a flexible cell culture membrane." Journal of Orthopaedic Research **12**(5): 709–719.

Schürmann, S., S. Wagner, S. Herlitze, C. Fischer, S. Gumbrecht, A. Wirth-Hücking, G. Prölß, L. Lautscham, B. Fabry and W. Goldmann (2016). "The IsoStretcher: An isotropic cell stretch device to study mechanical biosensor pathways in living cells." Biosensors & Bioelectronics **81**. 363–372.

Scuor, N., P. Gallina, H. V. Panchawagh, R. L. Mahajan, O. Sbaizero and V. Sergo (2006). "Design of a novel MEMS platform for the biaxial stimulation of living cells." Biomedical Microdevices **8**(3): 239–246.

Shimizu, K., A. Shunori, K. Morimoto, M. Hashida and S. Konishi (2011). "Development of a biochip with serially connected pneumatic balloons for cell-stretching culture." Sensors and Actuators. B, Chemical **156**(1): 486–493.

Smith, L. R., S. Cho and D. E. Discher (2018). "Stem cell differentiation is regulated by extracellular matrix mechanics." Physiology (Bethesda) **33**(1): 16–25.

Sun, J. Y., X. Zhao, W. R. Illeperuma, O. Chaudhuri, K. H. Oh, D. J. Mooney, J. J. Vlassak and Z. Suo (2012). "Highly stretchable and tough hydrogels." Nature **489**(7414): 133–136.

Tang, J., R. Peng and J. Ding (2010). "The regulation of stem cell differentiation by cell-cell contact on micropatterned material surfaces." Biomaterials **31**(9): 2470–2476.

Trepat, X., M. Grabulosa, F. Puig, G. N. Maksym, D. Navajas and R. Farré (2004). "Viscoelasticity of human alveolar epithelial cells subjected to stretch." American Journal of Physiology. Lung Cellular and Molecular Physiology **287**(5): L1025–1034.

Wang, J. D., N. J. Douville, S. Takayama and M. ElSayed (2012). "Quantitative analysis of molecular absorption into PDMS microfluidic channels." Annals of Biomedical Engineering **40**(9): 1862–1873.

Yang, T., F. Bragheri and P. Minzioni (2016). "A comprehensive review of optical stretcher for cell mechanical characterization at single-cell level." Micromachines (Basel) **7**(5): 90.

Yao, Z., C. C. Kwan and A. W. Poon (2020). "An optofluidic "tweeze-and-drag" cell stretcher in a microfluidic channel." Lab on a Chip **20**: 601–613.

Joanna Zemła, Yara Abidine, Claude Verdier
3.6 Microrheology

3.6.1 Oscillatory Rheology

When oscillatory strain γ deformation with amplitude γ_0 and angular frequency ω ($\omega = 2\pi f$) is applied, the stress σ will oscillate in time (t), but its oscillations will be phase-shifted by φ (Figure 3.6.1):

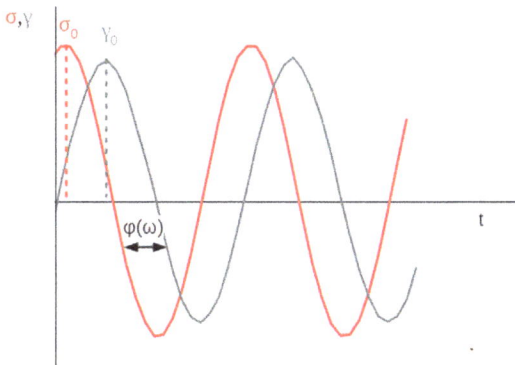

Figure 3.6.1: Stress σ and strain γ time evolution in oscillatory measurements. Stress and strain are shifted by angle φ. Stress and strain amplitudes are labeled with σ_0 and γ_0, respectively.

$$\gamma(t) = \gamma_0 \sin \omega t \tag{3.6.1}$$

$$\sigma(t) = \sigma_0 \sin(\omega t + \varphi) \tag{3.6.2}$$

The phase shift is always between 0° and 90°. In case of ideally elastic materials, the phase shift is 0°, while ideally viscous samples reveal $\varphi = 90°$. Materials with phase shift in between these values are viscoelastic, and their stress is expressed as a sum of storage modulus G' and loss modulus G'', parameters reflecting elastic and viscous properties of the sample, respectively:

Acknowledgments: Joanna Zemła acknowledges the financial support of the French Government and the French Embassy in Poland. Claude Verdier thanks the Nanoscience Foundation for financial support of the AFM, the LabeX Tec21 (Investissements d'Avenir, grant agreement no. ANR-11-LABX-0030), and the ANR "TRANSMIG" (grant no. 12-BS09-020-01).

Joanna Zemła, Institute of Nuclear Physics Polish Academy of Sciences, Krakow, Poland
Yara Abidine, University Grenoble Alpes, CNRS, LIPhy, Grenoble, France
Claude Verdier, University Grenoble Alpes, CNRS, LIPhy, Grenoble, France

https://doi.org/10.1515/9783110640632-012

$$\sigma(t) = \gamma_0 \left(G'' \cos(\omega t) + G' \sin(\omega t) \right) \tag{3.6.3}$$

These relations show that stress σ is proportional to strain or strain amplitude, which is true for all material types at low strains and is called the linear viscoelastic range (LVER). Thus, at the LVER G' and G'' are independent of stress or strain amplitudes. Usually, one has to perform a strain sweep in order to determine under which deformation (γ_0) moduli G' and G'' remain constant, that is, in the LVER.

3.6.2 AFM-Based Oscillatory Microrheology

3.6.2.1 Microrheological Measurements

Rheological measurements were performed with an atomic force microscope, Nanowizard II (JPK, Berlin, Germany), working in force modulation mode. The AFM system was integrated with inverted optical microscope Observer D1 (Carl Zeiss, Jena, Germany). Cells were measured in nonsupplemented medium at 37 °C, with silicon nitride MLCT cantilevers (Bruker, Germany). The MLCT (probe C) is a four-sided pyramid with nominal spring constant $k \sim 0.01$ N/m, which was calibrated using the thermal noise method (Butt and Jaschke, 1995). The loading operating force (200 pN) corresponded to an initial indentation depth δ_0 lower than 1 µm. The relationship between the force and indentation depth is described with Hertz–Sneddon contact mechanics model (Sneddon, 1965):

$$F = \frac{3}{4} \cdot \frac{E \cdot \tan\theta}{1 - v^2} \cdot \delta^2 \tag{3.6.4}$$

where E is the cell's Young's modulus, v is the Poisson's ratio, which for incompressible cells is roughly 0.5, and $\theta \sim 20°$ is the front half opening angle of the probing tip. When the low-amplitude oscillations at initial indentation depth δ_0 are superposed, then the complex shear modulus $G^*(\omega)$ can be determined (Alcaraz et al., 2003) using the formula:

$$G^*(\omega) = \frac{1 - v^2}{3\delta_0 \tan\theta} \cdot \frac{F^*(\omega)}{\delta^*(\omega)} \tag{3.6.5}$$

where $F^*(\omega)$ and $\delta^*(\omega)$ are the Fourier transforms of measured force and sample indentation depth, respectively, $\omega = 2\pi f$ is the angular frequency, and f is the frequency in Hz (Alcaraz et al., 2003, Abidine et al., 2015a, 2015b, 2018).

In addition, the complex shear modulus $G^*(\omega) = G'(\omega) + i\, G''(\omega)$, where $G'(\omega)$ is the storage modulus – a measure of the elastic energy stored and recovered per cycle of oscillations; and $G''(\omega)$ is the so-called loss modulus – a measure of the energy dissipated per cycle of sinusoidal oscillations. The ratio of $G''(\omega)$ and $G'(\omega)$

equals tan φ, a parameter also called the loss factor. If tan $\varphi \ll 1$, a solid-like behavior of the sample is assumed, and if loss factor $\gg 1$, a Newtonian fluid behavior is assumed.

3.6.2.2 Microrheological Models

Rheological models have been studied for years in the polymer community, since polymers can be subjected to shear deformations at very different frequencies which are important in the industry (rubbers, plastics, pastes, foods, foams, etc.). The basic models are the Maxwell and Voigt models (Verdier et al., 2009) which can be used in series or parallel. Such models contain different relaxation times, which can be used in association (Maxwell modes):

$$G' = \sum_{i=1}^{N} G_i \frac{\omega^2 \cdot \tau_i^2}{1 + \omega^2 \cdot \tau_i^2} \tag{3.6.6}$$

$$G'' = \sum_{i=1}^{N} G_i \frac{\omega \cdot \tau_i}{1 + \omega^2 \cdot \tau_i^2} \tag{3.6.7}$$

where τ_i is a relaxation time and G_i is the corresponding modulus.

These relations are quite useful but appear to be insufficient to describe a more complex rheology. Therefore, integral models (Baumgaertel et al., 1990) or fractal ones (Palade et al., 1996, Abidine et al., 2015b) have been used instead. These models are quite efficient for describing the whole range covering several decades in frequency f (Hz). More recently, Sollich et al. (1997) proposed an elegant model based on structural disorder and metastability, after introducing a mean-field temperature and proper statistical treatment. They were able to find various behaviors (yield stress, shear-thinning, glassy behavior) that encompass most of the rheology of soft materials, including cells and tissues. This model has been successfully used for describing adherent cell behaviors (Bursac et al., 2005).

Although complex, these models often seem to predict generic power-law behaviors (Alcaraz et al., 2003, Abidine et al., 2015a, 2018) in a specific range of frequencies. Therefore, it is sometimes useful to replace them with an easier model with fewer parameters to explain the dynamic behavior of moduli G' and G''. This will be shown in Section 3.6.4 for microrheology of cancer cells. The model parameters that are found can be used for describing cells or cell behavior. In particular, it was shown recently that cancer cells in a glassy state could remodel their actin microstructure quite rapidly in order to transmigrate through the endothelium during cancer metastasis (Abidine et al., 2018). Therefore, the present techniques can become quite efficient for predicting different behaviors, that is, normal versus sick cells and constitute a new powerful tool for characterizing/differentiating cells in vitro.

Finally, viscoelastic behaviors are one type of classical behavior at low deformation, but more recent studies show that viscoelastoplastic or poroelastic behaviors can also be encountered for cells (Moeendarbary et al., 2013) and tissues (Preziosi et al., 2010).

3.6.3 Oscillatory Shear Macrorheology

The basic oscillatory macrorheological experiments are called *frequency sweep* and *strain/stress amplitude sweep*. During frequency sweep experiment, the G' and G'' evolutions as a function of frequency are observed. This type of measurement gives information about the (micro)structure and dynamics of the system. Strain sweeps are oscillatory measurements at fixed frequency with increasing strain amplitude. In this approach, a LVER should be observed at low strain amplitudes (see Section 3.6.1). For most gel samples at higher strain values, G' and G'' will depend on the strain amplitude. Higher values of G' will be observed at low strains, while at high-stress amplitudes, G'' may exceed the storage modulus G'. This larger G'' reflects the breakage of the gel structure. Out of the LVER, storage and loss moduli are not well defined as the strain signals will not be sine functions, as they may contain different frequencies (Preziosi et al., 2010). However, Storm et al. (2005) have shown, for highly nonlinear material, that the error made in estimating the moduli by fitting such data using a sine function can be small.

Macro-rheological experiments were performed using a parallel plate rotational rheometer MRC302 (Anton Paar, Graz, Austria) at 22 °C with a plate diameter of 4 mm and a torque limit of 200 mNm. Measurements were conducted on muscle tissue samples ~1 mm thickness. The slices were placed between the plates, and the environmental chamber was closed. Amplitude sweeps were conducted at a frequency of 1 rad/s and shear range of 0.01–1%. Storage G' and loss G'' moduli values were calculated using the Anton Paar software.

3.6.3.1 Cells and Tissue Samples

Nonmalignant cancer cell of ureter cell line (HCV29) and four cancerous cell lines of different grades (RT112, G2; 5637, G2; T24, G2-3; and J82, G3) were examined. The cells were cultured in RPMI 1640 medium supplemented with 10% fetal calf serum (fetal bovine serum) at 37 °C in a humidified 5% CO_2 atmosphere. Cells were seeded on plastic (TPP Petri dishes, Switzerland) or glass substrates covered with FN with cell density allowing measurements of isolated cells after 24 and 48 h of growth.

Tibialis anterior (TA) muscles were explanted from wild-type B6 SCID mice and mdx SCID (mouse model for studying Duchenne's muscular dystrophy) during experiments on cell-transplantation-based treatment (Iyer et al., 2018). TA samples were placed in tubes containing Dulbecco's modified Eagle medium admixed with 10% DMSO (dimethyl sulfoxide) and stored in liquid nitrogen for additional experiments. Prior to the measurements, samples were defreezed to room temperature and subsequently cut using lancet and biopsy punch to obtain tissue samples of 1 mm in thickness and 4 mm in diameter. The samples were cut along the muscle fibers.

3.6.3.2 AFM-Based Characterization of Cell's Viscoelastic Properties

Rheology studies show how materials deform in time under an applied external force. Information on the viscous properties of the material can be provided by applying small-amplitude oscillatory strains or stresses to the sample. When oscillatory strain deformation is applied, the stress will also oscillate in time t, but its oscillations will be phase shifted.

Shroff et al. (1995) were the first to use an oscillating AFM probe to study the mechanical properties of rat atrial myocytes. They investigated mechanical changes of the cell during a single contraction and found a dynamic increase of cell stiffness proportional to its contraction. It has also been shown that environmental conditions (substrate rigidity, concentration of Ca^{2+} ions, and fixation) result in growth of the elasticity parameter. This approach was further developed by Alcaraz et al. (2003). The method they have introduced allows determination of the complex shear modulus $G^*(\omega)$ from oscillatory measurements over a chosen frequency range (see Section 3.6.2.1). What is important is that the method takes into account the probe-cell contact geometry as well as the viscous drag corrections in the microrheological model. Alcaraz et al. (2003) studied the microrheological properties of alveolar (A549) and lung epithelial cells (BEAS-2B) in the frequency range of 0.1–100 Hz. They showed that both G' and G'' change with frequency and that the rheology of lung epithelial cells resembles the one of soft glassy materials close to a glass transition, and assumed that structural disorder and metastability might be fundamental features of cell organization (Alcaraz et al., 2003).

3.6.4 Viscoelastic Characteristics of Bladder Cancer Cell Lines

The AFM-based microrheological technique was also applied to study viscoelastic properties of T24 bladder cancer cells (Abidine et al., 2015a, 2018). Measurements performed at different locations on the cell (over nucleus, at perinucleus, and the edge) showed that there exists a varying plateau elastic modulus, depending on the cell location: cell stiffens away from its nucleus (Figures 3.6.2 and 3.6.4A–C); thus, G_N^0 (nucleus) < G_N^0 (perinucleus) < G_N^0 (edge) (Abidine et al., 2015a), which is in agreement with the conclusions of Shroff et al. (1995).

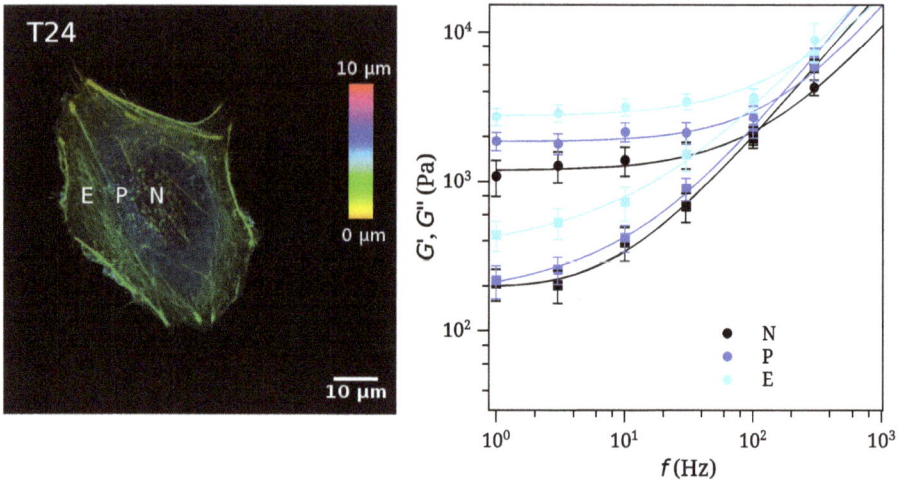

Figure 3.6.2: Left: Confocal images of a single T24 cell. A z-projection with a color scale corresponding to the height (yellow is the basal side of the cell, and red is on top of the cell). Indentations are made at three locations: nucleus (N), perinucleus (P), and edge (E). The cells are in a low migrating state. Right: Evolution of moduli G' (circle) and G'' (square) on the nucleus (black), perinucleus (blue), and edge of the cell (cyan); $N = 20$ and error bars represent SEM. Curves were fitted with the model (reprinted with permission from Abidine et al. (2015a)).

Additionally, the mechanical properties of bladder cancer cell lines of different malignancy have been compared, RT112, T24, and J82 bladder cancer cell lines, ordered by increasing malignant potential (Abidine et al., 2015a, 2018). The results indicate that $|G^*|$ decreases with invasiveness and cells become glassy (decrease of transition frequency f_T, being the crossover of G' and G''), which was shown by a decrease in storage modulus values obtained for RT112, T24, and J82 bladder cancer cell lines (Figure 3.6.3). This research also showed that viscoelastic properties of cells strongly depend on cytoskeleton organization. Bladder cancer cells treated with latrunculin A, a drug causing depolymerization of actin fibers, revealed a decrease in storage modulus at low frequencies (Abidine et al., 2015a). It has also been investigated whether substrate rigidity modifies the nanomechanical characteristics of cells. T24 cells were grown on polyacrylamide gels of rigidity of 5, 8, and 28 kPa (Figure 3.6.4), and an increase of $|G^*|$ was observed, showing a clear mechanosensitivity effect. It has already been reported that measurements of mechanical properties of cells grown on rigid substrates may result in overestimated Young's modulus (E) values; interestingly, data presented in Abidine et al. (2018) indicate that the elastic modulus E obtained for cells grown on soft substrates may result in an underestimation of E. This assumption is illustrated by comparison of G^* and G^*_{cor} data obtained for bladder cancer lines, where $|G^*|$ is the modulus calculated from eq. (3.6.4) above and $|G^*_{cor}|$ from eqs. (3.6.5) and (3.6.6) in Abidine et al. (2018).

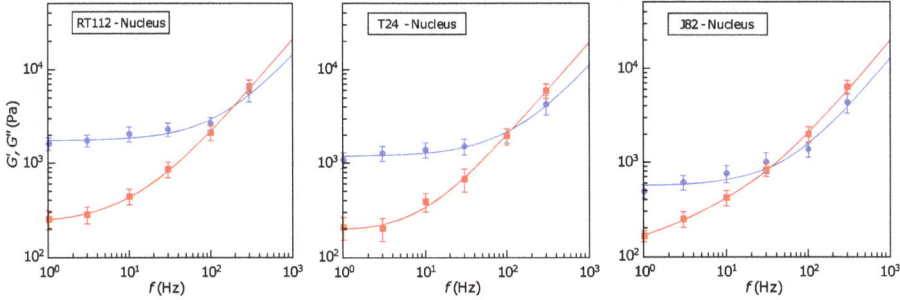

Figure 3.6.3: Evolution of G′ (blue circle) and G″ (red square) on the nucleus (N) of three cancer cell lines: RT112, T24 and J82 (left to right, respectively, N = 10, N = 20, N = 10; error bars represent mean ± SEM). Reprinted with permission from Abidine et al. (2015a).

Studies on characterization of viscoelastic properties of bladder cancer cell lines have been continued using nonmalignant ureter cancer, HCV29, and 5637 bladder cancer cell lines. The T24 cell line was also investigated as a reference, allowing to compare the obtained results with data presented by Abidine et al. (2015a). The measurements were performed using a similar setup and protocol as described in Section 3.6.2.1 over cell nuclei after 48 h of cell growth. The data have been fitted using a simplified model, where the cell elastic and viscous moduli are described with the following equations:

$$G'(\omega) = G_N^0 + k_1 \cdot \omega^a \tag{3.6.8}$$

$$G''(\omega) = k_0 + b \cdot k_1 \cdot \omega^a \tag{3.6.9}$$

With this approach, the angular transition frequency ω_T corresponding to the crossing of G′ and G″ can be simply calculated from solving the equation G′ = G″, which gives (since b > 1)

$$\omega_T = \left(\frac{G_N^0 - k_0}{k_1 \cdot (b-1)} \right)^{\frac{1}{a}} \tag{3.6.10}$$

The evolution of G′ and G″ as a function of angular frequency is presented in Figure 3.6.5D–F. The plateau modulus values obtained for the T24 cell line are in agreement with data published in Abidine et al. (2015a) (compare Figures 3.6.3 and 3.6.4). We find that benign cells are stiffer than the cancerous ones, as expected.

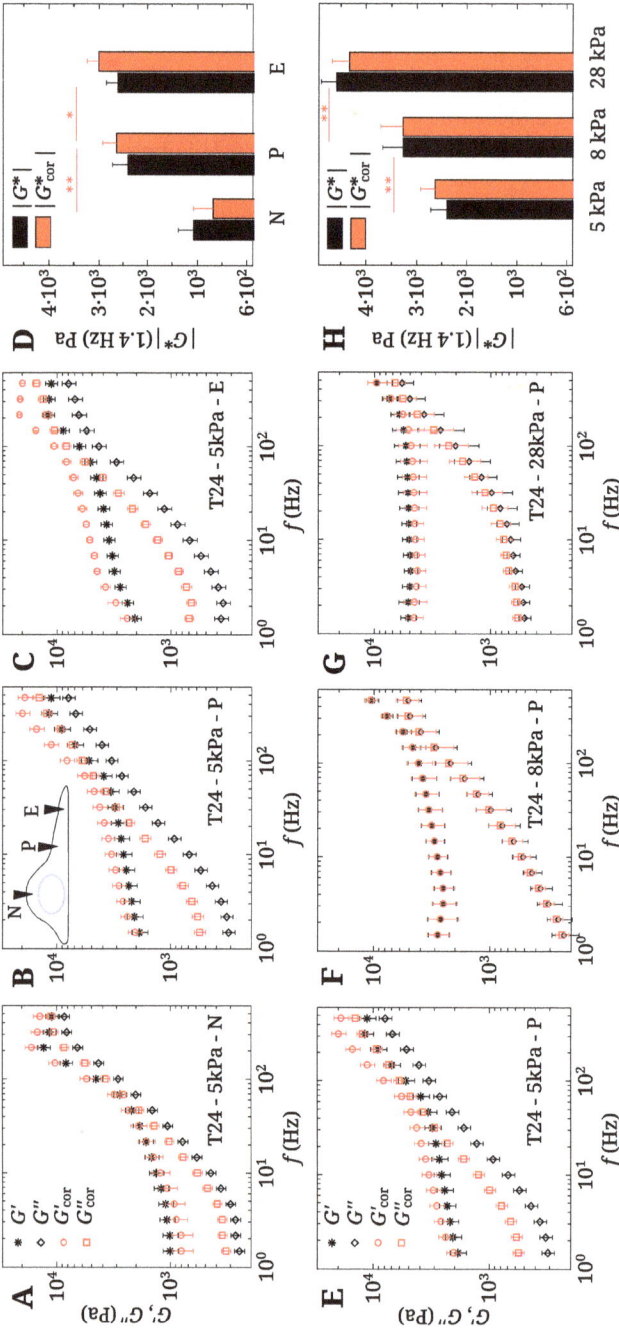

Figure 3.6.4: Raw and corrected viscoelastic moduli of T24 cells. (A–C) Data on a 5 kPa gel at three locations (nucleus (N), perinuclear region (P), and edge (E) are shown. (D) Modulus $|G^*|$ (1.4 Hz) at the three locations (N, P, and E) is shown. (E–G) Data measured in the perinuclear region (P) of cells on three gels (E2 = 5, 8, and 28 kPa). N = 5, and error bars represent the mean ± SE. (H) Modulus $|G^*|$ (1.4 Hz) for the three gels are shown (E2 = 5, 8, and 28 kPa). Statistical relevance is shown for corrected values of $|G^*|$. Reprinted with permission from Abidine et al. (2018).

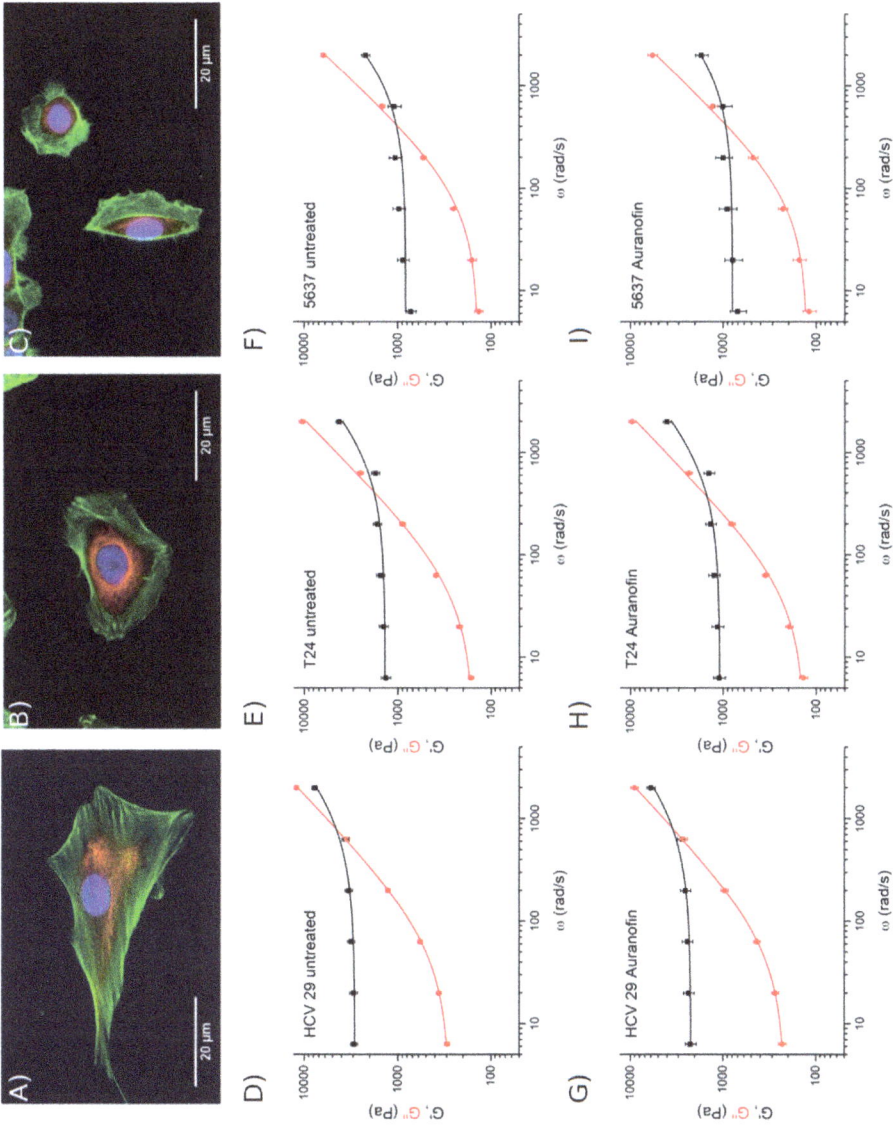

Figure 3.6.5: Cytoskeleton structure of HCV29, T24, and 5637 cell lines imaged with fluorescence microscopy. Actin filaments are stained in green, microtubules are in red, and nuclei in blue (A–C). G′ (black) and G″ (red) of untreated (D–F) and Auranofin-treated (G–I) bladder cancer cell lines. Data are presented as mean ± SEM. N(untreated) = 72 (HCV29), 58 (T24), 31 (5637), and N (Auranofin) = 44 (HCV29), 46 (T24), 24 (5637) (unpublished data, J Zemła, IFJPAN).

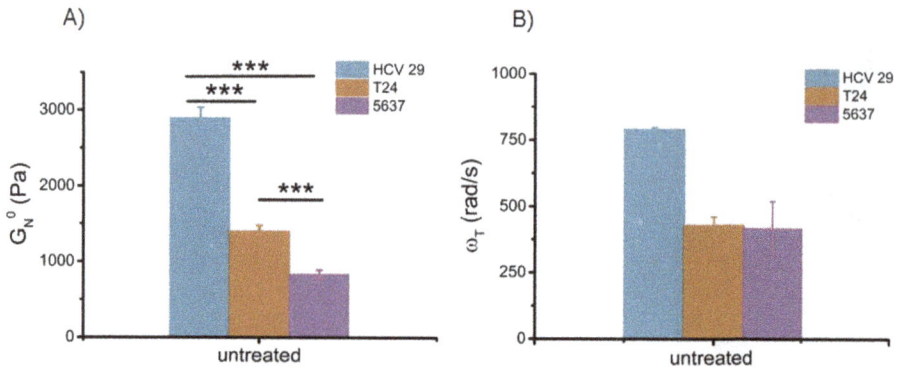

Figure 3.6.6: Mean values of plateau modulus $G_N{}^0$ (A) and transition angular frequency ω_T (B) obtained for nonmalignant (HCV29) and cancer cell lines (5637, G2 and T24, G2-3). The error bars are SEM, and statistical analysis was performed with a two-sample unpaired Student's t-test (*** refer to $p < 0.001$). N = 72 (HCV29), 58 (T24), 31 (5637) (unpublished data, J Zemła, IFJPAN).

Additionally, the plateau modulus of the 5637 cell line is lower than both HCV29 and T24 cell lines (Figure 3.6.6).

This effect may result from their cytoskeleton structure, which is more complex in the case of HCV29 cells (Figure 3.6.5A–C). Fluorescent staining reveals stress fibers in HCV29 as well as in T24 cell lines. The cytoskeleton of 5637 cells is very poorly developed. They seem to be unable to develop either stress fibers or microtubules. The obtained results allow to assume that the cytoskeleton structure is closely linked to the mechanical properties of cell.

It has been concluded that in the case of bladder cancer cell lines, changes of cell rigidity are correlated with their invasive potential (Abidine et al., 2015a, 2018). In Figure 3.6.6, plateau modulus mean values as well as transition angular frequency mean values calculated for the abovementioned bladder cancer cell lines are compared. The microrheological properties of the benign cell line are characterized by the highest values of $G_N{}^0$ and ω_T, which is in the agreement with the previous conclusions. Yet, investigation of Figure 3.6.6B reveals no difference (within error bars) between ω_T values obtained for T24, grades 2–3, and 5637, grade 2, cancer cell lines. These results may indicate that plateau modulus values are determined by cytoskeleton structure, and the transition frequency is the parameter strongly correlated with cell invasiveness, that is, its ability to remodel the cytoskeleton rapidly or not.

It has already been shown that exposing cells to chemicals modifying their cytoskeleton structure results in changes of microrheological properties (Abidine et al., 2015a). However, the intriguing question was if other types of anticancer drugs modify the cell's viscoelastic properties as well. HCV29, T24, and 5637 bladder cancer cell lines were exposed for 72 h to Auranofin (AF), a chemical causing hyperoxidation via mitochondria deregulation. A relatively low dose of 0.2 μM was

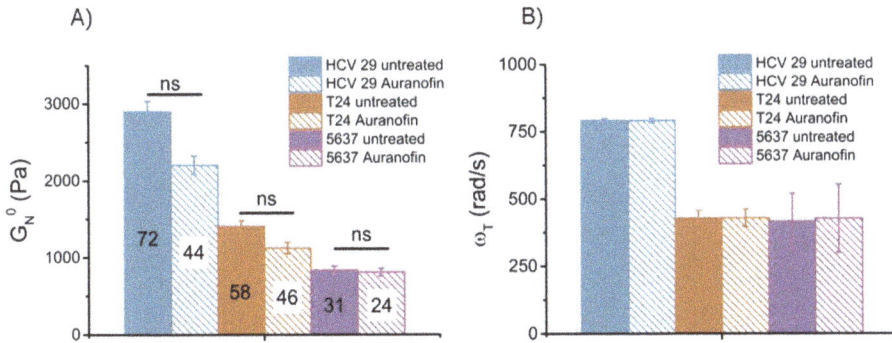

Figure 3.6.7: Comparison of microrheological model parameters calculated for HCV29, T24, and 5637 cell lines treated with Auranofin. G_N^0 and ω_T values obtained for reference samples are plotted as well. The error bars are SEM, and statistical analysis was performed with a two-sample unpaired Student's t-test (p > 0.05 refers to no statistical significance, ns). In bars, the amounts of measured cells are indicated (unpublished data, J Zemła, IFJPAN).

used. The concentration was chosen based on cell proliferation tests (MTS, Sigma Aldrich) performed at AF concentrations of 0.01, 0.1, 1.0, and 10 µM. The results were normalized to control samples, and no drug effect on bladder cells proliferation was observed at concentrations of 0.01 and 0.1 µM. About 1.0 µM AF did not affect HCV29 proliferation, while the number of T24 cells decreased by 30% and 5637 cells by 75%. The results indicate that cells respond to 0.2 µM AF by changes in proliferation rate.

Before the microrheological measurements, the drug-containing medium was replaced with supplemented RPMI 1640 medium. The cells were indented on top of their nuclei. The G' and G'' values were calculated using eq. (3.6.4). The data presented in Figures 3.6.5 and 3.6.7 show that this dose of AF has not influenced the mechanical properties of studied cells, which may result from either too low AF concentration or the fact that this chemical does not target cytoskeleton structures directly.

3.6.5 Viscoelastic Properties of Model Tissues

We first start to study model tissues. Such tissues contain extracellular matrix (ECM) and cells. They can be tested using classical rheology using a plate–plate rheometer (Iordan et al., 2010). This geometry is preferred in this case since it allows to compress the biological tissue into the rheometer without including too much prestress through normal forces. Several studies are discussed, where the concentration of the collagen is varied and the cell concentration is also changed. In our study, we used CHO cells (Chinese Hamster Ovary). Considering the collagen only, we find that the

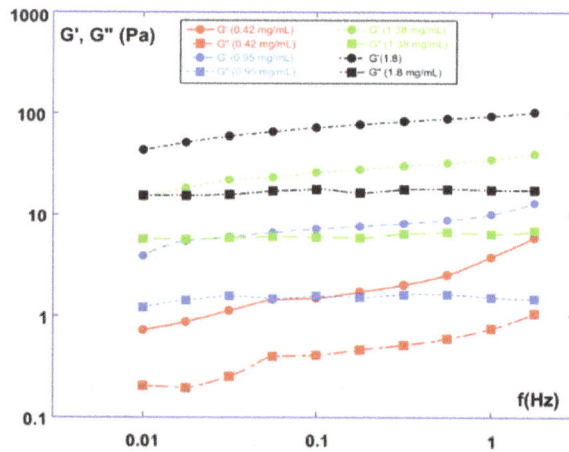

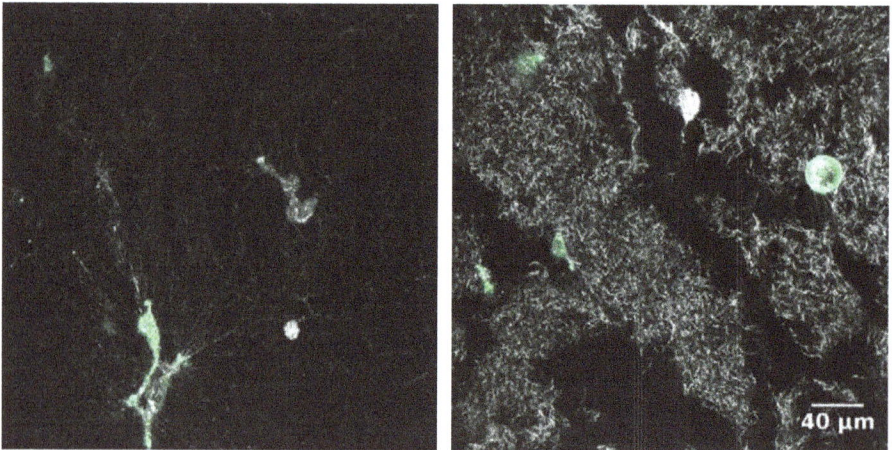

Figure 3.6.8: Left: Viscoelastic properties of collagen networks (G' and G") as a function of collagen concentration, from 0.42 to 1.8 mg/mL. Center and right: CHO cells embedded in collagen matrix (center 0.42 mg/mL, and right 1.8 mg/mL). Reprinted with permission from Iordan et al. (2010).

collagen network without cells is viscoelastic, as expected, with a slightly higher elastic modulus G' as compared to the loss modulus G'' (see Figure 3.6.8).

The plateau modulus $G_N{}^0$ can be obtained at the lowest frequency (0.01 Hz) and plotted in terms of the collagen concentration c and leads to a relationship of the kind $G_N{}^0 \sim c^{2.6}$, in agreement with the work of Vader et al. (2009). Note that the LVER lies below 5% strain in such gels.

Now we insert cells into the matrix (collagen network) and the results show again viscoelastic effects (Iordan et al., 2010). The amount of CHO cells that were included goes up to $1.8 \cdot 10^7$ cells/mL. There is an elasticity increase as cells are inserted into the gel (e.g., at 0.95 mg/mL G' increases from 3 to 8 Pa) but at collagen concentrations higher than 0.95 mg/mL, the inverse is observed and the elasticity decreases (e.g., from 50 Pa down to 10 Pa for the 1.8 mg/mL collagen concentration). Plotting again G_N^0 as a function of c leads to a smaller slope $G_N^0 \sim c^1$, raising the question of what governs such a dependence.

Obviously, interactions between cells and the matrix take place, which can be followed, thanks to confocal reflectance microscopy (Friedl et al., 2001). Figure 3.6.8 shows typical CHO cells embedded in two different collagen matrices (0.42 and 1.8 mg/mL). Their behavior changes as there is enough space at low collagen concentration to migrate and adhere to the fibers; therefore, they are more elongated. On the other hand, at higher concentrations, they remain round as there is no space. It was also shown that cells attract collagen as they move slowly into the matrix and carry it along, leaving holes or tunnels behind (see last figure on the right). We conclude that the interplay between collagen fibers and cells has two opposite effects:

a) Cells in low-density networks have space, and elongate, pull on fibers, modifying the collagen structure, and therefore its rheology;

b) cells in high-density matrix have no space so they interact more with collagen to remodel it. They can attract it, and therefore can dig tunnels through it. Hence, the collagen structure breaks down, and the relationship between the plateau modulus G_N^0 and the concentration shows a much smaller slope at high concentrations.

To conclude, ECM interactions with cells are complex and need be taken into account for proper understanding/modeling of tissues. Classical hyperelasticity laws for tissues can be used but they need to be refined, taking into account cell–cell and cell–ECM interactions in order to be more accurate, as proposed in recent works (Preziosi et al., 2010).

3.6.6 Macrorheological Properties of Brain, Liver, and Kidney Tissues

It is a known fact that many pathological reactions such as cancer (Caroline et al., 2014), liver and kidney fibrosis (Desmoulière et al., 2003), respiratory system diseases (Zemła et al., 2018b), or Duchenne muscular dystrophy (DMD) (Puttini et al., 2009, van Zwieten et al., 2014) manifest themselves by tissue/ECM remodeling. The remodeling results in changes of mechanical properties of tissues, which was revealed by Young's modulus determination (AFM measurements) of diseased (diseased and healthy) brain (Ciasca et al., 2016), breast (Plodinec et al., 2012, Ansardamavandi et al., 2016), liver (Tian et al., 2015), and cervical cancer biopsies (Cui et al., 2017), or asthmatic (Zemła et al., 2018b) and DMD (van Zwieten et al., 2014, Iyer et al., 2018) tissue samples, and also shown in model tissues (Section 5.2.3). At microscale, AFM measurements of tissue samples allow to distinguish between the pathological and normal tissue samples, although the complexity of the tissue structure results in wide E distributions (Ansardamavandi et al., 2016, Zemła et al., 2018b) and, in some cases, in the multimodal distributions of E (Tian et al., 2015, Iyer et al., 2018). This, however, does not discriminate AFM as a tool to study mechanical properties of tissues but it seems that the complexity of the system may require a different method to reveal their viscoelastic properties such as macrorheological measurements.

First, studies of macrorheological properties of biological samples were performed for brain tissue samples (Bilston et al., 1997, Bilston, 2003, Nicolle and Palierne, 2012, Mao et al., 2012, Pogoda et al., 2014, Mihai et al., 2015, Qing et al., 2019) and liver (Tan et al., 2013, Perepelyuk et al., 2016). The abovementioned oscillatory shear measurements revealed a narrow LVER of tissue samples at low strains (below 1%). Additionally, the issue of sample immobilization was also discussed. There are three common approaches:
(1) no modification of plates,
(2) gluing sandpaper disks to rheometer plates, and
(3) gluing the sample with a cyanoacrylate adhesive to the plates.

Nicolle and Palierne (2012) reviewed results obtained at these conditions and compared moduli values of porcine kidney tissue with respect to the tissue immobilization method. They found no difference in storage and loss moduli for glue and sandpaper immobilization type. The results, however, show that measurements performed without any additional tissue attachment resulted in decreased values of G' and G''. To avoid tissue sample slippage during oscillatory test, a preload, compressive strain exerted on a sample may be applied (Tan et al., 2013). It has to be mentioned, though, that measured G' and G'' moduli values depend on the preload. The higher compressive strain is applied, the higher storage and loss moduli we get (Tan et al., 2013, Pogoda et al., 2014, Mihai et al., 2015). These results suggest that for comparative studies, the loss factor $\tan(\varphi)$ may be a good parameter as G'/G'' ratio does not depend on sample immobilization protocol (Nicolle and Palierne, 2012).

Recently, macrorheological measurements have been performed to compare viscoelastic properties of normal and sclerosis complex (TSC) brain tissue (Qing et al., 2019). TSC is a genetic disorder with a high penetrance of autism spectrum disorders. Unfortunately, both AFM-nanomechanical characterization of the samples and macrorheological measurements did not exhibit differences between healthy and pathological tissue samples.

3.6.7 Macrorheological Characteristics of Duchenne Muscular Dystrophy

Strain amplitude sweep tests (Section 3.6.3) were performed to characterize the viscoelastic properties of normal and DMD tissue samples (Figure 3.6.9A). The samples were prepared according to the procedure described in Section 3.6.3.1. Macrorheological tests were performed at oscillations of 1 rad/s and at strain range of 0.01–2%. Higher storage modulus values were obtained for wild-type TA samples in comparison to mdx tissues (Figure 3.6.9).

We have observed an increase of the G' modulus with increasing shear strain. Interestingly, a similar behavior was shown for cross-linked biopolymer networks (Storm et al., 2005). Indeed, TA muscle structure resembles that of a cross-linked network in a way, as it is built of aligned muscle fibers interconnected with collagen fibers present in endomysium, which envelopes each muscle fiber (Beunk et al., 2019). In Figure 3.6.9B, loss factor values calculated for both types of samples studied are plotted, and regardless of strain amplitude G''_{mdx} to G'_{mdx} ratio is higher than G''_{wt} to G'_{wt} ratio, which indicates that the storage modulus contributes less to the overall mechanical characteristics of mdx tissues as compared to wt tissue samples.

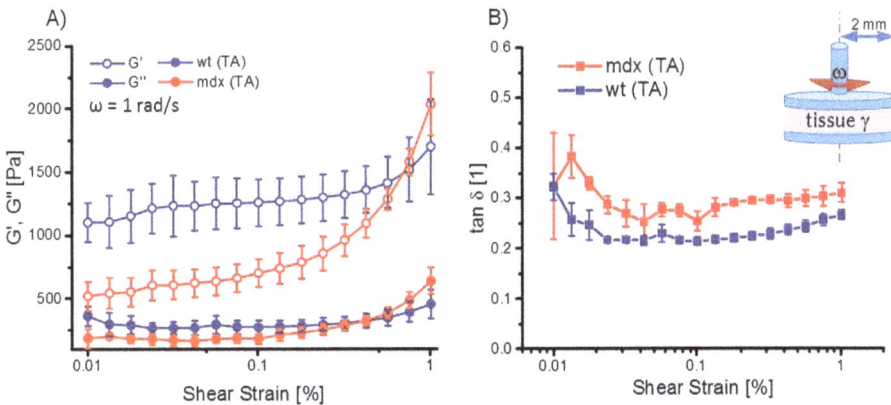

Figure 3.6.9: Comparison of storage (G′) and loss (G″) modulus of normal (wt) and Duchenne dystrophy (mdx) mice tibialis anterior (TA) muscles. (A) Loss factor calculated for wt and mdx tissues. In the inset, the scheme of the rheometer is presented. (B) Data are mean ± SEM, N = 5 (unpublished data, J Zemła, IFJ PAN).

3.6.8 Conclusions

Rheological tools, both at the micro- and macroscales, are important tools to help understand the properties of cells and tissues. In any case, they cannot alone predict or elucidate all properties. When coupled with complementary observation techniques (classical and confocal microscopy, new super-resolution microscopy, STED or STORM, ultrasound, X-rays, biology techniques), they allow to correlate the time-dependent microstructure of cells/tissues in order to understand the main mechanical features.

On the experimental side, there are many challenges to investigate the many different cell/tissue systems which still remain difficult to obtain, prepare, and characterize according to well-defined protocols needing an impressive expertise. Probably, this is the most important task that remains to be considered in the future, but it is very important, in particular since normal and insane tissues definitely need to be characterized and compared.

From a theoretical point of view, it is still difficult nowadays to come up with rheological models containing both viscoelastoplastic effects and the active nature of the cells, and include their interactions between themselves and the ECM. The new challenge for the next decades will be to use the available models and enrich them with such properties.

References

Abidine, Y., A. Constantinescu, V. M. Laurent et al. (2018) "Mechanosensitivity of cancer cells in contact with soft substrates using AFM". Biophysical Journal **114**: 1165–1175.

Abidine, Y., V. M. Laurent, R. Michel et al. (2013) "Microrheology of complex systems and living cells using AFM". Computer Methods in Biomechanics and Biomedical Engineering **16**: 15–16.

Abidine, Y., V. M. Laurent, R. Michel et al. (2015a) "Local mechanical properties of bladder cancer cells measured by AFM as a signature of metastatic potential". European Physical Journal Plus **130**: 202–215.

Abidine, Y., V. M. Laurent, R. Michel et al. (2015b) "Physical properties of polyacrylamide gels probed by AFM and rheology". Europhysics Letters **109**: 38003–38008.

Alcaraz, J., L. Buscemi, M. Grabulosa et al. (2003) "Microrheology of human lung epithelial cells measured by atomic force microscopy". Biophysical Journal **84**: 2071–2079.

Ansardamavandi, A., M. Tafazzoli-Shadpour, R. Omidvar and I. Jahanzad (2016). "Quantification of effects of cancer on elastic properties of breast tissue by Atomic Force Microscopy". Journal of the Mechanical Behavior of Biomedical Materials **60**: 234–242.

Baumgaertel, M., A. Schausberger, H.H. Winter (1990). "The relaxation of polymers with linear flexible chains of uniform length". Rheological Acta **29**:400–408.

Beunk, L., K. Brown, I. Nagtegaal et al. (2019) "Cancer invasion into musculature: Mechanics, molecules and implications". Seminars in Cell & Developmental Biology **93**: 36–45.

Bilston, L. E. (2003) "Brain tissue properties at moderate strain rates" IMECE2003-4. In: American Society of Mechanical Engineers, Bioengineering Division (Publication) BED55:3–4.

Bilston, L. E., Z. Liu and N. Phan-Thien (1997). "Linear viscoelastic properties of bovine brain tissue in shear". Biorheology **34**: 377–385.

Bursac, P., G. Lenormand, B. Fabry, M. Oliver, D. A. Weitz, V. Viasnoff, J. P. Butler and J. J. Fredberg (2005). "Cytoskeletal remodelling and slow dynamics in the living cell". Nature Materials **4**: 557–561.

Butt, H. J. and M. Jaschke (1995). "Calculation of thermal noise in atomic force microscopy". Nanotechnology **6**: 1–7.

Caroline, B., C. Jonathan and Z. Werb (2014). "Remodelling the extra-cellular matrix in development and disease". Nature Reviews. Molecular Cell Biology **15**: 786–801.

Ciasca, G., T. E. Sassun, E. Minelli et al (2016) "Nanomechanical signature of brain tumours". Nanoscale **8**: 19629–19643.

Cui, Y., X. Zhang, K. You et al (2017) "Nanomechanical characteristics of cervical cancer and cervical intraepithelial neoplasia revealed by atomic force microscopy". Medical Science Monitor **23**: 4205–4213.

Desmoulière, A., I. A. Darby and G. Gabbiani (2003). "Normal and pathologic soft tissue remodeling: role of the myofibroblast, with special emphasis on liver and kidney fibrosis". Lab Investig **83**: 1689–1707.

Florea, C., P. Tanska, M. E. Mononen et al (2017) "A combined experimental atomic force microscopy-based nanoindentation and computational modeling approach to unravel the key contributors to the time-dependent mechanical behavior of single cells". Biomechanics and Modeling in Mechanobiology **16**: 297–311.

Friedl, P., S. Borgann and E. B. Bröcker (2001). "Amoeboid leukocyte crawling through extra-cellular matrix: Lessons from the Dictyostelium paradigm of cell movement". Journal of Leukocyte Biology 491–509.

Iordan, A., A. Duperray, A. Gérard, A. Grichine and C. Verdier (2010). "Breakdown of cell-collagen networks through collagen remodeling". Biorheology **47**: 277–295.

Iyer, P. S., L. O. Mavoungou, F. Ronzoni et al (2018) "Autologous cell therapy approach for Duchenne muscular dystrophy using PiggyBac transposons and mesoangioblasts". Molecular Therapy **26**: 1093–1108.

Mao, J., S. Duan, A. Song et al (2012). "Macroporous and nanofibrous poly(lactide-co-glycolide)(50/50) scaffolds via phase separation combined with particle-leaching". Materials Science and Engineering C 1407–1414.

Mihai, L. A., L. K. Chin, P. A. Janmey and A. Goriely (2015). "A comparison of hyperelastic constitutive models applicable to brain and fat tissues". Journal of the Royal Society, Interface/The Royal Society **12**: 20150486–20150497.

Moeendarbary, E., L. Valon, M. Fritzsche et al (2013) "The cytoplasm of living cells behaves as a poroelastic material". Nature Materials **12**: 253–261.

Nicolle, S. and J. F. Palierne (2012). "On the efficiency of attachment methods of biological soft tissues in shear experiments". Journal of the Mechanical Behavior of Biomedical Materials **14**: 158–162.

Palade, L. I., V. Verney and P. Attané (1996). "A modified fractional model to describe the entire viscoelastic behavior of polybutadienes from flow to glassy regime". Rheologica Acta **35**: 265–273.

Perepelyuk, M., L. Chin, X. Cao et al (2016) "Normal and fibrotic rat livers demonstrate shear strain softening and compression stiffening: A model for soft tissue mechanics". PLoS One **11**: 1–18.

Plodinec, M., M. Loparic, C. A. Monnier et al (2012) "The nanomechanical signature of breast cancer". Nature Nanotechnology 7: 757–765.

Pogoda, K., L. Chin, P. C. Georges et al (2014) "Compression stiffening of brain and its effect on mechanosensing by glioma cells". New Journal of Physics 16: 075002.

Preziosi, L., D. Ambrosi and C. Verdier (2010). "An elasto-visco-plastic model of cell aggregates". Journal of Theoretical Biology 262: 35–47.

Puttini, S., M. Lekka, O. M. Dorchies et al (2009) "Gene-mediated restoration of normal myofiber elasticity in dystrophic muscles". Molecular Therapy 17: 19–25.

Qing, B., E. P. Canovic, A. S. Mijailovic et al (2019) "Probing Mechanical Properties of Brain in a Tuberous Sclerosis Model of Autism". Journal of Biomechanical Engineering 141: 031001.

Shroff, S. G., D. R. Saner and R. Lal (1995). "Dynamic micromechanical properties of cultured rat atrial myocytes measured by atomic force microscopy". The American Journal of Physiology 269: C286–C292.

Sneddon, I. N. (1965). "The relation between load and penetration in the axisymmetric Boussinesq problem for a punch of arbitrary profile". International Journal of Engineering Science 3: 47–57.

Sollich, P., F. Lequeux, P. Hébraud and M. E. Cates (1997). "Rheology of soft glassy materials". Physical Review Letters 78: 2020–2023.

Storm, C., J. J. Pastore, F. C. MacKintosh et al (2005) "Nonlinear elasticity in biological gels". Nature 435: 0–3.

Tan, K., S. Cheng, L. Jugé and L. E. Bilston (2013). "Characterising soft tissues under large amplitude oscillatory shear and combined loading". Journal of Biomechanics 46: 1060–1066.

Tian, M., Y. Li, W. Liu et al (2015) "The nanomechanical signature of liver cancer tissues and its molecular origin". Nanoscale 7: 12998–13010.

Vader, D., A. Kabla, D. Weitz and L. Mahadevan (2009). "Strain-induced alignment in collagen gels". PLoS One 4: e5902.

van Zwieten, R. W., S. Puttini, M. Lekka et al (2014) "Assessing dystrophies and other muscle diseases at the nanometer scale by atomic force microscopy". Nanomedicine 9: 393–406.

Verdier, C., J. Etienne, A. Duperray and L. Preziosi (2009). "Review: Rheological properties of biological materials". Comptes Rendus Physique 10: 790–811.

Zemła, J., J. Danilkiewicz, B. Orzechowska et al (2018a) "Atomic force microscopy as a tool for assessing the cellular elasticity and adhesiveness to identify cancer cells and tissues". Seminars in Cell & Developmental Biology 73: 115–124.

Zemła, J., T. Stachura, I. Gross-Sondej et al (2018b) "AFM-based nanomechanical characterisation of bronchoscopic samples in asthma patients". Journal of Molecular Recognition 31: e2752.

Chau Ly, Tae-Hyung Kim, Amy C. Rowat

3.7 Fluidic Approaches to Measure Cellular Deformability

Over the past decades, we have gained detailed insights into the fundamental mechanical properties of cells through pioneering methods such as micropipette aspiration (Mitchison et al., 1954, Hochmuth, 2000, Evans and Kukan, 1984, Needham and Nunn, 1990, Dahl et al., 2005), atomic force microscopy (AFM) (Hoh and Schoenenberger, 1994, Rotsch and Radmacher, 2000, Maloney et al., 2010, Rosenbluth et al., 2006, Müller and Dufrêne, 2011), and optical tweezers (Zhang and Liu, 2008, Dao et al., 2003, Guck et al., 2001). Such methods have advanced knowledge of the molecules that mediate cellular mechanotype. Actin and microtubules are well-established regulators of cellular mechanotype (Fletcher and Mullins, 2010, Murrell et al., 2015). Other specific mechanosignaling mediators and pathways such as nuclear protein lamin A and YAP/TAZ signaling also contribute to the regulation of cellular mechanotype (Roca-Cusachs et al., 2012, Cho et al., 2017, Discher et al., 2017, Engler et al., 2006, Swift et al., 2013, Nardone et al., 2017). While existing mechanotyping methods have advanced our understanding of cellular mechanotype, higher throughput methods could deepen our knowledge of diverse mechanisms that mediate cellular mechanotype and enable mechanotyping to be harnessed for translational applications.

In this chapter, we discuss fluidic approaches to measure cellular deformability, which harness fluid flow to deform cells. The filtration of cell suspensions has provided a straightforward approach to assay cellular deformability for decades and, more recently, is enabling higher throughput assays. Another class of fluidic methods enables rapid, single-cell measurements of cellular deformability. Fluidic approaches to measure cell deformability have exciting potential to deepen our fundamental knowledge of cell biology, to enhance the characterization of cells in diseased states, and to identify potential therapeutic interventions.

Acknowledgments: We are grateful to support from the National Science Foundation (BMMB-1906165 and BRITE Fellow Award CMMI-2135747 to A.C.R.), the Jonsson Comprehensive Cancer Foundation, the University of California Cancer Research Coordinating Committee (CRR-18-526901 to A.C.R.), and the Farber Family Foundation.

Chau Ly, Tae-Hyung Kim, Amy C. Rowat, University of California, Los Angeles

https://doi.org/10.1515/9783110640632-013

3.7.1 Filtration Assays to Measure Cellular Deformability

Cellular filtration methods have been inspired by the vascular system, whereby red blood cells with a diameter of 6–8 μm flow through capillaries as small as 3 μm (Downey et al., 1990, Buchan, 1980). Filtration relies on driving a suspension of cells to deform through a porous membrane; the resultant flow behavior can be detected by measuring flow rate or mass of filtrate (Downey and Worthen, 1988, Qi et al., 2015). Cellular filtration has revealed how factors impact the flow of red and white blood cells, such as blood viscosity and hemoglobin levels (Gregerson et al., 1967, Chien et al., 1971, Buchan, 1980, Worthen, et al., 1989, Downey et al., 1990). Filtration methods have since been developed for application to other cell types including cancer cells, fibroblasts, and stem cells (Qi et al., 2015, Gill et al., 2019b, Lipowsky et al., 2018).

Key components of a filtration assay include a membrane that contains micrometer-scale pores that are typically 0.25–0.75 the cell diameter and a method to drive a suspension of cells across the membrane, such as a perfusion pump (Chien et al., 1971) or air pressure (Qi et al., 2015). Cellular filtration also requires an approach to detect the flow of fluid and/or cells across the porous membrane, which depends on the number of occluded pores. The rate of filtration (Chien et al., 1971), final filtration volume (fluid mass or absorbance) (Qi et al., 2015, Jones et al., 1985, Gill et al., 2019b), or number of cells quantified by radiolabeling (Downey and Worthen, 1988) have all been used to quantify cellular filtration.

3.7.1.1 Theoretical Framework to Understand Filtration

The ability of cells to deform through micron-scale pores is determined by the pore size, applied pressure, cell density, filtration time, and cellular physical properties, including stiffness and size (Qi et al., 2015). To understand the physical mechanisms governing cell filtration, Darcy's law provides a simple model to describe the flow of fluid through a porous material (eq. (3.7.1), Figure 3.7.1; Darcy, 1856). The flow rate (Q) of a suspension of cells through micron-scale pores is described by the change in filtrate volume (V) over time (t). The flow rate depends on the applied pressure (P), flow area (A), membrane thickness (L), viscosity of the cell medium (μ), and membrane permeability (k), which decreases over time as cells occlude pores:

$$Q = \frac{dV}{dt} = \frac{\Delta PAk(t)}{L\mu} \qquad (3.7.1)$$

Filtration assays to measure cellular deformability

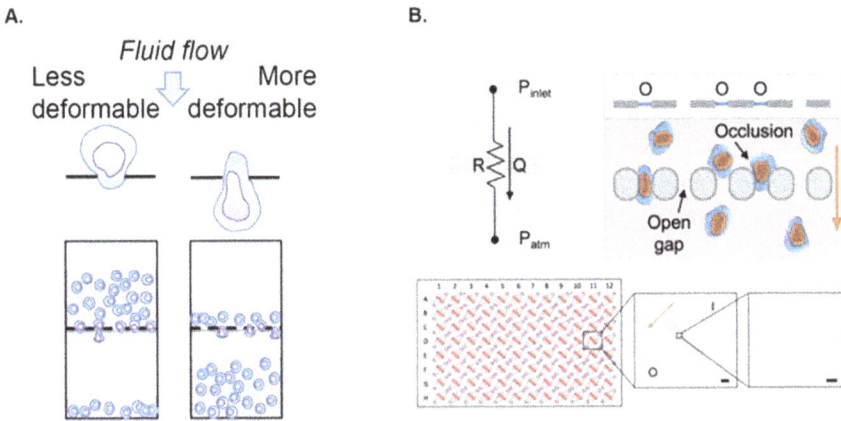

Single-cell fluidic assays to measure cellular deformability

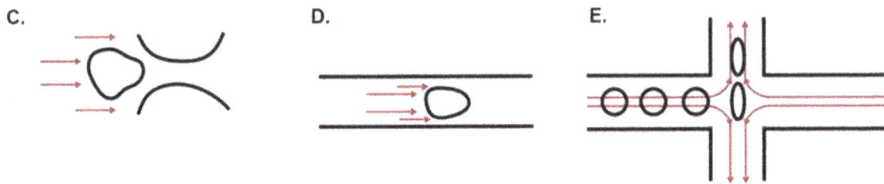

Figure 3.7.1: Overview of fluidic methods to probe cell deformability. (A) Cell filtration (Qi et al., 2015). The volume of filtrate in the bottom well or volume of retained fluid in the top well can be used as a readout for cellular deformability. (B) High-throughput filtration (Gill et al., 2019b). To simulate the filtration process, the filtration device can be considered as an electric circuit, where R is the fluidic resistance, Q is the flow, P_{inlet} is the driving pressure, and P_{atm} is atmospheric pressure. The schematic illustration shows how fluidic resistance is determined by the number of occluded (O) versus open gaps. Bottom row shows a plan view of an array of 96 filtration devices. Middle inset shows a single filtration device. Arrow indicates the direction of fluid flow from inlet (I) to outlet (O). Scale, 1 mm. Right inset shows an array of posts within a single filtration device through which cells need to deform in order to filter through the device. Scale, 100 µm. Single-cell methods to measure cell deformability include (C) quantitative deformability cytometry (q-DC). The deformation of cells through micron-scale constricted channels to measure transit time or elastic modulus based on the time-dependent changes in cell shape (Nyberg et al., 2017). (D) In real-time deformability cytometry (RT-DC), cell deformations are quantified by the changes in cell shape due to fluid shear stresses (Otto et al., 2015). (E). Inertial deformability cytometry (DC) subjects cells to the forces generated by inertial flow at a junction of channels in a microfluidic device (Gossett et al., 2012). Figure (B) reprinted with permission.

The decreasing membrane permeability over time is captured by $k(t)$ as cells occlude the pores over the timescale of filtration (Qi et al., 2015, Belfort et al., 1994).

More detailed mathematical models have been developed to explain the blockage of the pores, or fouling behavior that occurs during filtration. The pore blockage model describes how pores occluded by cells result in a localized reduction in flux; flowing cells can only pass through unblocked pores and/or deposit on top of occluded pores (Ho and Zydney, 2000, Duclos-Orsello et al., 2006). The cake filtration model accounts for cells that deposit on top of the membrane as pores become occluded and thereby increase the hydraulic resistance over time (Duclos-Orsello et al., 2006, Lu et al., 2001). These models to describe cellular filtration are not mutually exclusive (Duclos-Orsello et al., 2006).

3.7.1.2 High-Throughput Filtration

Recent advances have facilitated higher throughput measurements of cell filtration in a parallelized, multiwell format: our group developed a technique known as parallel microfiltration (PMF), which enables a 96-well array of individual cell samples to be simultaneously filtered through polycarbonate membranes with micrometer-sized pores by applying air pressure uniformly across wells (Qi et al., 2015). After pressure is applied on the timescale of seconds, the volume of filtrate collected provides a metric of cellular deformability: less deformable cells are more likely to occlude the pores, thereby preventing the cell medium from filtering through and reducing the filtrate volume (Figure 3.7.1B). The volume or mass of fluid retained above the membrane can equivalently be measured as an indicator of cellular deformability, whereby less deformable cells result in increased retention. When conducting PMF assays, it is important to note that both cell stiffness and size can impact filtration; cells that are larger and/or stiffer are more likely to occlude pores (Qi et al., 2015, Nyberg et al., 2017). While we have not observed any differences in cell size across numerous pharmacologic and genetic perturbations of various cell types (Gill et al., 2019a, Kim et al., 2016, 2019), cell size distributions should nonetheless be quantified alongside each filtration experiment; in the case of a strong positive correlation between cell size and filtration, more detailed studies of cell stiffness can be conducted, for example, using a method that enables quantification of cell size and deformability at the single-cell level such as deformability cytometry (Figure 3.7.1C–E; see Section 3.7.2) or AFM (Chapter 3.1).

PMF provides an accessible assay that can be used to quickly evaluate changes in cellular deformability. The PMF setup is straightforward to build and implement, since readouts require only access to a plate reader or balance (Table 3.7.1; Gill et al., 2017). Moreover, porous polycarbonate membranes typically used in filtration assays are commercially available in a range of pore sizes from 3 to 12 µm. However, the required user input to set up and operate the device can lead to variability between users.

Alternatively, microfabrication methods can be used to generate polydimethyl-siloxane (PDMS) membranes that contain arrays of 96–384 individual filtration devices; each device has a region of micropillars, where cells are forced to deform between in order to filter through the device (Figure 3.7.1B; Gill et al., 2019b). While the fabrication of PDMS membranes requires additional resources – such as a cleanroom facility – the operation of such devices requires less user input and may therefore be subject to less user variability. As in any high-throughput assay, the initial screen should identify conditions that elicit a change in cell phenotype, which can be investigated in follow-up studies. In this way, PMF is a valuable method for an initial screen to simply and quickly reveal changes or differences in cell filtration relative to other samples; more detailed studies of cellular deformability using higher resolution methods such as quantitative deformability cytometry (q-DC) or AFM can be used to obtain quantitative measurements of cell mechanical properties, such as elastic

Table 3.7.1: Overview of methods using fluidic approaches to measure cellular deformability.

Class of method	Methods	References	Advantages	Limitations
Cell filtration: Single-sample filtration method for bulk measurements of cell deformability	Filtration apparatus	– Downey and Worthen (1988)	– Straight-forward setup – Low-cost equipment	– Measures bulk suspension of cells; unable to account for population heterogeneity – Relative measurement of deformability
	Filtration apparatus	– Gregerson et al. (1967)		
High-throughput cell filtration: Filtration method enabling deformability; Measurements of multiple samples simultaneously	Parallel microfiltration (PMF)	– Qi et al. (2015) – Gill et al. (2019b)	– Straight-forward setup – Requires equipment that biological labs typically have access to (e.g., plate reader) – Can be performed in multiwell plate format – Amenable to screening	– Measures bulk suspension of cells; unable to account for population heterogeneity – Provides relative measurement of deformability

Table 3.7.1 (continued)

Class of method	Methods	References	Advantages	Limitations
Single-cell fluidic methods: Rapid assays for single-cell measurements	Quantitative deformability cytometry (q-DC)	– Nyberg et al. (2017)	– Can extract multiple parameters in a single experiment (cell size, cellular elastic modulus, fluidity, transit time) – Can account for differences in cell size and deformability at the single-cell level	– Computationally intensive and requires high-end data storage – Requires clean room for device fabrication and high-speed camera for single-cell imaging – Requires highly trained personnel
	Microconstriction arrays	– Lange et al. (2015)		
	Transit time analysis	– Rosenbluth et al. (2006) – Nyberg et al. (2016)	– Can extract multiple parameters in a single experiment (cell size, transit time, deformability) – Can account for differences in cell size and deformability at the single-cell level	
	Real-time fluorescence deformability cytometry (RT-FDC) and RT-DC	– Rosendahl et al. (2018) – Otto et al. (2015) – Toepfner et al. (2018)		
	Inertial deformability cytometry (DC)	– Gossett et al. (2012) – Tse et al. (2013)		
	Suspended microchannel resonator (SMR)	– Byun et al. (2013)		
	Deformability cytometry for the cell nucleus	– Hodgson et al. (2017)		
	Electrical-based deformability cytometry	– Adamo et al. (2012)		

modulus or viscosity. Such parallelized filtration assays have exciting potential to identify novel small molecules or proteins that regulate cellular deformability and to generate a comprehensive, systems-level knowledge of mechanotype.

3.7.2 Single-Cell Deformability Assays Using Fluidic Approaches

To achieve single-cell deformability measurements, the development of microfluidic devices paired with high-speed imaging enables tracking the shape changes of single cells through micron-scale channels on timescales of microseconds to milliseconds (Rosenbluth et al., 2008, Rowat et al., 2013). Such methods can rapidly gather deformability data for hundreds to thousands of cells, thereby enabling population analysis, which is essential to study subpopulations and phenotypic heterogeneity. The rich spatial and temporal data obtained from images of single-cell deformations also enables multiple parameters to be extracted. Multiparametric analysis can improve accuracy to classify cells (Lin et al., 2017) and to predict functionally relevant behaviors, such as cancer cell invasion (Nyberg et al., 2018).

3.7.2.1 Probing Cell Deformability by Measuring Passage Through Micron-Scale Constrictions

One class of single-cell fluidic assays relies on deforming cells as they flow through micron-scale constrictions that are smaller than the cell diameter. Such devices mimic the physiological gaps that a cell encounters on its journey through circulation. Quantifying the timescale of cell transit provides a measure of cellular deformability: cells with longer transit times tend to be less deformable (Rosenbluth et al., 2008). More detailed studies of microgel particles revealed that particles with increased stiffness have longer transit times (Nyberg et al., 2016). Such transit time measurements have shown differences in the mechanical properties of leukemia cells with chemotherapy treatment, as indicated by the increased transit time and occlusion of micron-scale gaps of microfluidic devices (Lam et al., 2007). By probing the transit of HL-60 cells with varying levels of lamin A, a transit time microfluidic assay also revealed that the cell nucleus rate limits the transit of cells through channels with a gap size that is smaller than the nucleus (Rowat et al., 2013).

While transit time analysis provides a straightforward metric for analysis of cellular deformability, more advanced image analysis has enabled quantification of cell mechanical properties. Recording the shape changes of single cells as they deform through micron-scale constrictions makes it possible to analyze time-dependent deformations (Figure 3.7.1C); such data can be fitted with existing models to describe cells as materials, for example, using a power law rheology model (Fabry et al., 2001, Nyberg et al., 2017). However, the exact stresses at the constriction of a microfluidic device can vary between experiments due to variability in channel dimensions generated using soft lithography. By using gel particles with a well-defined stiffness, the mechanical stresses on a particle deforming through a micron-scale gap of the device can be quantified, thereby enabling calibrated measurements of cell's mechanical properties; this technique is known as q-DC. Achieving calibrated measurements of cell mechanical properties is essential to compare measurements over time and also across independent laboratories. While such calibrated and quantitative measurements of cellular mechanical properties are valuable, the occlusion of channels over time can limit the duration and total number of cells assayed in a single q-DC or transit time experiment.

3.7.2.2 Measuring Deformations of Cells in Response to the Forces of Fluid Flows

To achieve even higher throughput deformability measurements, the real-time DC (RT-DC) method provides an elegant approach: cells are flowed through a straight-flow channel geometry and the cellular shape changes that occur in response to fluid shear stresses are monitored using a high-speed camera (Otto et al., 2015; Figure 3.7.1D). Shape changes are quantified as cell deformation, D, where $D = 1 -$ circularity. Circularity, C, for a perfect circle is 1 and can be calculated using $C = \left(2\sqrt{\pi A} \right)/l$ with A representing the cell area and l representing cell perimeter. Cells exhibiting larger changes in shape have higher values of deformation. The RT-DC approach enables real-time analysis at rates of 100 cells/s, for larger populations of ~100,000 cells. With such rapid measurement capabilities, RT-DC has more recently provided measurements of up to 1 mL of diluted whole blood at rates of 1,000 cells/s (Toepfner et al., 2018). Subjecting cells to the inertial forces of flow in a microfluidic device can also be used to extract single-cell deformability measurements (Gossett et al., 2012; Figure 3.7.1E).

While single-cell fluidic methods enable label-free quantification of single-cell deformability (Otto et al., 2015, Gossett et al., 2012), the addition of fluorescence detection with mechanical phenotyping can enable multiparameter measurements of mechanotype together with genetic or protein biomarkers (Rosendahl et al., 2018). Efforts to parallelize the imaging of single cells – for example, using a beam splitter to facilitate simultaneous imaging of multiple fluid channels – is another promising direction (Muñoz et al., 2018).

3.7.3 What Determines the Deformability of Cells Through Micron-Scale Gaps on Timescales of Seconds to Minutes?

Measurements of cellular mechanotype are highly dependent on both timescales and length scales. Fluidic-based methods enable a range of deformation length scales on the micron scale, which is roughly equivalent to 30–60% strain (Nyberg et al., 2017, Otto et al., 2015). The timescale of fluidic-based deformability measurements ranges from seconds for filtration to microseconds to milliseconds for single-cell fluidic methods. By contrast, AFM typically deforms cells over nanometer to micrometer-length scales over timescales of seconds to minutes (Kim et al., 2016, Dufrêne et al., 2017).

Both cytoskeletal and nuclear components determine the deformation of cells through micron-scale gaps on the timescales and length scales of fluidic-based methods. Actin and microtubules are major determinants of filtration across cell types (Gill et al., 2019b, Kim et al., 2016, Qi et al., 2015). The filamentous (F) actin-perturbing drug cytochalasin D results in increased filtration (Qi et al., 2015), consistent with the decreased stiffness of cells lacking F-actin (Ting-Beall et al., 1995, Nyberg et al., 2017). Fluidic-based deformability assays are also sensitive to changes in the organization of F-actin: the decreased filtration of cancer cells treated with the βAR agonist, isoproterenol, was associated with increased density of cortical actin (Kim et al., 2016; Figure 3.7.2).

Single-cell fluidic assays are also sensitive to actin organization: cytochalasin D treatment results in increased deformability of HL-60 cells as measured by RT-DC (Otto et al., 2015) and q-DC (Nyberg et al., 2017). Treatment of cancer cells and fibroblasts with drugs that stabilize microtubules (paclitaxel) or inhibit microtubule polymerization (colchicine) results in decreased filtration (Gill et al., 2019a, Kim et al., 2016, Qi et al., 2015, Gill et al., 2019b).

Fluidic methods have also revealed that the nucleus – which is typically the stiffest and largest organelle (Rowat et al., 2008, 2013, Harada et al., 2014, Wolf et al., 2013) – is a key contributor to the ability of cells to deform through narrow gaps on timescales from milliseconds to hours. Our previous work showed that the nucleus rate limits the deformation of neutrophil-type HL-60 cells through 3 and 5 μm gaps that were ~30% the median cell size on the timescale of seconds (Rowat et al., 2013). For a 3 μm gap size, even cytochalasin D treatment had no significant effect on the timescale of cell transit, despite overexpression of lamin A, which is a major determinant of cellular and nuclear deformability (Rowat et al., 2013, Swift et al., 2013). Interestingly, lamin A-overexpressing cells were also less likely to migrate through 3 μm pores in a transwell migration assay (Rowat et al., 2013), suggesting that fluidic-based assays that probe cell deformability on timescales of milliseconds can provide insight into the mechanisms of active cell migration. Nuclear structure also impacts cellular filtration as measured by PMF. Mouse embryo fibroblasts (MEFs) that are lacking lamin A/C

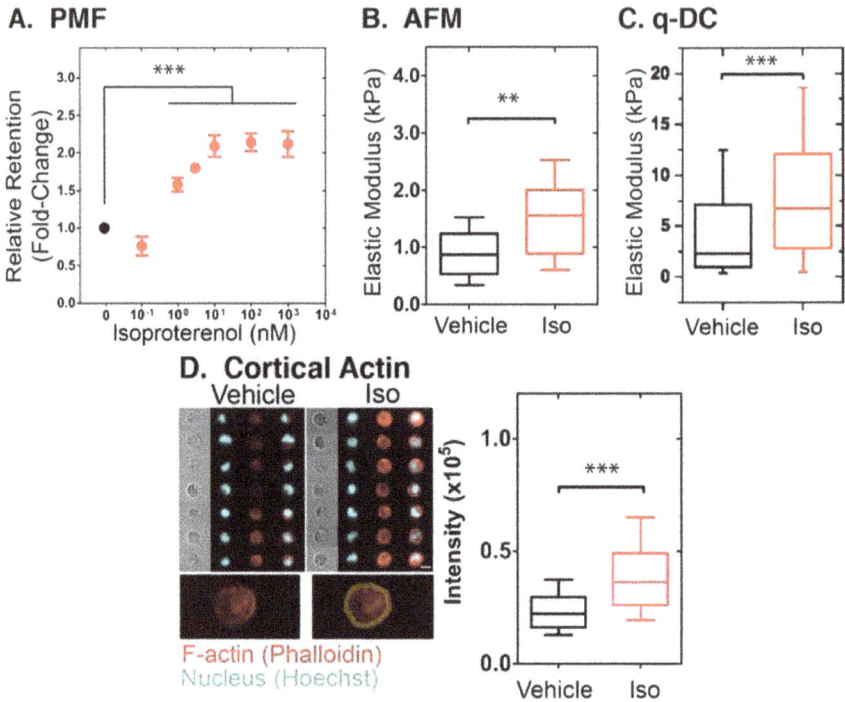

Figure 3.7.2: Comparison of mechanotyping methods. Human breast cancer cells (MDA-MB-231) treated with β-adrenergic agonist, isoproterenol (Iso), were measured using various fluidic and standard methods for mechanotyping: (A) PMF; (B) AFM; (C) q-DC; (D) in this example showing effects of ßAR activation, the increased cell stiffness and retention is associated with increased levels of cortical actin. Statistical significance is determined using a one-way ANOVA with Tukey's multiple comparison test for PMF data (as in A) and Mann–Whitney U-test for nonparametrically distributed data as in (B)–(D). ** $p < 0.01$, *** $p < 0.001$. Data from Kim et al. (2016).

or lamin B1 exhibit increased filtration compared to wild-type controls (Gill et al., 2019a). Decondensing chromatin by histone deacetylase inhibitor trichostatin A treatment results in increased filtration of human ovarian cancer cell lines (Gill et al., 2019b). While the nucleus is a major contributor to cellular deformability on timescales of milliseconds to hours, the measurement timescale may impact the relative contributions of specific proteins to cellular mechanical properties (Swift et al., 2013).

The interaction of cells with their microenvironment also contributes to cellular mechanotype via reciprocal mechanical feedback between cells and the surrounding extracellular matrix (Engler et al., 2006). Harvesting cells into a suspended state prior to fluidic-based deformability measurements results in cytoskeletal remodeling on the timescale of seconds to minutes (Sen and Kumar, 2009). Major changes in cellular signaling pathways may also occur when cells are lifted into a suspended state, such as the activation of Hippo signaling, which inhibits YAP activity (Zhao et al., 2012). For these reasons, an equilibration time of ~30 min is recommended prior to any filtration measurement (Barnes et al., 2012). How does the deformability of a cell in a suspended state compare to an adhered state? Certain components are known to regulate the deformability of cells in both suspended and adhered states. For example, cells in a suspended state with increased lamin A levels exhibit increased transit time (Rowat et al., 2013) and retention (Gill et al., 2019a); the stiffness of adhered cells also scales with levels of lamin A (Swift et al., 2013). We also find that βAR activation consistently increases the stiffness of MDA-MB-231 cells, as indicated by the increased elastic modulus of adhered cells measured using AFM, and of suspended cells as determined by q-DC; consistent with these findings, we also discovered that βAR activation increases the retention of suspended cells measured by PMF (Figure 3.7.2; Kim et al., 2016). However, there are also contexts where the mechanical properties of adhered and suspended cells differ. Different trends in the deformability of cells that are in adhered versus suspended states may be attributed to differences in the higher order structure and dynamics of cytoskeletal and nuclear architecture. Previous reports show that inhibition of myosin II using a pharmacologic inhibitor such as blebbistatin or the ROCK inhibitor Y27632 decreases the cytoplasmic stiffness of adhered cells (Nagayama et al., 2004, Nijenhuis et al., 2014, Ayala et al., 2017, Sbrana et al., 2008). The same treatments to inhibit myosin II activity in suspended cells have shown both similar softening effects (Gabriele et al., 2009, Cartagena-Rivera et al., 2016, Thuet et al., 2011, Smith et al., 2018); yet other studies show opposite effects, whereby cells in suspension become stiffer with myosin II inhibition (Chan et al., 2015). The differential effects of blebbistatin on adhered versus suspended cells may be attributed to differences in cytoskeletal remodeling when cells are in suspension and lack the cell–matrix and cell–cell adhesions that promote generation of intracellular tension. Variations in the magnitude of deformation across methods may also contribute to the varied results: the stiffness of the cortical region may dominate deformations of 1–3 μm, and a decreased turnover of actin due to blebbistatin treatment could result in increased cortical stiffness (Cartagena-Rivera et al., 2016, Nie et al., 2015). By contrast, the nucleus tends to dominate larger deformations of 5–6 μm, and inhibition of myosin II activity could decrease intracellular tension and thereby result in a more deformable nucleus (Kim and Wirtz, 2015). Differences in cell type, culture conditions, and the passage number of cultured cells may also contribute to discrepancies across measurements (Targosz-Korecka et al., 2013).

3.7.4 Physiological and Disease Applications of Fluidic-Based Deformability Assays

3.7.4.1 Fluidic Methods to Measure Cell Deformability Mimic Physiological and Disease Contexts That Require Cells to Deform Through Narrow Gaps

Blood cells routinely flow and deform through narrow gaps of the vasculature. Stiffening of red and white blood cells contributes to physiological regulation of leukocyte trafficking (Fay et al., 2016) as well as pathologies from sickle cell disease (Higgins et al., 2007) to chronic lymphocytic leukemia (Zheng et al., 2015). Circulating tumor cells (CTCs) are required to deform through micron-scale constrictions in order to disseminate to distant sites; on the other hand, CTCs must also survive the shear stresses of fluid flow (Barnes et al., 2012) and occlude capillary beds to seed secondary tumors (Zeidman, 1961).

Fluidic-based mechanotyping methods can also provide key biological insights into the behaviors of other cell types, such as fibroblasts or epithelial cells, which tend to adhere to their surrounding cells and matrix rather than flow as single cells in a suspended state. The transit of cells through narrow gaps can simulate the requirements for the migration and invasion of adhered cells through the extracellular matrix, such as the invasion of cancer cells, which is required for metastasis (Geho et al., 2005). The migration of cells through confined microfluidic channels also mimics the physical constraints that immune cells face during their invasion through dense collagen matrices (Raab et al., 2016). Indeed, the increased deformability of cancer and immune cells is associated with their invasion and motility (Xu et al., 2012, Swaminathan et al., 2011, Nyberg et al., 2018, Kim et al., 2016, 2019) and is linked to transitions from epithelial to mesenchymal to amoeboid phenotypes (Qi et al., 2015, Reis-Sobreiro et al., 2018). Fluidic-based methods can thus provide assays to assess how cells behave during diverse physiologically and disease-relevant deformations.

3.7.4.2 Mechanotyping Using Fluidic Approaches Provides Mechanistic Insights into Cell Biology

Understanding the changes in cell deformability that occurs in response to pharmacologic or genetic perturbations can provide insight into underlying molecular mechanisms that regulate cellular behaviors such as motility and mechanosensing, which are often regulated by shared mediators of mechanotype. Using PMF, we established that stress hormone cues – activation of βAR signaling by treatment with the agonist isoproterenol – alter the stiffness and motility of breast cancer cells through a RhoA-ROCK axis (Kim et al., 2016). Cell filtration revealed that activation of βAR signaling decreases the deformability of macrophages and alters the actin cytoskeleton to favor branched, Arp2/3-nucleated filament structures (Kim et al., 2019). Cell filtration also revealed that dermal fibroblasts from dystonia patients are more deformable than normal fibroblasts (Figure 3.7.3B, 3.7.3C; Gill et al., 2019a); the increased deformability of DYT1 fibroblasts was associated with increased cell death in response to mechanical stretching, suggesting that DYT1 dystonia is characterized by altered cellular mechanobiology. Since the molecular mechanisms of dystonia pathophysiology still remain poorly understood, mechanotyping could provide a promising novel approach to develop effective therapeutics.

3.7.4.3 Deformability as Biomarker

The translational potential of fluidic methods to measure cell deformability is already being realized as a diagnostic and prognostic tool for blood disorders such as sepsis (Crawford et al., 2018) and shows promise to enhance cancer prognosis and diagnosis (Tse et al., 2013, Byun et al., 2013). The ability of high-throughput DC to identify malignant cells from patient's pleural effusions based on their increased deformability could complement traditional visual examination by cytopathologists (Tse et al., 2013); rapid screening of cellular deformability could thus be a valuable screening tool to eliminate samples negative for malignant cells from hospital workflow. Fluidic-based methods can also be applied to a range of normal and pathological cell types, including immune cells, fibroblasts, and cancer cells (Figures 3.7.2 and 3.7.3; Tse et al., 2013, Gill et al., 2019a, Kim et al., 2016), and thus have exciting translational potential.

3.7.5 Future Outlook

The higher throughput capacity of fluidic methods to measure cell deformability is enabling advances in basic research and translational applications. The ability to conduct high-throughput screens has exciting potential to provide a systems-level understanding of the molecules and pathways that regulate mechanotype including cellular deformability and contractility; ultimately this will enable us to define the "mechanome" – the set of genes, proteins, and pathways that regulate cellular mechanotype. High-throughput mechanotyping could enable screening large libraries of pharmacologically active compounds and CRISPR/Cas9 knockouts to drive the discovery of molecular mediators of mechanotype. Conducting high-throughput screens across libraries of small molecules based on cellular deformability will further broaden our understanding of the mechanisms through which cells translate

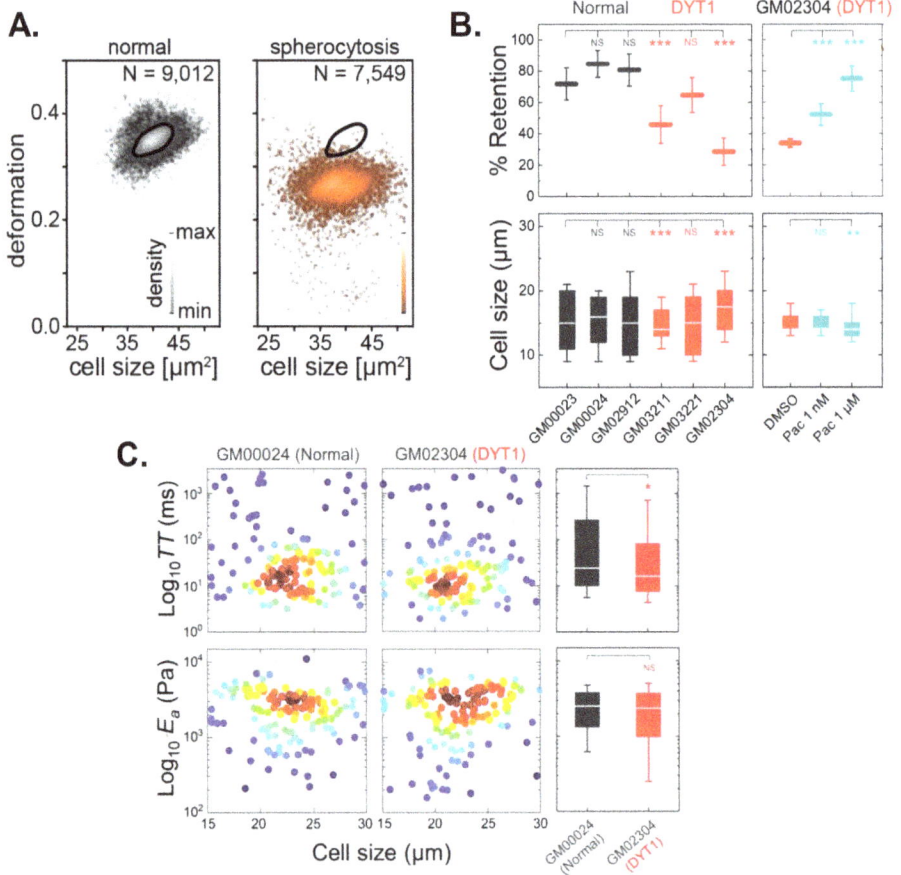

Figure 3.7.3: Application of fluidic-based methods to measure cell deformability for physiological and disease applications. (A) Red blood cells from patients with spherocytosis are discriminated from normal red blood cells using RT-DC (Toepfner et al., 2018); (B) dermal fibroblasts from patients with DYT1 dystonia (GM02304) compared to fibroblasts from normal individuals (GM00024) using PMF; (C) transit time (TT) and elastic modulus (E_a) are measured using q-DC (Gill et al., 2019a).

diverse signals from soluble cues into changes in cellular mechanotype; ultimately such knowledge will deepen our understanding of cellular homeostasis and identify points of leverage to intervene and control cellular mechanotype, including cellular deformability and cell–matrix interactions. Despite advances in defining the genome and epigenome, there is still a lack of understanding of genome-to-phenome relationships, or the molecular signatures that describe complex phenotypes such as mechanotype; a systems-level understanding of the mechanome should help to fill these knowledge gaps.

Mechanotyping also has exciting potential to be applied to identify compounds with therapeutic value that could not be found using existing methods. Screening cancer cells treated with small molecules could be used to identify novel therapeutic molecules based on their ability to change cell deformability (Gill et al., 2019b). In contrast to many existing screening methods that rely on detecting a specific gene or protein target, cellular mechanotyping enables screening assays based on whole-cell mechanotype. Since cellular mechanotype is altered across broad spectrum of disease contexts from cancer to malaria to diabetes (Otto et al., 2015, Bow et al., 2011, Shin and Ku, 2005), mechanotype screening could thus trigger discoveries of mechanisms of pathophysiology across diverse fields in biomedical research and offer potential therapeutic value across a range of human diseases.

References

Adamo, A., A. Sharei, L. Adamo, B. Lee, S. Mao and K. F. Jensen (2012). "Microfluidics-based assessment of cell deformability." Analytical Chemistry **84**(15): 6438–6443.

Ayala, Y. A., B. Pontes, B. Hissa, A. C. M. Monteiro, M. Farina, V. Moura-Neto, N. B. Viana and H. M. Nussenzveig (2017). "Effects of cytoskeletal drugs on actin cortex elasticity." Experimental Cell Research **351**(2): 173–181.

Barnes, J. M., J. T. Nauseef and M. D. Henry (2012). "Resistance to fluid shear stress is a conserved biophysical property of malignant cells." PLoS One **7**(12): e50973.

Belfort, G., R. H. Davis and A. L. Zydney (1994). "The behavior of suspensions and macromolecular solutions in crossflow microfiltration." Journal of Membrane Science **96**(1–2): 1–58.

Bow, H., I. V. Pivkin, M. Diez-Silva, S. J. Goldfless, M. Dao, J. C. Niles, S. Suresh and J. Han (2011). "A microfabricated deformability-based flow cytometer with application to malaria." Lab on a Chip **11**(6): 1065–1073.

Buchan, P. C. (1980). "Evaluation and modification of whole blood filtration in the measurement of erythrocyte deformability in pregnancy and the newborn." British Journal of Haematology **45** (1): 97–105.

Byun, S., S. Son, D. Amodei, N. Cermak, J. Shaw, J. H. Kang, V. C. Hecht, M. M. Winslow, T. Jacks, P. Mallick and S. R. Manalis (2013). "Characterizing deformability and surface friction of cancer cells." Proceedings of the National Academy of Sciences **110**(19): 7580–7585.

Cartagena-Rivera, A. X., J. S. Logue, C. M. Waterman and R. S. Chadwick (2016). "Actomyosin cortical mechanical properties in nonadherent cells determined by atomic force microscopy." Biophysical Journal **110**(11): 2528–2539.

Chan, C. J., A. E. Ekpenyong, S. Golfier, W. Li, K. J. Chalut, O. Otto, J. Elgeti, J. Guck and F. Lautenschläger (2015). "Myosin II activity softens cells in suspension." Biophysical Journal **108**(8): 1856–1869.

Chien, S., S. A. Luse and C. A. Bryant (1971). "Hemolysis during Filtration through Micropores - Scanning Electron Microscopic and Hemorheologic Correlation." Microvascular Research **3**(2): 183–203.

Cho, S., J. Irianto and D. E. Discher (2017). "Mechanosensing by the nucleus: From pathways to scaling relationships." Journal of Cell Biology **216**(2): 305–315.

Chowdhury, F., Y. Li, Y. C. Poh, T. Yokohama-Tamaki, N. Wang and T. S. Tanaka (2010). "Soft substrates promote homogeneous self-renewal of embryonic stem cells via downregulating cell-matrix tractions." PLoS One 5(12): e15655.

Crawford, K., A. DeWitt, S. Brierre, T. Caffery, T. Jagneaux, C. Thomas, M. Macdonald, H. Tse, A. Shah, D. Di Carlo and H. R. O'Neal (2018). "Rapid biophysical analysis of host immune cell variations associated with sepsis." American Journal of Respiratory and Critical Care Medicine 198(2): 280–282.

Cross, S. E., J. Kreth, L. Zhu, R. Sullivan, W. Shi, F. Qi and J. K. Gimzewski (2007). "Nanomechanical properties of glucans and associated cell-surface adhesion of Streptococcus mutans probed by atomic force microscopy under in situ conditions." Microbiology 153(9): 3124–3132.

Dahl, K. N., S. M. Kahn, K. L. Wilson and D. E. Discher (2004). "The nuclear envelope lamina network has elasticity and a compressibility limit suggestive of a molecular shock absorber." Journal of Cell Science 117(20): 4779–4786.

Dahl, K. N., P. Scaffidi, M. F. Islam, A. G. Yodh, K. L. Wilson and T. Misteli (2006). "Distinct structural and mechanical properties of the nuclear lamina in Hutchinson–Gilford progeria syndrome." Proceedings of the National Academy of Sciences 103(27): 10271–10276.

Dao, M., C. T. Lim and S. Suresh (2003). "Mechanics of the human red blood cell deformed by optical tweezers." Journal of the Mechanics and Physics of Solids 51(11–12): 2259–2280.

Darcy, H. (1856). The public fountains of the city of Dijon: Exposition and application of principles to follow and formulas to use in questions of water distribution. Paris, Dalmont.

Dahl, K. N., A. J. Engler, J. D. Pajerowski and D. E. Discher (2005). "Power-law rheology of isolated nuclei with deformation mapping of nuclear sub-structures." Biophysical Journal 89:2855–2864.

Discher, D. E., L. Smith, S. Cho, M. Colasurdo, A. J. García and S. Safran (2017). "Matrix mechanosensing: From scaling concepts in 'omics data to mechanisms in the nucleus, regeneration, and cancer." Annual Review of Biophysics 46: 295–315.

Downey, G. P., D. E. Doherty, B. Schwab, 3rd, E. L. Elson, P. M. Henson and G. S. Worthen (1990). "Retention of leukocytes in capillaries: Role of cell size and deformability." Journal of Applied Physiology 69(5): 1767–1778.

Downey, G. P. and G. S. Worthen (1988). "Neutrophil retention in model capillaries: Deformability, geometry, and hydrodynamic forces." Journal of Applied Physiology 65(4): 1861–1871.

Dufrêne, Y. F., T. Ando, R. Garcia, D. Alsteens, D. Martinez-Martin, A. Engler, C. Gerber and D. J. Müller (2017). "Imaging modes at atomic force microscopy for application in molecular and cell biology." Nature Nanotechnology 12(4): 295.

Duclos-Orsello, C., W. Li and C. Ho (2006). "A three mechanism model to describe fouling of microfiltration membranes." Journal of Membrane Science 280(1-2): 856–866.

Engler, A. J., S. Sen, H. L. Sweeney and D. E. Discher (2006). "Matrix elasticity directs stem cell lineage specification." Cell 126(4): 677–689.

Evans, E. and B. Kukan (1984). "Passive material behavior of granulocytes based on large deformation and recovery after deformation tests." Blood 64(5): 1028–1035.

Fabry, B., G. N. Maksym, J. P. Butler, M. Glogauer, D. Navajas and J. J. Fredberg (2001). "Scaling the microrheology of living cells." Physical Review Letters 87(14): 148102.

Fay, M. E., D. R. Myers, A. Kumar, C. T. Turbyfield, R. Byler, K. Crawford, R. G. Mannino, A. Laohapant, E. A. Tyburski, Y. Sakurai, M. J. Rosenbluth, N. A. Switz, T. A. Sulchek, M. D. Graham and W. A. Lam (2016). "Cellular softening mediates leukocyte demargination and trafficking, thereby increasing clinical blood counts." Proceedings of the National Academy of Sciences 113(8): 1987–1992.

Fletcher, D. A. and R. D. Mullins (2010). "Cell mechanics and the cytoskeleton." Nature 463(7280): 485.

Gabriele, S., A. M. Benoliel, P. Bongrand and O. Théodoly (2009). "Microfluidic investigation reveals distinct roles for actin cytoskeleton and myosin II activity in capillary leukocyte trafficking." Biophysical Journal **96**(10): 4308–4318.

Geho, D. H., R. W. Bandle, T. Clair and L. A. Liotta (2005). "Physiological mechanisms of tumor-cell invasion and migration." Physiology **20**(3): 194–200.

Gill, N. K., C. Ly, P. H. Kim, C. A. Saunders, L. G. Fong, S. G. Young, G. W. Luxton and A. C. Rowat (2019a). "DYT1 dystonia patient-derived fibroblasts have increased deformability and susceptibility to damage by mechanical forces." Frontiers in Cell and Developmental Biology **7**: 103.

Gill, N. K., C. Ly, K. D. Nyberg, L. Lee, D. Qi, B. Tofig, M. Reis-Sobreiro, O. Dorigo, J. Rao, R. Wiedemeyer, B. Karlan, K. Lawrenson, M. R. Freeman, R. Damoiseaux and A. C. Rowat (2019b). "A scalable filtration method for high throughput screening based on cell deformability." Lab on a Chip **19**(2): 343–357.

Gill, N. K., D. Qi, T. H. Kim, C. Chan, A. Nguyen, K. D. Nyberg and A. C. Rowat (2017). A protocol for screening cells based on deformability using parallel microfiltration. Nature Protocol Exchange. https://protocolexchange.researchsquare.com/article/nprot-6167/v1

Gossett, D. R., T. K. Henry, S. A. Lee, Y. Ying, A. G. Lindgren, O. O. Yang, J. Rao, A. T. Clark and D. Di Carlo (2012). "Hydrodynamic stretching of single cells for large population mechanical phenotyping." Proceedings of the National Academy of Sciences **109**(20): 7630–7635.

Gregersen, M. I., C. A. Bryant, W. E. Hammerle, S. Usami and S. Chien (1967). "Flow Characteristics of Human Erythrocytes through Polycarbonate Sieves." Science **157**(3790): 825–827.

Guck, J., R. Ananthakrishnan, H. Mahmood, T. J. Moon, C. C. Cunningham and J. Käs (2001). "The optical stretcher: A novel laser tool to micromanipulate cells." Biophysical Journal **81**(2): 767–784.

Harada, T., J. Swift, J. Irianto, J. W. Shin, K. R. Spinler, A. Athirasala, R. Diegmiller, P. D. P. Dingal, I. L. Ivanovska and D. E. Discher (2014). "Nuclear lamin stiffness is a barrier to 3D migration, but softness can limit survival." Journal of Cell Biology **204**(5): 669–682.

Higgins, J. M., D. T. Eddington, S. N. Bhatia and L. Mahadevan (2007). "Sickle cell vasoocclusion and rescue in a microfluidic device." Proceedings of the National Academy of Sciences **104**(51): 20496–20500.

Ho, C. C. and A. L. Zydney (2000). "A combined pore blockage and cake filtration model for protein fouling during microfiltration." Journal of Colloid and Interface Science **232**(2): 389–399.

Hochmuth, R. M. (2000). "Micropipette aspiration of living cells." Journal of Biomechanics **33**(1): 15–22.

Hodgson, A. C., C. M. Verstreken, C. L. Fisher, U. F. Keyser, S. Pagliara and K. J. Chalut (2017). "A microfluidic device for characterizing nuclear deformations." Lab on a Chip **17**(5): 805–813.

Hoh, J. H. and C. A. Schoenenberger (1994). "Surface morphology and mechanical properties of MDCK monolayers by atomic force microscopy." Journal of Cell Science **107**(5): 1105–1114.

Jones, J. G., B. M. Holland, J. Humphrys, R. Quew and C. A. Wardrop (1984). "Evaluation of the contribution of red and white cells to the flow of suspensions of washed blood cells through 3 μm Nuclepore membranes." British Journal of Haematology **57**(3): 457–466.

Jones, J. G., B. M. Holland, J. Humphrys, R. Quew and C. A. Wardrop (1985). "Evaluation of the Contribution of Red and White Cells to the Flow of Suspensions of Washed Blood-Cells through 3 Mu-M Nuclepore Membranes." British Journal of Haematology **57**(3):457–466.

Kaleridis, V., G. Athanassiou, D. Deligianni and Y. Missirlis (2010). "Slow flow of passive neutrophils and sequestered nucleus into micropipette." Clinical Hemorheology and Microcirculation **45**(1): 53–65.

Kim, T. H., N. K. Gill, K. D. Nyberg, A. V. Nguyen, S. V. Hohlbauch, N. A. Geisse, C. J. Nowell, E. K. Sloan and A. C. Rowat (2016). "Cancer cells become less deformable and more invasive with activation of β-adrenergic signaling." Journal of Cell Science **129**(24): 4563–4575.

Kim, T. H., C. Ly, A. Christoulides, C. J. Nowell, P. W. Gunning, E. K. Sloan and A. C. Rowat (2019a). "Stress hormone signaling β-adrenergic receptors regulates macrophage mechanotype and function." The FASEB Journal **33**: 3997–4006.

Kim, D. H. and D. Wirtz (2015). "Cytoskeletal tension induces the polarized architecture of the nucleus." Biomaterials **48**: 161–172.

Lam, W. A., M. J. Rosenbluth and D. A. Fletcher (2007). "Chemotherapy exposure increases leukemia cell stiffness." Blood **109**(8): 3505–3508.

Lammerding, J., L. G. Fong, J. Y. Ji, K. Reue, C. L. Stewart, S. G. Young and R. T. Lee (2006). "Lamins A and C but not lamin B1 regulate nuclear mechanics." Journal of Biological Chemistry **281**(35): 25768–25780.

Lange, J. R., J. Steinwachs, T. Kolb, L. A. Lautscham, I. Harder, G. Whyte and B. Fabry (2015). "Microconstriction arrays for high-throughput quantitative measurements of cell mechanical properties." Biophysical Journal **109**(1): 26–34.

Lu, W. M., K. L. Tung, S. M. Hung, J. S. Shiau and K. J. Hwang (2001). "Constant pressure filtration of mono-dispersed deformable particle slurry." Separation Science and Technology **36**(11): 2355–2383.

Lin, J., D. Kim, T. T. Henry, P. Tseng, L. Peng, M. Dhar, S. Karumbayaram and D. Di Carlo (2017). "High-throughput physical phenotyping of cell differentiation." Microsystems & Nanoengineering **3**: 17013.

Lipowsky, H. H., D. T. Bowers, B. L. Banik and J. L. Brown (2018). "Mesenchymal stem cell deformability and implications for microvascular sequestration." Annals of Biomedical Engineering **46**(4): 640–654.

Maloney, J. M., D. Nikova, F. Lautenschläger, E. Clarke, R. Langer, J. Guck and K. J. Van Vliet (2010). "Mesenchymal stem cell mechanics from the attached to the suspended state." Biophysical Journal **99**(8): 2479–2487.

Mitchison, J. M. and M. M. Swann (1954). "The mechanical properties of the cell surface. I. The cell elastimeter." Journal of Experimental Biology **31**: 443–460

Murrell, M., P. W. Oakes, M. Lenz and M. L. Gardel (2015). "Forcing cells into shape: The mechanics of actomyosin contractility." Nature Reviews. Molecular Cell Biology **16**(8): 486.

Müller, D. J. and Y. F. Dufrêne (2011). "Atomic force microscopy: A nanoscopic window on the cell surface." Trends in Cell Biology **21**(8): 461–469.

Muñoz, H. E., M. Li, C. T. Riche, N. Nitta, E. Diebold, J. Lin, K. Owsley, M. Bahr, K. Goda and D. Di Carlo (2018). "Single-cell analysis of morphological and metabolic heterogeneity in Euglena gracilis by fluorescence-imaging flow cytometry." Analytical Chemistry **90**(19): 11280–11289.

Nardone, G., J. Oliver-De La Cruz, J. Vrbsky, C. Martini, J. Pribyl, P. Skládal, M. Pešl, G. Caluori, S. Pagliari, F. Martino, Z. Maceckova, M. Hajduch, A. Sanz-Garcia, N. M. Pugno, G. B. Stokin and G. Forte (2017). "YAP regulates cell mechanics by controlling focal adhesion assembly." Nature Communications **8**: 15321.

Nagayama, M., H. Haga, M. Takahashi, T. Saitoh and K. Kawabata (2004). "Contribution of cellular contractility to spatial and temporal variations in cellular stiffness." Experimental Cell Research **300**(2): 396–405.

Needham, D. and R. S. Nunn (1990). "Elastic deformation and failure of lipid bilayer membranes containing cholesterol." Biophysical Journal **58**(4): 997–1009.

Nie, W., M. T. Wei, H. D. Ou-Yang, S. S. Jedlicka and D. Vavylonis (2015). "Formation of contractile networks and fibers in the medial cell cortex through myosin-II turnover, contraction, and stress-stabilization." Cytoskeleton **72**(1): 29–46.

Nijenhuis, N., X. Zhao, A. Carisey, C. Ballestrem and B. Derby (2014). "Combining AFM and acoustic probes to reveal changes in the elastic stiffness tensor of living cells." Biophysical Journal **107** (7): 1502–1512.

Nyberg, K. D., S. L. Bruce, A. V. Nguyen, C. K. Chan, N. K. Gill, T. H. Kim, T. H. Kim, E. K. Sloan and A. C. Rowat (2018). "Predicting cancer cell invasion by single-cell physical phenotyping." Integrative Biology **10**(4): 218–231.

Nyberg, K. D., K. H. Hu, S. H. Kleinman, D. B. Khismatullin, M. J. Butte and A. C. Rowat (2017). "Quantitative deformability cytometry: Rapid, calibrated measurements of cell mechanical properties." Biophysical Journal **113**(7): 1574–1584.

Nyberg, K. D., M. B. Scott, S. L. Bruce, A. B. Gopinath, D. Bikos, T. G. Mason, J. W. Kim, H. S. Choi and A. C. Rowat (2016). "The physical origins of transit time measurements for rapid, single cell mechanotyping." Lab on a Chip **16**(17): 3330–3339.

Otto, O., P. Rosendahl, A. Mietke, S. Golfier, C. Herold, D. Klaue, S. Girardo, S. Pagliara, A. Ekpenyong, A. Jacobi, M. Wobus, N. Töpfner, U. F. Keyser, J. Mansfeld, E. Fischer-Friedrich and J. Guck (2015). "Real-time deformability cytometry: On-the-fly cell mechanical phenotyping." Nature Methods **12**(3): 199–202.

Pajerowski, J. D., K. N. Dahl, F. L. Zhong, P. J. Sammak and D. E. Discher (2007). "Physical plasticity of the nucleus in stem cell differentiation." Proceedings of the National Academy of Sciences **104**(40): 15619–15624.

Qi, D., N. K. Gill, C. Santiskulvong, J. Sifuentes, O. Dorigo, B. Taylor-Harding, W. R. Wiedemeyer and A. C. Rowat (2015). "Screening cell mechanotype by parallel microfiltration." Scientific Reports **5**: 17595.

Raab, M., M. Gentili, H. de Belly, H. R. Thiam, P. Vargas, A. J. Jimenez, F. Lautenschlaeger, R. Voituriez, A. M. Lennon-Duménil, N. Manel and M. Piel (2016). "ESCRT III repairs nuclear envelope ruptures during cell migration to limit DNA damage and cell death." Science **352** (6283): 359–362.

Reis-Sobreiro, M., J. F. Chen, T. Novitskaya, S. You, S. Morley, K. Steadman, N. K. Gill, A. Eskaros, M. Rotinen, C. Y. Chu, L. W. Chung, H. Tanaka, W. Yang, B. S. Knudsen, H. R. Tseng, A. C. Rowat, E. M. Posadas, A. Zijlstra, D. Di Vizio and M. R. Freeman (2018). "Emerin deregulation links nuclear shape instability to metastatic potential." Cancer Research **78**(21): 6086–6097.

Roca-Cusachs, P., T. Iskratsch and M. P. Sheetz (2012). "Finding the weakest link–exploring integrin-mediated mechanical molecular pathways." Journal of Cell Science **125**(13): 3025–3038.

Rosenbluth, M. J., W. A. Lam and D. A. Fletcher (2006). "Force microscopy of nonadherent cells: A comparison of leukemia cell deformability." Biophysical Journal **90**(8): 2994–3003.

Rosenbluth, M. J., W. A. Lam and D. A. Fletcher (2008). "Analyzing cell mechanics in hematologic diseases with microfluidic biophysical flow cytometry." Lab on a Chip **8**(7): 1062–1070.

Rosendahl, P., K. Plak, A. Jacobi, M. Kraeter, N. Toepfner, O. Otto, C. Herold, M. Winzi, M. Herbig, Y. Ge, S. Girardo, K. Wagner, B. Baum and J. Guck (2018). "Real-time fluorescence and deformability cytometry." Nature Methods **15**(5): 355.

Rotsch, C. and M. Radmacher (2000). "Drug-induced changes of cytoskeletal structure and mechanics in fibroblasts: An atomic force microscopy study." Biophysical Journal **78**(1): 520–535.

Rowat, A. C., J. Lammerding, H. Herrmann and U. Aebi (2008). "Towards an integrated understanding of the structure and mechanics of the cell nucleus." Bioessays **30**(3): 226–236.

Rowat, A. C., D. E. Jaalouk, M. Zwerger, W. L. Ung, I. A. Eydelnant, D. E. Olins, A. L. Olins, H. Herrmann, D. A. Weitz and J. Lammerding (2013). "Nuclear envelope composition determines the ability of neutrophil-type cells to passage through micron-scale constrictions." Journal of Biological Chemistry **288**(12): 8610–8618.

Sbrana, F., C. Sassoli, E. Meacci, D. Nosi, R. Squecco, F. Paternostro, B. Tiribilli, S. Zecchi-Orlandini, F. Francini and L. Formigli (2008). "Role for stress fiber contraction in surface tension development and stretch-activated channel regulation in C2C12 myoblasts." American Journal of Physiology – Cell Physiology **295**(1): C160–C172.

Schmid-Schönbein, G. W., K. L. Sung, H. Tözeren, R. Skalak and S. Chien (1981). "Passive mechanical properties of human leukocytes." Biophysical Journal **36**(1): 243–256.

Sen, S. and S. Kumar (2009). "Cell–matrix de-adhesion dynamics reflect contractile mechanics." Cellular and Molecular Bioengineering **2**(2): 218–230.

Shin, S. and Y. Ku (2005). "Hemorheology and clinical application: Association of impairment of red blood cell deformability with diabetic nephropathy." Korea-Australia Rheology Journal **17**(3): 117–123.

Smith, A. S., R. B. Nowak, S. Zhou, M. Giannetto, D. S. Gokhin, J. Papoin, I. C. Ghiran, L. Blanc, J. Wan and V. M. Fowler (2018). "Myosin IIA interacts with the spectrin-actin membrane skeleton to control red blood cell membrane curvature and deformability." Proceedings of the National Academy of Sciences **115**(19): E4377–E4385.

Stewart, M. P., Y. Toyoda, A. A. Hyman and D. J. Müller (2012). "Tracking mechanics and volume of globular cells with atomic force microscopy using a constant-height clamp." Nature Protocols **7**(1): 143–154.

Swaminathan, V., K. Mythreye, E. T. O'Brien, A. Berchuck, G. C. Blobe and R. Superfine (2011). "Mechanical stiffness grades metastatic potential in patient tumor cells and in cancer cell lines." Cancer Research **71**(15): 5075–5080.

Swift, J., I. L. Ivanovska, A. Buxboim, T. Harada, P. D. P. Dingal, J. Pinter, J. D. Pajerowski, K. R. Spinler, J. W. Shin, M. Tewari, F. Rehfeldt, D. W. Speicher and D. E. Discher (2013). "Nuclear lamin-A scales with tissue stiffness and enhances matrix-directed differentiation." Science **341**(6149): 1240104.

Targosz-Korecka, M., G. D. Brzezinka, K. E. Malek, E. Stępień and M. Szymonski (2013). "Stiffness memory of EA. hy926 endothelial cells in response to chronic hyperglycemia." Cardiovascular Diabetology **12**(1): 96.

Thuet, K. M., E. A. Bowles, M. L. Ellsworth, R. S. Sprague and A. H. Stephenson (2011). "The Rho kinase inhibitor Y-27632 increases erythrocyte deformability and low oxygen tension-induced ATP release." American Journal of Physiology-Heart and Circulatory Physiology **301**(5): H1891–H1896.

Ting-Beall, H. P., A. S. Lee and R. M. Hochmuth (1995). "Effect of cytochalasin D on the mechanical properties and morphology of passive human neutrophils." Annals of Biomedical Engineering **23**(5): 666–671.

Toepfner, N., C. Herold, O. Otto, P. Rosendahl, A. Jacobi, M. Kräter, J. Stächele, L. Menschner, M. Herbig, L. Ciuffreda, L. Ranford-Cartwright, M. Gryzbek, U. Coskun, E. Reithuber, G. Garriss, P. Mellroth, B. Henriques-Normark, N. Tregay, M. Suttorp, M. Bornhauser, E. R. Chilvers, R. Berner and J. Guck (2018). "Detection of human disease conditions by single-cell morpho-rheological phenotyping of blood." Elife **7**: e29213.

Trickey, W. R., G. M. Lee and F. Guilak (2000). "Viscoelastic properties of chondrocytes from normal and osteoarthritic human cartilage." Journal of Orthopaedic Research **18**(6): 891–898.

Tse, H. T. K., D. R. Gossett, Y. S. Moon, M. Masaeli, M. Sohsman, Y. Ying, K. Mislick, R. P. Adams, J. Rao and D. Di Carlo (2013). "Quantitative diagnosis of malignant pleural effusions by single-cell mechanophenotyping." Science Translational Medicine **5**(212): 212ra163–212ra163.

Worthen, G. S., B. Schwab, 3rd, E. L. Elson and G. P. Downey (1989). "Mechanics of stimulated neutrophils: cell stiffening induces retention in capillaries." Science **245**(4914): 183–186.

Wolf, K., M. Te Lindert, M. Krause, S. Alexander, J. Te Riet, A. L. Willis, R. M. Hoffman, C. G. Figdor, S. J. Weiss and P. Friedl (2013). "Physical limits of cell migration: Control by ECM space and nuclear deformation and tuning by proteolysis and traction force." Journal of Cell Biology **2013** (7): 1069–1084.

Wyss, H. M., T. Franke, E. Mele and D. A. Weitz (2010). "Capillary micromechanics: Measuring the elasticity of microscopic soft objects." Soft Matter **6**(18): 4550–4555.

Xu, W., R. Mezencev, B. Kim, L. Wang, J. McDonald and T. Sulchek (2012). "Cell stiffness is a biomarker of the metastatic potential of ovarian cancer cells." PLoS One **7**(10): e46609.

Zeidman, I. (1961). "The fate of circulating tumor cells: I." Passage of Cells through Capillaries. Cancer Research **21**(1): 38–39.

Zhang, H. and K. K. Liu (2008). "Optical tweezers for single cells." Journal of the Royal Society Interface **5**(24): 671–690.

Zhao, B., L. Li, L. Wang, C. Y. Wang, J. Yu and K. L. Guan (2012). "Cell detachment activates the Hippo pathway via cytoskeleton reorganization to induce anoikis." Genes & Development **26**(1): 54–68.

Zheng, Y., J. Wen, J. Nguyen, M. A. Cachia, C. Wang and Y. Sun (2015). "Decreased deformability of lymphocytes in chronic lymphocytic leukemia." Scientific Reports **5**: 7613.

Sébastien Janel, Vincent Dupres, Lorena Redondo-Morata, and Frank Lafont

3.8 Progresses in Correlative Microscopy and Techniques for AFM in Biomechanics

Since its introduction in life sciences, research activities using atomic force microscopy (AFM) aim at investigating the living matter with a clear identification of the many components of complex systems at several scales, from molecules to organisms. Identification was first based on structural fiducial parameters such as DNA grooves or cytoskeletal filaments. As soon as AFM was performed on cells, the question arose on the identity of the components sensed by the tip of the cantilever, especially during indentation studies. Hence, the need to establish a physical connection to optical microscopes in order to go along with correlative analysis was present already. However, it required some time before the combination really worked in a user-friendly and useful way and at the same metric range, with the development of super-resolution (SR) photonic microscopes. Actually, this was one of the main development lines that the AFM field experienced these last years, the other one being the faster acquisition rate. Most of the studies to date concern using a fluorescence optical microscope with an AFM, but some more original coupling already proved to be useful. In this chapter, we cover how and to which extent correlative approaches combined with AFM can help answering biomechanical questions.

3.8.1 Controlling the AFM

3.8.1.1 Setup

The first atomic force microscopes, custom made like the one from Binnig and Rohrer, or commercial like the MultiMode (Bruker), did not have any optical microscopic capabilities as the ones available nowadays (Figure 3.8.1). Having a compact design was a

Acknowledgments: We apologize to colleagues whose works were not cited herein owing to space limitation. We thank Dr. Michka Popoff for sharing data of Figure 3.8.2E. This work was supported by grants from the ANR (10-EQPX-04-01, 15-CE18-0016-01, and 16-CE15-00003-03), FEDER (12001407), and EU-ITN Phys2Biomed 812772.

Sébastien Janel, Vincent Dupres, Lorena Redondo-Morata, Frank Lafont, Cellular Microbiology and Physics of Infection Group, University of Lille, CNRS, Inserm, CHU Lille, Institut Pasteur Lille, Center for Infection and Immunity of Lille, Lille, France

https://doi.org/10.1515/9783110640632-014

key to achieve low noise and high resolution in the AFM sense, and no optical integration was present. As a consequence, setting up the microscope for the alignment of the optical lever detection can be difficult for inexperienced users. Therefore, the first benefit of having an AFM coupled to an optical microscope is the ability to easily set up the instrument by seeing through oculars or a camera. It helps positioning the laser in a reproducible manner, which is important for the calibration of the cantilever.

3.8.1.2 Positioning

Cultured cells tend to grow in a random way, and it is important for the user to be able to see where the measurement is going to be performed. Working on living cells, apparent elasticity measured by force–distance curves can vary if recorded on edge, in the perinuclear, or nuclear region, because of the heterogeneous intracellular composition and height differences.

On a bigger scale, tissues are also not homogeneous by nature, and precise positioning must be achieved for the mechanical measurement.

3.8.1.3 Manipulation

Another important application where optical microscopy is needed for the achievement of an AFM experiment is the manipulation of the object. The first one is the picking of cells on a tipless cantilever. This can be performed for nonadherent cells or for the study of cell mechanics considering cells as a whole body, using, for instance, wedge-AFM cantilevers (Stewart et al., 2013). The other application is the picking of in-medium colloids for batch mechanical measurement on cells using the FluidFM technology (Guillaume-Gentil et al., 2014). Bacteria can also be picked up and placed on living cells for the study of not only the adhesion process but also the cell mechanical response (Ciczora et al., 2019). Another more indirect possibility is to coat a cantilever with fibronectin and monitor the cell response upon contact, stimulation, stretching, or pushing (Colombelli et al., 2009).

3.8.1.4 Side-View Imaging

Numerous works have shown how much side-view imaging can be an important technique when performing cell mechanics experiments. This can be achieved through a dedicated apparatus giving direct images (e.g., with a 45° mirror) or by computing a z-stack reconstruction from confocal imaging. This proved to be key to monitor the cell profile (i.e., indentation depth and contact area) during AFM indentation (Harris and Charras, 2011). It also demonstrated how cells are pushed inside a soft gel during AFM indentation (Rheinlaender et al., 2020), and how the soft substrate effect must be taken into account for the measurement of cell cortical elasticity. Side-view confocal imaging was used to monitor the tip–nucleus contact, nuclear compression, and the fate of the cell during indentation, using either bead or conical tips (Krause et al., 2013). Gonnermann et al. measured stiffnesses of non-blebbing and blebbing membranes of *Xenopus* cranial neural crest cells with bead cantilevers. Side view allowed to precisely position the bead on the center top of the membrane (Gonnermann et al., 2015). Haase and Pelling (2013) also used this technique to monitor the recovery of the cell shape (actin and nucleus) following mechanical perturbation, using time-lapse imaging. Rianna and Radmacher (2017) measured elasticities and imaged different indentations of normal, cancer, and metastatic cells on soft polyacrylamide gels (3 kPa). As another example, Staunton et al. (2016) studied the force response of metastatic breast cancer cells embedded in 3D extracellular matrix by combining AFM, axial confocal laser scanning microscopy (CLSM) imaging, and finite element analysis. One of the limitations of side view is the poor optical resolution although Beicker et al. (2018) set up a vertical light sheet enhanced side-view imaging to nicely see how cytoplasm and nucleus are deformed upon colloidal tip indentation, opening the field to plenty more applications.

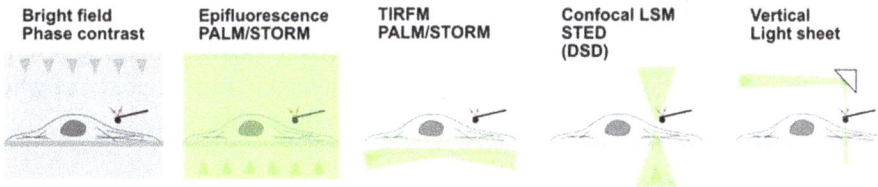

Figure 3.8.1: Types of illumination patterns used in current AFM/optical correlative microscopy.

3.8.2 Identifying the Sample

3.8.2.1 Photonic Microscopy

We cover biomechanical studies, from tissue down to molecules, using optical microscopy technologies coupled with AFM to date. That includes wide field such as bright field, phase contrast, fluorescence, total internal fluorescence (TIRF), photoactivated localization microscopy (PALM), and stochastic optical reconstruction microscopy (STORM). It also includes scanning technologies: confocal (CLSM), stimulated emission depletion (STED), near-field scanning optical microscope (SNOM), and differential spinning disk (DSD). There is no differential interference contrast (DIC) compatibility to commercial AFM microscopes to date because of the AFM head perturbation of the optical path, even though a prototype has been developed in the past (Lugmaier et al., 2005).

3.8.2.1.1 Tissues

Tissues have been investigated by AFM, after biomolecules and living cells. Tissues were not ideal candidates because of the difficult immobilization and the high variation in height compared to the scanning size of the AFM piezoelectric elements. When a tissue is quite homogeneous, indirect correlation can be performed, either with photonic or electron microscopy (EM, here on the cartilage; Stolz et al., 2009). But correlative microscopy seems mandatory when tissues are more heterogeneous, in order to identify which part of the tissue is being examined. One of the seminal articles on the matter (Plodinec et al., 2012) applied hematoxylin and eosin and immunohistochemistry histologic staining imaging after the AFM has been performed on a biopsy. It was probably impossible to image the AFM scan area because the sample was too high to achieve good optical resolution. Later, Calò et al. associated higher elasticity values with collagen-enriched domains of a human liver tissue section. Force–volume measurements were correlated with bright-field images and correlated with corresponding histologically stained adjacent sections (Calò et al., 2020). Jang et al. (2016) correlated AFM topography/elasticity with bright field, X-ray, and strain mapping on interradicular alveolar bone. In the future, correlative AFM and photonic microscopy on the tissue will probably be performed on thin slices (micrometers range). It will first eliminate the uncertainty of performing AFM measurements on the surface of tissue while correlating it to a deeper histologic section. Furthermore, it will be essential to help deciphering why and how elasticities vary from one area to another on the same tissue.

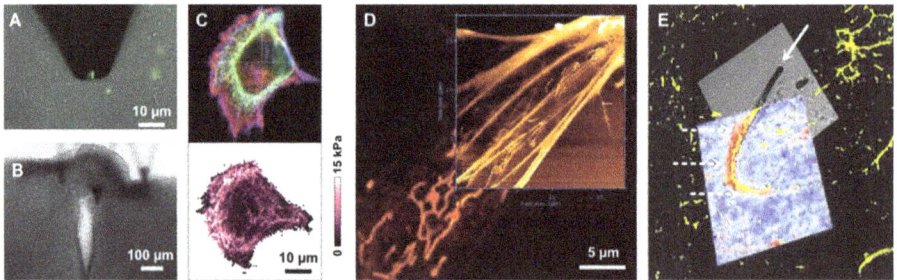

Figure 3.8.2: Examples of correlative microscopy used from microbes to tissues. (A) Manipulating live *Shigella flexneri AfaE GFP* bacteria on a tipless cantilever. (B) Brain tissue slice probed for elasticity by MLCT-SPH-DC cantilever. (C) Correlation between the cytoskeletal organization of retinal pigment epithelial (RPE1) cells as seen in multicolor CLSM for actin (blue), tubulin (red), and vimentin (green) (top) and elasticity (bottom). (D) Fluorescent mitochondria using MitoTracker inside *Potorous tridactylus* Kidney (PtK2) cells and AFM piezo-height image showing both mitochondria and actin cytoskeleton (top-right corner). (E) *Shigella flexneri*'s actin tail comet (dashed arrows) inside fixed PtK2 cells is 3D visualized in PALM with a 30 nm localization precision (black and gold), AFM elasticity mapping (10 μm², blue to red), and TEM sectioning (gray), where the bacterium is visible (white arrow).

3.8.2.1.2 Cell Identification and Selection

One key and frequent application of light microscopy is to sort out which cells can be explored with the AFM. For instance, transfection of cells with siRNA or DNA is not 100% efficient and a fluorescent control is often required. That can be a co-transfected agent or the expression of a fluorescent protein. Fluorescence can also help finding out specific cells in a mixed population. Plodinec et al. (2011) employed a bright field to assess the viability of Rat2 fibroblasts and selected the transfected ones with GFP-desmin by epifluorescence. It helped to compare the effects of desmin variants on cell elasticity and demonstrated that alterations in the intermediate filaments influence the nanomechanical properties of living cells, both at the large and nanometer scale. Staunton et al. (2016) studied how metastatic adenocarcinoma cells become stiffer when invading 3D collagen hydrogels compared to cells staying on the surface. CLSM gave the position of membrane-stained cells in 3D while AFM performed piecewise-z elasticity measurements. Stiffening was determined to be Rho/ROCK-dependent by adding inhibitors directly in the medium.

3.8.2.1.3 Cellular Structures

A considerable amount of work has been produced on the study of the cellular cytoskeleton by correlating AFM and photonic microscopy. Undeniably, microfilaments are considered to be the main participant in cell mechanical properties, together with intermediate filaments and microtubules, and imaging and quantifying them help understanding variations measured with the AFM. Haga et al. (2000) performed CLSM imaging on living fibroblasts and found out that actin and vimentin are correlated with high-elasticity regions while microtubules are correlated with low-elasticity regions. Li et al. (2008) imaged actin and measured elasticity for the study of breast cancer cells. Celik et al. (2013) studied the effect of both mechanical indentation by an AFM tip and shear flow on microtubules network of Chinese Hamster Ovary cells imaged by CLSM. They show displacement of the fibers, and disruption and polymerization of microtubules. Abidine et al. used confocal microscopy to identify the region where they performed rheology measurements (1–300 Hz): edge, perinucleus, and nucleus of the bladder cancer cell. They correlated their measurements with actin fluorescence intensity and found out some different behaviors at the cell periphery (Abidine et al., 2015). Gavara and Chadwick (2016) found out that actomyosin's amount greatly modulates cytoskeletal stiffness by imaging and computing the fiber amount. Grady et al. measured elastic properties of various cell types: chondrocytes, fibroblasts, human umbilical vein endothelial cells, hepatocellular carcinoma cells, and fibrosarcoma cells. They used cytoskeletal-affecting drugs such as cytochalasin D (affecting actin) and nocodazole (affecting microtubules). They correlate elasticity with actin amount in normal cells but also that cancer cells show stiffness variability when subjected to nocodazole (Grady et al., 2016). As the last example of many, Schierbaum et al. correlated viscoelastic measurements of normal and cancerous breast cells using both force clamp force mapping and imaging of the actin cytoskeleton. They show normal cells can increase their stiffness and reduce their fluidity when they become confluent, which is not the case for cancerous cells (Schierbaum et al., 2017).

TIRF microscopy was early coupled with AFM. The AFM was able to image the apical topography of the cell and probe its mechanical properties while TIRF depicted focal contacts. Mechanical stress was also performed with the AFM tip, and consequent effects at the basal level were monitored by fluorescence (Mathur et al., 2000). Trache and Meininger (2005) also set up early a TIRF-AFM and were able to show the rearrangement of focal adhesions of vascular muscle cells upon mechanical stimulation in real time. Labernadie et al. (2010) correlated topography and elasticity dynamics with actin and vinculin staining of podosomes grown on micropatterned fibrinogen. In the field of infection, intracellular *Plasmodium falciparum* inside erythrocytes was detected by staining, allowing to select cells according to the development step of the parasite. The surface structure called knobs, enriched in adhesion protein coming from the parasite, were characterized based on the parasite's genotype (Nacer et al., 2011). Bhat et al. (2018) demonstrated how anthropogenic chemical (2,4-D) effects can be monitored on model cells (*E. coli*, *C. albicans*, HEK 293) using cell topography, cell elasticity measurements, and multicolor confocal microscopy following the increase of the ROS signal and the disorganization of the microtubules network. SR was a breakthrough in photonic microscopy, as it increased the resolution by order of magnitude, and gave access to a clearer image of the cell surface and internal structure. It was early adapted to AFM with the STED modality, where the actin filaments could be seen and their elasticity assessed (Harke et al., 2012).

Atomic force, SR photonic, and electron microscopies were achieved on the same sample (Janel et al., 2017) to validate the stiffness tomography analysis on biological samples (Janel et al., 2019). It demonstrated the stiffening of intracellular mitochondria upon drug treatment in living cells. No PALM/STORM biomechanical experiments have been published to date, partly because of the complex sample preparation and the long acquisition time. An example of such coupling has been performed on fixed samples (Dahmane et al., 2019, Hirvonen and Cox, 2018), looking at the topography. Odermatt et al. (2015) achieved AFM/PALM imaging of living mammalian cells. We show in Figure 3.8.2E an example of PALM and elasticity of bacterial actin tail comet inside PtK2 cells, adding EM section for the ultrastructure imaging of the bacteria. PALM/STORM is better suited for slow biological processes or on fixed samples, as illustrated, while STED is more relevant for fast acquisition.

3.8.2.1.4 Biomolecules

Miranda et al. optimized fluorescence light and spinning disk noise to carry out simultaneous measurements of DSD fluorescence optical sectioning microscopy and quantitative imaging nanomechanical mapping on DOPC/DOPS-supported bilayers (Miranda et al., 2015).

High-speed AFM (HS-AFM) has been one of the best late innovations for AFM in biology, especially to see biomolecules at work. It has been applied to analyze cell mechanics at new frequencies (Rigato et al., 2017) and to map the viscoelastic properties of a live fibroblast cell (Schächtele et al., 2018). Up to now, there are few correlative microscopic works because of the somewhat optical limitations of current microscopes. Some new original setups like HS-AFM-SNOM have achieved high-speed correlation at 3 s/frame and 39 nm fluorescence resolution on biomolecules (Umakoshi et al., 2020).

3.8.2.2 Electron Microscopy

Soon after the application of AFM on biological samples, there has been a will to compare the possibilities of this high-resolution technique to the golden standard of high-resolution imaging, EM. One of the first works investigated the morphology of freeze-fracture replicas of rat atrial tissue (Kordylewski et al., 1994). As their conclusion pointed out, AFM has the advantage of being able to image biological samples in their native environment. Indeed, the two techniques differ greatly in terms of preparation of the sample as the AFM does not need much, while samples for EM have to be prepared with several unkind treatments such as dehydration and staining or metalization. Stukalov et al. (2008) showed by AFM height measurement how uranyl acetate staining shrank the height of *S. oneidensis* bacteria height by a factor of 2, and pleaded for cryotechniques for preparing samples. AFM and field emission scanning electron microscopy (SEM) were reported on human saliva exosomes (Sharma et al., 2010). The authors performed deformation measurements, EM imaging, and force spectroscopy with anti-CD63 IgG-functionalized AFM tips to find out the endosomal origin of the exosomes. Transmission electron microscopy (TEM) can give the best insight into the internal organization of cells. Yamada et al. (2017) developed correlative HS-AFM and TEM preparing a dedicated gold patterned surface. The demonstration was done by imaging actin filaments and immuno-EM of actin antibody with colloidal gold. The latest development of so-called correlative light atomic force electron microscopy shows how TEM or SEM can be deployed for ultrastructure identification after both AFM (topography, cell mechanics) and fluorescence imaging (selection, identification) have been carried out on fixed and living cells (Janel et al., 2017, 2019). The added value of EM is for the high resolution of identified (as by fluorescence) or unidentified structures, and possibly affecting cell mechanics locally.

3.8.2.3 Raman Spectroscopy

Raman spectroscopy gives the possibility to determine the internal molecular composition of the biological sample. Therefore, coupling to AFM can help understand why some variations in the viscoelastic properties of cells or tissues are observed. McEwen et al. (2013) linked chemical component analysis of the cancer cell membrane with cell's biophysical properties. Bright-field imaging, Raman spectroscopy, and then AFM (elasticity and adhesion) were applied on three cancer cell lines (A549, MDA-MB-435 +/– BRMS1). Maase et al. (2019) combined Raman spectroscopy for biochemical analysis with AFM nanomechanics of the endothelium of native ex vivo aortas. They linked an increased cortical elasticity with intracellular lipid content for dysfunctional endothelium (ApoE/LDLR–/–). Raman mapping (3,030–2,800 cm^{-1}) and AFM were performed sequentially using a top access Raman spectrometer. There are no correlative tip-enhanced Raman spectroscopy (TERS) publications on cell mechanics yet, as most of the studies were performed in air or on fixed samples. Hermelink et al. (2017) developed a noteworthy TERS/EM correlative approach for the study of the tobacco mosaic virus. TEM was executed first and AFM/TERS afterward. For similar sample preparation reasons, established spectroscopic techniques such as AFM infrared and time-of-flight secondary ion mass spectrometry did not give any insights into biomechanical studies with AFM to date.

3.8.3 Correlative Techniques

3.8.3.1 Micropillars

The cell environment in living organisms is quite different from the standard plastic or glass Petri dishes used in AFM biomechanics experiments. Micropillars are substrate protrusions of a given size that help understand how living cells adapt to their environment by modifying their shape. Dimensions (shape, width, and height) can be defined, and stiffness adjusted by changing the composition (PDMS, PLLA) during the microfabrication process. Badique et al. (2013) used micropillars to force the deformation of the nucleus of different osteosarcoma cells, and measured their elastic properties while monitoring the cytoskeleton reorganization by fluorescence microscopy. They correlated deformation capacity and higher stiffness as measured by AFM. This was further linked to the cell phenotype (thick layer of actin filaments around the nucleus, dense perinuclear microtubules, and vimentin filament networks) and not to the substrate chemistry nor elasticity. Palankar et al. (2016) carried out AFM elasticity experiments on stem-cell-derived cardiomyocytes grown on micropillars and planar surfaces. Differences were explained by the remodeling of sarcomeres and clustering of integrins around the pillars, leading to modification of calcium signaling.

3.8.3.2 Traction Force Microscopy

Traction force microscopy (TFM) is another means of exploring cell mechanics, well complementary to AFM. Such a correlation can reveal a link between viscoelastic properties and the contractile prestress of living cells in their native environment. Al-Rekabi and Pelling (2013) studied the interplay between substrate elasticity and nanomechanical stimulation, by observing a rapid increase in the traction of muscle precursor cells when stimulated by loading of a 10 nN local force. Cell traction was highest on gels with similar elasticities (64–89 kPa), and appeared to be transient (30 s). van Helvert and Friedl (2016) monitored the strain stiffening of biopolymer networks at the leading edge of normal and tumor cells with the AFM nanoindentation and confocal reflection microscopy. Schierbaum et al. (2019) employed the same AFM-TFM correlative microscopy to further describe the viscoelastic properties and contractile prestress of fibroblast and epithelial cells. Stiffer cells had lower fluidity and larger prestress than softer cells. Actomyosin machinery seems to be the dominant factor influencing such behaviors.

3.8.3.3 Scanning Ion-Conductance Microscopy

Scanning ion-conductance microscopy (SICM) is a younger and challenger technique to AFM, as it scans the samples from the top. Here the cantilever/tip is replaced by a glass or silicon pipette, through which ion current is flowing. Resolution is governed by the aperture size of the pipette. By using the current signal as feedback in a hopping mode, SICM can produce noncontact topographic images of very soft samples such as neurons in less than a minute (Watanabe et al., 2019). It can additionally measure surface charges and establish mechanical mapping. Pellegrino et al. (2012) coupled SICM with an inverted AFM setup to understand forces produced by ion current, and applied this to measure elastic properties of fibroblasts. Ushiki et al. (2012) compared AFM, SICM, and TEM from biomolecules to living cells and the luminal surface of rat trachea, and show how SICM can be relevant in the case of samples with steep slopes. Rheinlaender et al. (2015) studied the mechanics of hemostasis by comparing thrombin-induced and cytochalasin D-mediated elasticity changes of human platelet. Another approach developed by Ossola et al. (2015) is a completely integrated AFM-SICM setup using the FluidFM technology, where a microchannel inside the AFM cantilever acts as the pipette.

3.8.4 Correlation Method

3.8.4.1 Hardware Setup

The first requirement to achieve is the mechanical integration of the two techniques. AFM manufacturers provide stages for the main optical inverted microscope companies. The goal of the AFM stage is to provide mechanical stability to the AFM head, to align the tip with the field of view, and to move the sample. The sample on the stage can be motionless (tip scanning) or moving (sample scanning). Some setups provide tip and sample scanning. This is very important, depending on the technique being added to the AFM: wide-field imaging is better achieved with tip scanning, but a stable and perfect correlation can be achieved between a LASER beam and AFM with sample scanning. All this is valid for transparent substrate (i.e., glass). When working on an opaque sample, the solution is to work with a shuttle stage that will go from the upright optical microscope to the AFM stage with a positioning accuracy of a few µm (e.g., Bruker BioMat).

3.8.4.2 Software Integration

To have two techniques integrated, meaning scanning the same sample at the same time, there must be a unique software handling both microscopes or two software that work synchronously. AFM controlling being the more demanding task, AFM manufacturers tend to implement optical acquisition on their software. It usually consists in importing the image from a camera or a shared folder automatically. To do so, the first step is to calibrate the optical image with respect to the precise AFM x–y movement. The image is then imported, corrected, and displayed in the AFM field of the scan. The user can decide to AFM scan the region of interest. This video integration in the AFM software is useful; nevertheless, it has many disadvantages: there is a limited number of cameras with software drivers available, there is no microscope control, and there is no image processing. That means advanced microscopy acquisition must be performed by another software and controller. The triggering of these acquisitions and image transfer can still be automatized on some systems like AFM/STED.

For better correlation of the two data, the user can use fiducial markers such as TetraSpeck or FluoSpheres that will both be visible in fluorescence and AFM, with small sizes (hundreds of nm). An offline software will then correct the optical image to match the position of the spheres as detected by the AFM scan. An AFM stage or head with x–y sensors should give the most accurate dimensions.

3.8.4.3 Repositioning

To perform EM microscopy correlation for cell mechanics, the sample needs to be moved away from the AFM'stage, so the location of the cell is lost. One of the solutions is to prepare the sample on photo-etched grid coverslips, like the ones pictured in Figure 3.8.3.

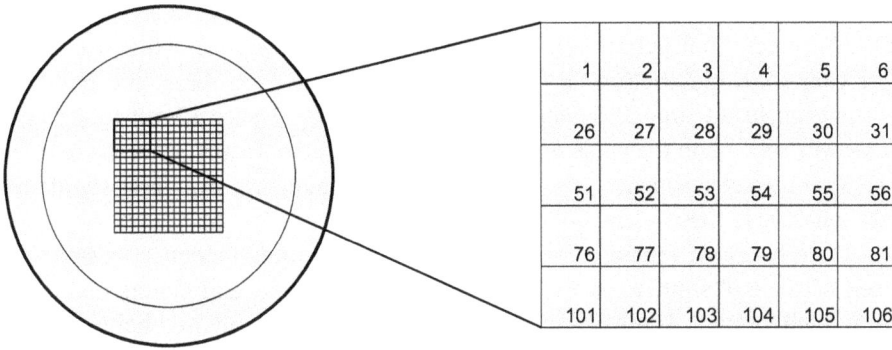

1	2	3	4	5	6
26	27	28	29	30	31
51	52	53	54	55	56
76	77	78	79	80	81
101	102	103	104	105	106

Figure 3.8.3: Photo-etched coverslip used for repositioning AFM/Optics with subsequent electron microscopy.

The user can note the position of the AFM scan from bright-field imaging and then detect it on the EM preparation to find the same cell. The different methods of EM preparation will often slightly deform the sample: ethanol dehydration may shrink the sample in both TEM and SEM, and TEM sectioning might also compress the sample in the direction of the cut. As a result, what is seen in AFM and EM might not be exactly of the same dimensions and so the EM image will have to be offline corrected to match the AFM image. Correlation can be performed with the existing software such as ImageJ or Icy (ec-CLEM; Paul-Gilloteaux et al., 2017).

3.8.5 Limitations

Can we perform experiments with the best sensibility of the two coupled techniques at the same time, or are there any issues when integrating AFM with another technique? Taking the example of AFM/inverted fluorescence microscope setup, we will list here the cross-talk happening between the two techniques.

3.8.5.1 Noise Affecting the AFM Measurements

3.8.5.1.1 Mechanical Noise

The mechanical noise comes from several parts. The first source is ambient noise coming from cooling fans of cameras, controllers, and LASER sources. This might make very sensitive experiments difficult to achieve. We observed a 100% increase of noise on the force curve baseline when using the internal fan of an Andor iXon EMCCD camera (not published). To circumvent this, the user can either use a fan-less camera, water cooling, or disable the fan, isolate controllers, and build an acoustic enclosure around the microscope. The amount of noise affecting the cantilever will depend on the noise frequency and the resonant frequency of the cantilever. As a nice consequence, a short, high-resonant frequency cantilever should be less sensitive to these sources of noise.

The other main source of mechanical noise comes from the optical microscope itself because of shutters, scanning mirrors, moving disks, and motors that automate the light path. These effects are important, so it is best advised not to change any light path during AFM scanning, as this can induce lines on topographic images, noise on force curves. It can even damage the tip, which is unwanted as it is crucial to keep a constant tip geometry during mechanical testing of the sample.

3.8.5.1.2 Photothermal Noise and Radiation Pressure Force

There are two causes of interference of the illumination on the AFM tip/cantilever. First, photothermal effects will make the cantilever bend because of the different thermal expansion coefficients of the bimetallic reflective coatings, and secondly, radiation pressure forces acting on the tip or the cantilever. The levels of cantilever deflection will depend on the optical setup and power used. Even simple bright-field illumination has an effect on the cantilever deflection. We recorded up to 10 pN noise on Olympus AC40 and Bruker MLCT-BIO-DC cantilevers with 50% power illumination from a standard Zeiss LED source. The noise increased up to 60 pN with 100% power, with the temperature staying constant in the medium. It is best to either keep the bright-field illumination at constant low power or even turn it off during AFM scanning. In the case of fluorescence imaging with a continuous source, the energy can be much higher and produce huge cantilever deflections. Also when working with fluorescence, the cantilever is bent, and so the deflection sensitivity calibration might not hold. Fernandes et al. (2020) described these effects in detail and were able to perform correlative AFM-fluorescence lifetime imaging microscopy on supported lipid bilayers by using low excitation power (120 mW) and specific cantilevers. Finally, illumination can heat up the medium (Figure 3.8.4C), which will consequently deflect the cantilever if metal-coated. We recorded a 0.2 °C increase for 10% illumination of a Lumencore Sola fluorescent source (not published). Besides deflection of the cantilever, temperature rise can have an effect on the sample mechanical properties itself. As shown by Sunnerberg et al. (2019), cortical neurons get softer and bigger when increasing temperature. Authors relate this to the dynamics of the cytoskeleton. It is therefore important to keep the temperature as stable and homogeneous as possible and thus reduce illumination. When using a confocal (or STED) microscopy, the illumination will be restricted in the x–y plane but also higher in terms of power. As a consequence, the amount of noise greatly depends on the position of the cantilever with respect to the LASER spot in x-, y-, and z planes. With the focus made on the tip, we noticed a 20–200 nm vertical deflection change using a 20 mW 485 nm confocal illumination in a medium. This was in the same order of magnitude when using a 1 W pulsed 775 nm STED illumination. It is worth mentioning that TIRF imaging should produce much less interference as the illumination is limited to a few hundreds of nanometers above the glass and so only the end of the tip should be affected.

All these effects can be summed up in Figure 3.8.4D, where the microscope automatically acquires a bright-field and fluorescence time-lapse movie. It seems unlikely to be able to perform low-noise, reliable mechanical measurements while changing the optical microscope setup as in a standard acquisition performed by biologists on living cells. As a consequence, the user must be extremely cautious when manipulating any optical microscopy during AFM mechanical measurement. The safest way is to acquire fluorescence and AFM sequentially, but one ingenious alternative for biomechanical experiments would be to perform fluorescence imaging while the cantilever is in the retracted position of a force curve cycle. This could be automatized in a pixel-by-pixel or line-by-line manner. Similar work synchronizing force and fluorescence have already been reported, following calcium response and force response of T cells and macrophages when stimulated with the AFM (Cazaux et al., 2016).

3.8.5.2 AFM Affecting Optical Microscopy

3.8.5.2.1 AFM Laser

The main issue when integrating an AFM on another microscope is the LASER of the optical beam detection system. It is often IR, not to interfere with fluorescence wavelengths (350–800 nm). The beam goes through the sample and reaches the camera, polluting the optical image (see Figure 3.8.4B). This is true for wide-field and scanning optical microscopes. AFM manufacturers provide a laser-blocking filter that seats after the objective turret, but without the best blocking properties. It is advised to change it to a more efficient one. Another possibility is to turn off the AFM LASER while acquiring the optical image.

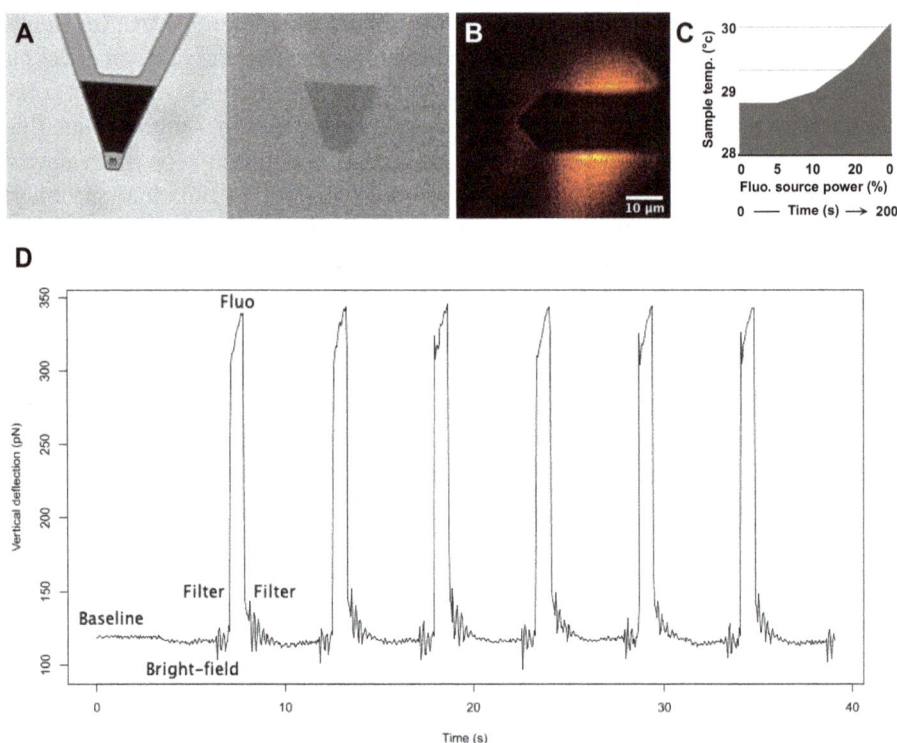

Figure 3.8.4: Some limitations of AFM/optical-integrated microscopes. (A) MLCT-BIO-DC cantilever seen in bright field (left) and its luminescence in the fluorescence channel (right). (B) AFM 850 nm IR laser as seen on a STED avalanche photodiode detector (APD). (C) Rise of the sample temperature when increasing wide-field standard fluorescent light source. (D) Perturbations of the cantilever deflection when operating an inverted wide-field fluorescence microscope.

3.8.5.2.2 Cantilever

One possible source of the problem is the cantilever itself. It can both slightly deviate the path of the transmitted light and be a source of luminescence (Figure 3.8.4A) or reflection if coated with a reflective layer (Au and Ti). If this is a problem, the user can either try uncoated cantilevers, higher tips to move the cantilever away from the surface, or retract the cantilever during acquisition. Another possibility is to synchronize both AFM and optical scanning and perform optical pixel acquisition when the tip is retracted in a force-triggered mode.

3.8.6 Perspectives

In the past years, there have been considerable developments in all three kinds of microscopy (scanning probe, photonic, and electronic), hallmarked by the Nobel Prizes in Physics in 1986 (AFM and EM) and Chemistry in 2014 (SR) and 2017 (cryo-EM). The rise of HS-AFM made a breakthrough with a huge increase in the scanning speed, bringing new insights into biological processes. Also, new imaging modes brought more qualitative and quantitative information on the sample, together with an easier scanning on biological samples. On the photonic microscopy side, the resolution has greatly progressed from hundreds of nanometers to tens of nanometers (STED, PALM/STORM), and in 2020 reached more or less the resolution of the AFM with MINFLUX technology (Gwosch et al., 2020). There is, therefore, a convergence of both scanning speed and resolution in both techniques. Together with the advent of cryoelectron microscopy, this opens a lot of possibilities for correlative microscopy for higher spatial and temporal resolutions.

Also, the field of optical microscopy has already embraced automation in order to be able to scan much more samples than a human user could be able to perform by hand on a microscope. Actually, high-content screening allows to scan complete drug libraries or gene banks to find specific targets in diseases, using automatic confocal fluorescence imaging and automatic image analysis. Automation has already been achieved in AFM on prokaryotes (Dujardin et al., 2019), and there is no doubt that this will be developed further in the future. It will make AFM more relevant to tackle important biological questions, for instance, by profiling mechanical or adhesive properties of samples with corresponding fluorescence markers.

What AFM-correlative microscopy must therefore achieve in the future is an adaptation to both faster scanning and more qualitative and quantitative information on more biological samples. This will demand better integration of the techniques, meaning synchronization of the measurements and handling of the data.

References

Abidine, Y., V. M. Laurent, R. Michel, A. Duperray and C. Verdier (2015). "Local mechanical properties of bladder cancer cells measured by AFM as a signature of metastatic potential". European Physical Journal Plus **130**: 202.

Al-Rekabi, Z. and A. E. Pelling (2013). "Cross talk between matrix elasticity and mechanical force regulates myoblast traction dynamics". Physical Biology **10**: 066003.

Badique, F., D. R. Stamov, P. M. Davidson, M. Veuillet, G. Reiter, J.-N. Freund, C. M. Franz and K. Anselme (2013). "Directing nuclear deformation on micropillared surfaces by substrate geometry and cytoskeleton organization". Biomaterials **34**: 2991–3001.

Beicker, K., E. T. O'Brien, M. R. Falvo and R. Superfine (2018). "Vertical light sheet enhanced side-view imaging for AFM cell mechanics studies". Scientific Reports **8**: 1504.

Bhat, S. V., T. Sultana, A. Körnig, S. McGrath, Z. Shahina and T. E. S. Dahms (2018). "Correlative atomic force microscopy quantitative imaging-laser scanning confocal microscopy quantifies the impact of stressors on live cells in real-time". Scientific Reports **8**: 8305.

Calò, A., Y. Romin, R. Srouji, C. P. Zambirinis, N. Fan, A. Santella, E. Feng, S. Fujisawa, M. Turkekul, S. Huang, A. L. Simpson, M. D'Angellica, W. R. Jarmagin and K. Manova-Todorova (2020). "Spatial mapping of the collagen distribution in human and mouse tissues by force volume atomic force microscopy". Scientific Reports **10**: 15664.

Cazaux, S., A. Sadoun, M. Pélicot-Biarnes, M. Martinez, S. Obeid, P. Bongrand, L. Limozin and P. H. Puech (2016). "Synchronizing atomic force microscopy force mode and fluorescence microscopy in real time for immune cell stimulation and activation studies". Ultramicroscopy **160**: 168–181.

Celik, E., M. H. Abdulreda, D. Maiguel, J. Li and V. T. Moy (2013). "Rearrangement of microtubule network under biochemical and mechanical stimulations". Methods **60**: 195–201.

Ciczora, Y., S. Janel, M. Soyer, M. Popoff, E. Werkmeister and F. Lafont (2019). "Blocking bacterial entry at the adhesion step reveals dynamic recruitment of membrane and cytosolic probes". Biology of the Cell/under the Auspices of the European Cell Biology Organization **111**: 67–77.

Colombelli, J., A. Besser, H. Kress, E. G. Reynaud, P. Girard, E. Caussinus, U. Haselmann, J. V. Small, U. S. Schwarz and E. H. K. Stelzer (2009). "Mechanosensing in actin stress fibers revealed by a close correlation between force and protein localization". Journal of Cell Science **122**: 1665–1679.

Dahmane, S., C. Doucet, A. Le Gall, C. Chamontin, P. Dosset, F. Murcy, L. Fernandez, D. Salas, E. Rubinstein, M. Mougel, M. Nollmann and P.-E. Milhiet (2019). "Nanoscale organization of tetraspanins during HIV-1 budding by correlative dSTORM/AFM". Nanoscale **11**: 6036–6044.

Dujardin, A., P. De Wolf, F. Lafont and V. Dupres (2019). "Automated multi-sample acquisition and analysis using atomic force microscopy for biomedical applications". PLoS One **14**: e0213853.

Fernandes, T. F. D., O. Saavedra-Villanueva, E. Margeat, P.-E. Milhiet and L. Costa (2020). "Synchronous, crosstalk-free correlative AFM and confocal microscopies/spectroscopies". Scientific Reports **10**: 7098.

Gavara, N. and R. S. Chadwick (2016). "Relationship between cell stiffness and stress fiber amount, assessed by simultaneous atomic force microscopy and live-cell fluorescence imaging". Biomechanics and Modeling in Mechanobiology **15**: 511–523.

Gonnermann, C., C. Huang, S. F. Becker, D. R. Stamov, D. Wedlich, J. Kashef and C. M. Franz (2015). "Quantitating membrane bleb stiffness using AFM force spectroscopy and an optical sideview setup". Integrative Biology **7**: 356–363.

Grady, M. E., R. J. Composto and D. M. Eckmann (2016). "Cell elasticity with altered cytoskeletal architectures across multiple cell types". Journal of the Mechanical Behavior of Biomedical Materials **61**: 197–207.

Guillaume-Gentil, O., E. Potthoff, D. Ossola, C. M. Franz, T. Zambelli and J. A. Vorholt (2014). "Force-controlled manipulation of single cells: From AFM to FluidFM". Trends in Biotechnology **32**: 381–388.

Gwosch, K. C., J. K. Pape, F. Balzarotti, P. Hoess, J. Ellenberg, J. Ries and S. W. Hell (2020). "MINFLUX nanoscopy delivers 3D multicolor nanometer resolution in cells". Nature Methods **17**: 217–224.

Haase, K. and A. E. Pelling (2013). "Resiliency of the plasma membrane and actin cortex to large-scale deformation: Resiliency of the Plasma Membrane". Cytoskeleton **70**: 494–514.

Haga, H., S. Sasaki, K. Kawabata, E. Ito, T. Ushiki and T. Sambongi (2000). "Elasticity mapping of living fibroblasts by AFM and immunofluorescence observation of the cytoskeleton". Ultramicroscopy **82**: 253–258.

Harke, B., J. V. Chacko, H. Haschke, C. Canale and A. Diaspro (2012). "A novel nanoscopic tool by combining AFM with STED microscopy". Optical Nanoscopy **1**: 1.

Harris, A. R. and G. T. Charras (2011). "Experimental validation of atomic force microscopy-based cell elasticity measurements". Nanotechnology **22**: 345102.

van Helvert, S. and P. Friedl (2016). "Strain stiffening of fibrillar collagen during individual and collective cell migration identified by AFM nanoindentation". ACS Applied Materials & Interfaces **8**: 21946–21955.

Hermelink, A., D. Naumann, J. Piesker, P. Lasch, M. Laue and P. Hermann (2017). "Towards a correlative approach for characterising single virus particles by transmission electron microscopy and nanoscale Raman spectroscopy". Analyst **142**: 1342–1349.

Hirvonen, L. M. and S. Cox (2018). "STORM without enzymatic oxygen scavenging for correlative atomic force and fluorescence superresolution microscopy". Methods and Applications in Fluorescence **6**: 045002.

Janel, S., E. Werkmeister, A. Bongiovanni, F. Lafont and N. Barois (2017). "CLAFEM: Correlative light atomic force electron microscopy". Methods in Cell Biology **140**: 165–185.

Janel, S., M. Popoff, N. Barois, E. Werkmeister, S. Divoux, F. Perez and F. Lafont (2019). "Stiffness tomography of eukaryotic intracellular compartments by atomic force microscopy". Nanoscale **11**: 10320–10328.

Jang, A., R. Prevost and S. P. Ho (2016). "Strain mapping and correlative microscopy of the alveolar bone in a bone–periodontal ligament–tooth fibrous joint". Proceedings of the Institution of Mechanical Engineers Part H **230**: 847–857.

Kordylewski, L., D. Saner and R. Lal (1994). "Atomic force microscopy of freeze-fracture replicas of rat atrial tissue". Journal of Microscopy **173**: 173–181.

Krause, M., J. Te Riet and K. Wolf (2013). "Probing the compressibility of tumor cell nuclei by combined atomic force-confocal microscopy". Physical Biology **10**: 065002.

Labernadie, A., C. Thibault, C. Vieu, I. Maridonneau-Parini and G. M. Charriere (2010). "Dynamics of podosome stiffness revealed by atomic force microscopy". PNAS **107**: 21016–21021.

Li, Q. S., G. Y. H. Lee, C. N. Ong and C. T. Lim (2008). "AFM indentation study of breast cancer cells". Biochemical and Biophysical Research Communications **374**: 609–613.

Lugmaier, R. A., T. Hugel, M. Benoit and H. E. Gaub (2005). "Phase contrast and DIC illumination for AFM hybrids". Ultramicroscopy **104**: 255–260.

Maase, M., A. Rygula, M. Z. Pacia, B. Proniewski, L. Mateuszuk, M. Sternak, A. Kaczor, S. Chlopicki and K. Kusche-Vihrog (2019). "Combined Raman- and AFM-based detection of biochemical and nanomechanical features of endothelial dysfunction in aorta isolated from ApoE/LDLR−/− mice". Nanomedicine: Nanotechnology, Biology and Medicine **16**: 97–105.

Mathur, A. B., G. A. Truskey and W. Monty Reichert (2000). "Atomic Force and Total Internal Reflection Fluorescence Microscopy for the Study of Force Transmission in Endothelial Cells". Biophysical Journal **78**: 1725–1735.

McEwen, G. D., Y. Wu, M. Tang, X. Qi, Z. Xiao, S. M. Baker, T. Yu, T. A. Gilbertson, D. B. DeWald and A. Zhou (2013). "Subcellular spectroscopic markers, topography and nanomechanics of human lung cancer and breast cancer cells examined by combined confocal Raman microspectroscopy and atomic force microscopy". Analyst **138**: 787–797.

Miranda, A., M. Martins and P. A. A. De Beule (2015). "Simultaneous differential spinning disk fluorescence optical sectioning microscopy and nanomechanical mapping atomic force microscopy". Review of Scientific Instruments **86**: 093705.

Nacer, A., E. Roux, S. Pomel, C. Scheidig-Benatar, H. Sakamoto, F. Lafont, A. Scherf and D. Mattei (2011). "Clag9 is not essential for PfEMP1 surface expression in non-cytoadherent Plasmodium falciparum parasites with a chromosome 9 deletion". PLoS One **6**: e29039.

Odermatt, P. D., A. Shivanandan, H. Deschout, R. Jankele, A. P. Nievergelt, L. Feletti, M. W. Davidson, A. Radenovic and G. E. Fantner (2015). "High-resolution correlative microscopy: Bridging the gap between single molecule localization microscopy and atomic force microscopy". Nano Letters **15**: 4896–4904.

Ossola, D., L. Dorwling-Carter, H. Dermutz, P. Behr, J. Vörös and T. Zambelli (2015). "Simultaneous scanning ion conductance microscopy and atomic force microscopy with microchanneled cantilevers". Physical Review Letters **115**: 238103.

Palankar, R., M. Glaubitz, U. Martens, N. Medvedev, M. von der Ehe, S. B. Felix, M. Münzenberg and M. Delcea (2016). "3D micropillars guide the mechanobiology of human induced pluripotent stem cell-derived cardiomyocytes". Advanced Healthcare Materials **5**: 335–341.

Paul-Gilloteaux, P., X. Heiligenstein, M. Belle, M. C. Domart, B. Larijani, L. Collinson, G. Raposo and J. Salamero (2017). "EC-CLEM: Flexible multidimensional registration software for correlative microscopies". Nature Methods **14**: 102–103.

Pellegrino, M., M. Pellegrini, P. Orsini, E. Tognoni, C. Ascoli, P. Baschieri and F. Dinelli (2012). "Measuring the elastic properties of living cells through the analysis of current–displacement curves in scanning ion conductance microscopy". Pflügers Archiv: European Journal of Physiology **464**: 307–316.

Plodinec, M., M. Loparic, R. Suetterlin, H. Herrmann, U. Aebi and C.-A. Schoenenberger (2011). "The nanomechanical properties of rat fibroblasts are modulated by interfering with the vimentin intermediate filament system". Journal of Structural Biology **174**: 476–484.

Plodinec, M., M. Loparic, C. A. Monnier, E. C. Obermann, R. Zanetti-Dallenbach, P. Oertle, J. T. Hyotyla, U. Aebi, M. Bentires-Alj, R. Y. H. Lim and C.-A. Schoenenberger (2012). "The nanomechanical signature of breast cancer". Nature Nanotechnology **7**: 757–765.

Rheinlaender, J., S. Vogel, J. Seifert, M. Schächtele, O. Borst, F. Lang, M. Gawaz and T. E. Schäffer (2015). "Imaging the elastic modulus of human platelets during thrombininduced activation using scanning ion conductance microscopy". Thrombosis and Haemostasis **113**: 305–311.

Rheinlaender, J., A. Dimitracopoulos, B. Wallmeyer, N. M. Kronenberg, K. J. Chalut, M. C. Gather, T. Betz, G. Charras and K. Franze (2020). "Cortical cell stiffness is independent of substrate mechanics". Nature Materials **19**: 1019–1025.

Rianna, C. and M. Radmacher (2017). "Influence of microenvironment topography and stiffness on the mechanics and motility of normal and cancer renal cells". Nanoscale **9**: 11222–11230.

Rigato, A., A. Miyagi, S. Scheuring and F. Rico (2017). "High-frequency microrheology reveals cytoskeleton dynamics in living cells". Nature Physics **13**: 771–775.

Schächtele, M., E. Hänel and T. E. Schäffer (2018). "Resonance compensating chirp mode for mapping the rheology of live cells by high-speed atomic force microscopy". Applied Physics Letters **113**: 093701.

Schierbaum, N., J. Rheinlaender and T. E. Schäffer (2017). "Viscoelastic properties of normal and cancerous human breast cells are affected differently by contact to adjacent cells". Acta Biomaterialia **55**: 239–248.

Schierbaum., N., J. Rheinlaender and T. E. Schäffer (2019). "Combined atomic force microscopy (AFM) and traction force microscopy (TFM) reveals a correlation between viscoelastic material properties and contractile prestress of living cells". Soft Matter **15**: 1721–1729.

Sharma, S., H. I. Rasool, V. Palanisamy, C. Mathisen, M. Schmidt, D. T. Wong and J. K. Gimzewski (2010). "Structural-mechanical characterization of nanoparticle exosomes in human saliva, using correlative AFM, FESEM, and force spectroscopy". ACS Nano **4**: 1921–1926.

Staunton, J. R., B. L. Doss, S. Lindsay and R. Ros (2016). "Correlating confocal microscopy and atomic force indentation reveals metastatic cancer cells stiffen during invasion into collagen I matrices". Scientific Reports **6**: 19686.

Stewart, M. P., A. W. Hodel, A. Spielhofer, C. J. Cattin, D. J. Müller and J. Helenius (2013). "Wedged AFM-cantilevers for parallel plate cell mechanics". Methods **60**: 186–194.

Stolz, M., R. Gottardi, R. Raiteri, S. Miot, I. Martin, R. Imer, U. Staufer, A. Raducanu, M. Düggelin, W. Baschong, A. U. Daniels, N. F. Friedrich, A. Aszodi and U. Aebi (2009). "Early detection of aging cartilage and osteoarthritis in mice and patient samples using atomic force microscopy". Nature Nanotechnology **4**: 186–192.

Stukalov, O., A. Korenevsky, T. J. Beveridge and J. R. Dutcher (2008). "Use of atomic force microscopy and transmission electron microscopy for correlative studies of bacterial capsules". AEM **74**: 5457–5465.

Sunnerberg, J. P., P. Moore, E. Spedden, D. L. Kaplan and C. Staii (2019). "Variations of elastic modulus and cell volume with temperature for cortical neurons". Langmuir **35**: 10965–10976.

Trache, A. and G. A. Meininger (2005). "Atomic force-multi-optical imaging integrated microscope for monitoring molecular dynamics in live cells". Journal of Biomedical Optics **10**: 064023.

Umakoshi, T., S. Fukuda, R. Iino, T. Uchihashi and T. Ando (2020). "High-speed near-field fluorescence microscopy combined with high-speed atomic force microscopy for biological studies". Biochimica Biophysica Acta – General Subject **1864**: 129325.

Ushiki, T., M. Nakajima, M. Choi, M. S.-J. Cho and F. Iwata (2012). "Scanning ion conductance microscopy for imaging biological samples in liquid: A comparative study with atomic force microscopy and scanning electron microscopy". Micron **43**: 1390–1398.

Watanabe, S., S. Kitazawa, L. Sun, N. Kodera and T. Ando (2019). "Development of high-speed ion conductance microscopy". Review of Scientific Instruments **90**: 1–15.

Yamada, Y., H. Konno and K. Shimabukuro (2017). "„Demonstration of correlative atomic force and transmission electron microscopy using actin cytoskeleton". Biophysics and Physicobiology **14**: 111–117.

Subject Index (Volume 1 & Volume 2)

This is a merged subject index for volume 1 (Biomedical Methods) and volume 2 (Biomedical Applications). The bold number in front of the page references represents the volume number.

https://doi.org/10.1515/9783110640632-015

www.ingramcontent.com/pod-product-compliance
Lightning Source LLC
Chambersburg PA
CBHW082109220326
41598CB00066BA/5938